Chagas Disease

María-Jesús Pinazo Delgado
Joaquim Gascón
Editors

Chagas Disease

A Neglected Tropical Disease

Springer

Editors
María-Jesús Pinazo Delgado
Barcelona Institute for Global Health
Barcelona
Spain

Joaquim Gascón
Barcelona Institute for Global Health
Barcelona
Spain

ISBN 978-3-030-44056-5 ISBN 978-3-030-44054-1 (eBook)
https://doi.org/10.1007/978-3-030-44054-1

This Springer imprint is published by the registered company Springer Nature Switzerland AG
The registered company address is: Gewerbestrasse 11, 6330 Cham, Switzerland

Contents

Abbreviations

ACE	Angiotensin-converting enzyme
ACTC1	Alfa cardiac actin-1 gene
AHA	American Heart Association
AIDS	Acquired immune deficiency syndrome
ARB	Angiotensin receptor blockers
ASP-1	Amastigote-stage surface protein-1
AT	Agglutination tests
AT CL-ELISA	Antigen trypomastigote chemiluminescent ELISA
AUC_{0-24}	Area under the curve
BNP	B Natriuretic peptide
BNZ	Benznidazole
BOTOX	Botulinum toxin
c-ELISA	Conventional ELISA
CCC	Chronic Chagas cardiomyopathy
CCD	Chronic Chagas disease
CD	Chagas disease
CDC	Center for Diseases Control
CEChagas disease	Center of Excellence for Chagas Disease
CHF	Congestive heart failure
Chunap	Chagas urine nanoparticle assay
CLIA	Indirect chemiluminescent immunoassays
Cmax	Maximum (or peak) serum concentration
CMIA	Chemiluminescent microparticle immunoassay
CMR	Cardiac magnetic resonance
CNS	Central nervous system
CoML	Complement-mediated lysis test
CRT	Cardiac resynchronization therapy
CSF	Cerebrospinal fluid
CT	Computed tomography
CTL	Cytotoxic $CD8^+$ T cells
CYP51	Cytochrome P-450 sterol C14α-demethylase
dd	Digital droplet
DNA	Deoxyribonucleic acid
DNDi	Drugs for Neglected Diseases Initiative

DTU	Discrete typing unit
ε	Two-dimensional strain
E-wave	Mitral annulus E-wave
EBI	Ergosterol biosynthesis inhibitors
ECG	Electrocardiogram/electrocardiographic
EGD	Esophageal gastric junction
ELISA	Enzyme-linked immunosorbent assay
ESA	Enzyme strip assay
EXO	Exo-antigens
FDA	U.S. Food and Drug Administration
Fg	Fentograms
FINDECHAGAS	Federation of Associations of People Affected by Chagas Disease
GEB	Guanidine EDTA-blood
GI	Gastrointestinal tract
HAART	Highly active antiretroviral therapy
HAI	Indirect hemagglutination assay
HF	Heart failure
HIV	Human immunodeficiency virus
HLI	Healthy Living Initiative
ICD	Implantable cardiac defibrillator
ICT	Immunochromatographic test
IFAT	Immunofluorescence antibody test
IP	Immunoprecipitation
ISGlobal	Barcelona Institute for Global Health
ITT	Intention to treat
kDNA	Kinetoplastid DNA
KMP-11	Kinetoplastid membrane protein-11
LA	Left atrium
LAMP	Loop isothermal amplification assay
LES	Lower esophageal sphincter
LGE	Late gadolinium enhancement
LIC	Low-income countries
LMA	Liquid bead microarray
LV	Left ventricle
LVEF	Left ventricle ejection fraction
MAP	Microtubule-associated protein
MASPs	Mucin-associated surface proteins
MHC I	Histocompatibility complex class I
MNR	Magnetic resonance
MoH	Ministry of Health
MSF	Doctors Without Borders
NFX	Nifurtimox
NGO	Nongovernmental organization
NIH	National Institute of Health

NK	Natural killer cells
NNN	Novy-MacNeal-Nicolle agar
NO	Nitric oxide
nPCR	Nested PCR
NTD	Neglected tropical disease
NYHA	New York Heart Association
OR	Odds ratio
PaGIA	Particle gel immunoassay
PAHO	Pan American Health Organization
PCR	Polymerase chain reaction
PHC	Primary health care
PK/PD	Pharmacokinetic/pharmacodynamic
PLHIV	People living with HIV
PMA	Propidium monoazide
POC	Point-of-care
POEM	Per oral endoscopic myotomy
PP	Per protocol
PRV	Priority review voucher
qPCR	Quantitative PCR
RA	Right atrium
RDTs	Rapid diagnostic tests
RIPA	Radioimmunoprecipitation assay
RPA	Recombinase polymerase amplification
rRNA	Ribosomal RNA
rtPCR	Real-time PCR
RT	Room temperature
RT3DE	Real-time three-dimensional echocardiography
RV	Right ventricle
SAPA	Shed acute-phase antigen
satDNA	Satellite DNA
SDG	Sustainable development goals
SELEX	Systematic Evolution of Ligands by Exponential Enrichment
SHD	Structural heart disease
SLE	Systemic lupus erythematosus
STI	Speckle tracking imaging
T. cruzi	*Trypanosoma cruzi*
T_{CM}	Central memory cells
T_{EM}	Effector memory T cells
TESA	Trypomastigote excreted/secreted antigens
TG6 IEC	Technical Group No. 6 on Information, Education, and Communication
Th	T helper
$T_{NAÏVE}$	T naïve cells
TPP	Target product profile
TR	Tandem repeats

T. rangeli	*Trypanosoma rangeli*
Treg	Regulatory T cells
TS	Trans-sialidase
TSA-1	Trypomastigote-form surface antigen-1
UFG	Universidade Federal de Goiás
USA	United States of America
VT	Ventricular tachycardia
WB	Western blotting
WHO	World Health Organization

Chagas Disease: An Unknown and Neglected Disease

1

Ricardo Brandt de Oliveira, Cristina Ballart, Alba Abràs, Montserrat Gállego, and J. Antonio Marin-Neto

1.1 Description of the Parasite and Taxonomy

Trypanosoma cruzi (*T. cruzi*) is a flagellated protozoan parasite with the following taxonomic classification [1, 2]: kingdom: Protista, phylum: Euglenozoa, class: Kinetoplastea, order: Trypanosomatida, family: Trypanosomatidae, genus: *Trypanosoma*, subgenus: *Schyzotrypanum*, species: *T.* (*S.*) *cruzi*.

The whole-genome sequence of *T. cruzi* was published in 2005 [3], providing important information about the biology of the parasite. *T. cruzi* DNA is distributed among two structures, the nucleus and the kinetoplast [4]. The nuclear DNA is located inside the cellular nucleus, enveloped by the typically porous eukaryotic membrane, and containing a high number of repetitive sequences [4, 5]. The class Kinetoplastea is characterized by the presence of the so-called kinetoplast, an organelle with an extracellular DNA network corresponding to the mitochondrial genome of the parasite [6]. Kinetoplast DNA (kDNA) represents around 30% of the whole cellular DNA [4, 7] and is composed by two types of circular molecules, maxicircles and minicircles [6].

R. B. de Oliveira (✉) · J. A. Marin-Neto
Department of Internal Medicine, Medical School of Ribeirão Preto, University of São Paulo, São Paulo, Brazil
e-mail: rbdolive@fmrp.usp.br

C. Ballart · M. Gállego
Parasitology Section, Department of Biology, Healthcare and the Environment, Faculty of Pharmacy and Food Sciences, Universitat de Barcelona, Barcelona, Spain

Barcelona Institute for Global Health, Hospital Clínic - Universitat de Barcelona, Barcelona, Spain
e-mail: cristina.ballart@isglobal.org

A. Abràs
Department of Biology, Universitat de Girona, Girona, Spain

M.-J. Pinazo Delgado, J. Gascón (eds.), *Chagas Disease*,
https://doi.org/10.1007/978-3-030-44054-1_1

T. cruzi presents various morphological, biochemical, and genetic changes along its life cycle that results in the differentiation of several morphological forms when it is hosted in the vertebrate and invertebrate hosts. Non-replicative forms are trypomastigotes, present in the bloodstream of mammalian hosts and in the vector's gut, while epimastigotes and amastigotes are the replicative forms found in the insect midgut and mammalian cells. Intermediate forms during metacyclogenesis occur [8]. When stained, the bloodstream trypomastigotes, of about 20 μm of length, adopt a curved C or S shape, with a sharply pointed posterior end containing a large subterminal kinetoplast and a long and undulating membrane, which forms a free flagellum in the anterior region [9]. The nucleus is located centrally in the epimastigote and trypomastigote stages, while the flagellum and kinetoplast are repositioned from the anterior to the posterior region of the cell body along metacyclogenesis. Distinct kDNA topologies exist between the different stages of the parasite [8].

Natural populations of *T. cruzi* have a high level of genetic diversity. The taxonomy of the parasite has been always difficult and the risk of misidentification has increased due to the lack of standardized molecular typing methods and the use of alternative nomenclatures [10]: biodemes [11], zymodemes [12, 13], clones [14, 15], schizodemes [16], and discrete typing units (DTUs) [17], among others. This led to a great confusion in the denomination of the proposed groups [18].

In 1999, an international consensus was finally reached on the designation of two lineages of *T. cruzi*, named as TcI and TcII [19]. More recently, TcII was divided into five subgroups (TcIIa–TcIIe) [20, 21], which were then re-nominated by the current classification of the six DTUs: TcI–TcVI [22]. Although this latest revision did not include subdivisions for TcI, a great genetic variability within this lineage has been reported [23, 24]. *T. cruzi* genotypes present different distribution in endemic regions and in transmission cycles [25]. Subsequently, a seventh DTU with a single molecular pattern was first isolated from bats in Brazil and was named as Tcbat [26]. The concept of DTU refers to a set of stocks genetically closer to each other than to any other stock that are identifiable by common genetic, molecular, or immunological markers [17]. Thus, strains included in a DTU should not be considered as a single clone but as multiple clones or families of clones closely related because they share determinate patterns with groups of molecular markers but usually can be distinguished using higher resolution markers [27, 28].

1.2 History of Chagas Disease

The first evidence of *T. cruzi* infection in humans comes from 9000-year-old Chinchorro mummies from Pre-Columbian Andean countries [6, 29, 30]. Chinchorros were the first civilization to establish themselves along the South American coastal region of the Atacama Desert, in southern Peru and northern Chile [31, 32]. The description of lesions compatible with Chagas disease with PCR-positive results demonstrates the antiquity of the infection and disease in humans [6, 29, 31, 33, 34]. Further evidence of American trypanosomiasis in Pre-Columbian times comes from Peruvian ceramics dated to the thirteenth to sixteenth centuries showing possible representations of Chagas disease [35]. Beginning in the sixteenth century, and coinciding with the Spanish conquest, patients with symptoms compatible with Chagas disease began to be described [36, 37]. There were also stories about the insect vector, long before discovering its role in the transmission of the disease, such as the one written in 1835 in the diary of Charles Darwin (1809–1882) during his trip aboard the Beagle [37].

In 1908 the Brazilian hygienist and bacteriologist Carlos Ribeiro Justiniano das Chagas (1879–1934), during one of his campaigns against malaria in Lassance (State of Minas Gerais, Brazil), was warned by the workers who were building the Brazilian Central Railway of the presence of large blood-sucking insects that bite them at night. Observing preparations of the insects in the microscope, he discovered the presence of a flagellated protozoan in his feces. After consulting his teacher, the doctor, bacteriologist, and epidemiologist Oswaldo Gonçalves Cruz (1872–1917), he inoculated the parasite in healthy monkeys and later observed the protozoan in the blood. He also managed to isolate the parasite from other laboratory animals and named it *Schizotrypanum cruzi*, known today as *T. cruzi*, in honor of his mentor Oswaldo Cruz. Thus, the classic postulates of an infectious disease were fulfilled: isolating the parasite, identifying the vector, associating with disease in inoculated animals [37, 38].

Chagas was convinced that he had discovered a pathogenic organism that caused a disease in humans, but he did not know what pathology it was. In 1909, he extracted blood from numerous people in Lassance in search of the parasite and, finally, managed to find it in the blood of a 2-year-old girl, named Berenice, who had fever, facial edema (later called Romaña's sign), hepatosplenomegaly, and swollen lymph nodes [39, 40]. Chagas communicated the disease and it was soon known by the name of its discoverer, Chagas disease [41]. In 1912, Chagas identified the parasite in an armadillo and gradually it was found in other wild reservoirs, evidencing the zoonotic nature of the disease. However, other researchers have also contributed to related findings [42, 43]. Carlos Chagas was nominated for the Nobel Prize in 1913 and 1921 but was never awarded, possibly due to the hard opposition of some of his contemporary physicians and researchers [37, 40].

1.3 Natural History of Chagas Disease

Chagas disease presents a zoonotic character, in which the parasite *T. cruzi* circulates between triatomine insects and sylvatic and domestic mammal reservoirs as well as humans [44, 45]. Human infection with *T. cruzi* occurs through contact with the vectors, that transmit the parasites when defecating while biting their hosts, or through their feces that contaminate juices or other foods that are ingested [46, 47]. Other transmission routes are by congenital transplacental infestation and blood transfusion, and less frequently by organ transplants, laboratory accidents, handling of infected animals, ingestion of uncooked meat from infected animals, or it has even mentioned the sexual route [47, 48]. After an initial blood phase, the survival of the parasite depends on its ability to invade nucleated cells of the host [49].

Chagas disease occurs naturally in Latin-American countries where the disease is still considered mostly rural, harbored by people with lower socioeconomic status and poor domestic hygiene habits, wherever the kissing bugs could find the best conditions to live [50, 51]. Nevertheless, migratory movements and globalization have made it present in other regions, mainly in the United States and Spain in Europe [52, 53]. The possibility of the other transmission routes has allowed the presence of autochthonous cases of Chagas disease in these non-endemic countries where the disease can occur in urban settings [54].

The disease occurs in two clinical phases, acute and chronic; the latter may present a long initial asymptomatic period, characterized as the indeterminate form of Chagas disease [51, 54, 55]. The routes of entry of the parasite can influence the host immune response, the incubation period, and the mechanisms of the natural history of the disease [49]. After an incubation period lasting from 1 to 2 weeks, an acute phase takes place, with abundant parasites present in bloodstream [50, 51]. When symptomatic, unspecific signs such as fever, headache, lymphadenopathy, shortness of breath, myalgia, swelling, and abdominal or chest pain could be present [49]. If the parasite enters through the ocular mucosae, the typical Romaña sign with conjunctivitis, unilateral palpebral edema, and preauricular satellite adenopathy can be seen [51]. If not treated, patients undergo a chronic phase in which, usually after a long asymptomatic period, they can develop cardiac symptoms or, less frequently, digestive symptoms which show regional differences [48]. Infants, immunosuppressed patients, and patient acquiring infection by oral route and organ transplantation route could manifest more severe symptoms including neurological alterations [50, 53, 56, 57]. The pathology during the chronic phase is associated with the invasion and persistence of the parasites in the cardiac or digestive tissues where the typical mega-syndromes occur. In the chronic cardiac form, the complications include high-grade conduction blocks, arrhythmia and ventricular aneurysm, thromboembolic complications, heart failure, and sudden death. In the digestive form alterations of the motility, secretion, and absorption occur in the digestive tract, especially the esophagus and the colon [51, 52]. Variations in the disease's severity may be the effect of genetic polymorphism or the action of *T. cruzi* molecules and of host cell adhesion molecules and inflammatory immune response [49, 51, 58]. However, it has not yet been possible to determine a cause-effect association between the parasite genotype and the clinical stage in the chronic phase [45].

The knowledge of epidemiological background together with the physical examination of a patient with suspected Chagas disease and the availability of screening methods are crucial for the diagnosis, and therefore also for the assessment of prognosis, decisions about forms of treatment and to evaluate the efficiency of the epidemiological control measures [44, 59, 60]. Laboratory diagnosis of Chagas disease depends on the phase of the disease, with parasites being detected by parasitological methods in blood during the acute phase, while during the chronic phase serological methods are of choice [61, 62]. Molecular methods are also available [63, 64].

The etiologic treatment of Chagas disease has not changed over the years. It is based on administering nifurtimox and benznidazole, although there are other compounds under study in clinical trials as well as new therapeutic regimens of the older drugs are being tested [65, 66]. These drugs are effective during the acute phase of the disease and their efficacy to reduce the severity of symptoms during the chronic phase remains controversial and seems to be of greater effectiveness in younger patients [65, 67].

1.4 Pathogenesis and Pathophysiology of Cardiac Involvement in Chagas Disease

Although the pathogenesis of Chagas cardiomyopathy has not been completely elucidated, most of its pathophysiological features can be understood on the basis of the main pathogenesis mechanisms operating in acute and chronic phases of the disease.

1.4.1 Pathophysiologic Alterations in Acute Chagas Disease

Organ damage during the acute phase is associated with high-grade parasitemia, intense direct tissue parasitism, and the immuno-inflammatory response to the parasite. From experimental models of *T. cruzi* infection it is known that most organs may harbor the parasites, but the most typically affected sites include the muscle system in the heart, esophagus and colon, and the central nervous system [68–70]. A study using an "in vitro" experimental model to observe *T. cruzi* passage through the vascular barrier showed that the parasite does not always cause disruption of the endothelial monolayer, but uses a special transmigration process that is facilitated by bradykinin and CCL2 chemokine [71]. Histopathology demonstrates intense tissue parasitism, with prominent inflammatory changes near ruptured infected cells [72, 73]. Myocarditis is intense and diffuse with myocyte necrosis, vasculitis and capillary dilation, and mononuclear and polymorphonuclear infiltration [68–70, 72, 73]. Marked cardiomegaly and pericardial effusion leading to heart failure have been reported by investigators using echocardiography [74]. Lymphadenopathy and splenomegaly accompany the widespread immunologic reaction that on the one hand acts to control parasite multiplication but on the other hand may exacerbate tissue damage. Various host innate mechanisms act to detect and control parasite tissue invasion, including the activation of $CD4^{+}$, $CD8^{+}$ T, and B cells that induce direct antitrypanosoma cytotoxicity, as well as cytokine secretion and production of specific antibodies against the parasite [75]. Death resulting from myocarditis and/or meningo-encephalitis is uncommon in patients infected through the classical vectorial sting, but mortality rates are higher in cases of infection through the oral route, with ingestion of massive amounts of parasites [74].

The neuronal depopulation of the Meissner and Auerbach plexuses and nerve degeneration that occur in esophageal and colon tissues during the acute phase are a key factor favoring the appearance of megaesophagus and megacolon in the chronic phase [76, 77]. Direct damage of smooth muscle may also be a contributory factor, but this has been much less studied in humans and experimental models of Chagas disease.

1.4.2 Pathophysiologic Alterations in Chronic Phase of Chagas Cardiomyopathy (CCC)

Four main mechanisms are thought to contribute to the pathogenesis of chronic Chagas heart disease: neurogenic disturbances; microvascular derangements; parasite-dependent inflammation; and immune-mediated injury [78].

The first two mechanisms are considered to play only an ancillary role in the development of the cardiac lesions and clinical complications observed in patients with CCC, as opposed to the other two mechanisms, that most investigators now believe to be critically dependent on a previously neglected, but that is in fact a crucial factor for the installation and progression of chronic Chagas myocarditis: the parasite persistence.

1.4.2.1 Disturbances of the Cardiac Autonomic Control

Impaired cardiac autonomic function and neuronal depopulation have been clearly shown in patients with Chagas disease [76, 79, 80]. Parasympathetic dysfunction is an early finding, preceding the occurrence of left ventricular dysfunction [79–81]. In fact, systematic studies carried out in the 1950s showed striking depopulation of intramural parasympathetic cardiac cells in deceased patients with Chagas disease; these findings led to a postulated neurogenic ("parasympathicopriva") theory [82], implying that a long-lasting autonomic imbalance would lead to a catecholamine-induced cardiomyopathy characterized by myocardial hypertrophy and cardiac dilation. Other studies have shown that because of the dysautonomia most Chagas disease patients are deprived of the tonic inhibitory action normally exerted by the parasympathetic system on the sinus node and they also lack the vagally mediated fast chronotropic control mechanism to respond to transient changes in blood pressure or venous return [83]. However, several conceptual obstacles and clinical findings have challenged the neurogenic theory and, despite some anatomical and functional alterations that are deemed to accompany dysautonomia caused by Chagas disease, no clear clinical syndrome has been reported in the context [78]. In fact, there is considerable subtleness and variability of the intensity of cardiac denervation in CCC patients. Also, there is no correlation between the severity of the autonomic derangement and the extent of left ventricular dysfunction and no correlation has yet been found between the degree of autonomic impairment and any prognostic indicator [84]. Moreover, sympathetic denervation has also been shown at the sinus node level and in myocardial regions during early stages of disease [78, 85] and impaired sympathetic innervation at the myocardial level is involved in the genesis of complex ventricular arrhythmia [86]. The complex dysautonomia that occurs in patients with Chagas disease is further compounded by the possibility that the parasympathetic impairment may also trigger malignant arrhythmia and sudden death, disturb the control of the coronary microcirculation, and/or exacerbate the mechanical consequences of ventricular dyssynergy [78]. It is indeed plausible that malignant ventricular arrhythmia may result from the combination of parasympathetic impairment and adrenergically denervated myocardium identified by using MIBG-I123 scintigraphy [86]. The extent of cardiac sympathetic denervation is also correlated with the occurrence of severe ventricular arrhythmias and scattered areas of myocardial fibrosis [87].

Finally, a pathophysiological link between impaired cardiac parasympathetic control and abnormal neuroimmunomodulatory regulation has been recently suggested on the basis of the protective effects of pyridostigmine, a cholinesterase inhibitor, in a murine model of chronic experimental *T. cruzi* infection. A reduction of myocardial inflammation, fibrosis, and hypertrophy was reported in association with a decrease in serum levels of IFN-gamma, and no change in IL-10 levels, following the cholinergic stimulation treatment; these investigators postulated that the autonomic dysregulation caused by *T. cruzi* infection could cause an impairment of neuroimmunomodulatory anti-inflammatory role normally played by the parasympathetic nervous system [88].

1.4.2.2 Microvascular Derangements

Various coronary microvascular abnormalities, including increased platelet activation, microthrombi, microvascular spasm, and endothelial dysfunction have been reported in animal models of *T. cruzi* infection [89] and also in pathological and clinical studies in humans [85, 90–92].

These abnormalities indicate the presence of vascular endothelial cell damage caused by inflammation directly related to the *T. cruzi* persistence or mediated by immune effector cells [89, 92]. It has been suggested that endothelin-1, a potent vasoconstrictor, which is also a pro-inflammatory cytokine, may be involved in causing the microvascular dysfunction [93]. Abnormal epicardial coronary artery reactivity to vasodilating and vasoconstricting stimuli has also been reported in Chagas disease, but the role of such disturbance in causing myocardial damage has not been determined [94]. In contrast, it is quite likely that microvascular derangements contribute to the exacerbation of myocardial ischemic damage and subsequent development of fibrosis. These abnormalities may participate in the genesis of ischemic-like symptoms exhibited by 30–50% of patients, electrocardiographic changes, and perfusion defects described in CCC patients with angiographically normal coronary arteries [90, 95]. In support to this theory is the longitudinal observation in CCC patients using Tc-99m-sestamibi myocardial perfusion scans which showed that the deterioration of LV systolic function over time was associated with both the presence of reversible ischemic defects at the initial assessment and an increase in the extent of irreversible perfusion defects during long-term follow-up [91]. It is also a plausible hypothesis that coalescent micro-infarctions may evolve into the typical Chagas disease-related ventricular aneurysms occurring in watershed coronary areas [96]. Also, recent investigations using the *T. cruzi*-infected hamster have shown that in the chronic phase of Chagas cardiomyopathy myocardial resting perfusion abnormalities are associated with left ventricular areas of inflammatory changes and contractile dysfunction, in the absence of significant degrees of regional fibrosis [97, 98]. In summary, there is convincing evidence for the occurrence of microcirculatory derangements producing ischemic myocardial damage [95]; in addition, current ongoing research in humans employing therapeutic interventions to antagonize the microvascular abnormalities may be expected to exert a positively impact on the natural history of Chagas cardiomyopathy [99].

1.4.2.3 Parasite-Dependent Inflammation

CCC is a dilated form of cardiomyopathy characterized by sparse inflammatory infiltrates, minimal parasitemia, low-grade tissue parasitism, and intense and extensive reparative and reactive fibrosis [100]. Classical histologic techniques usually fail to detect the parasite but *T. cruzi* antigens have been identified, using immunohistochemistry and PCR based methods, in inflammatory foci in biopsy and autopsy specimens of patients with chronic Chagas disease [101, 102]. A recent consensus has grown among most investigators that parasite persistence directly causing cell death and parasite-driven immune response are the key mechanisms for the development of CCC [103–105]. The low-grade, but virtually incessant myocarditis is the pivotal mechanism for the late appearance of biventricular heart failure (because of myocardial loss), with widespread fibrosis (the main trigger of malignant arrhythmias and sudden death), and systemic and pulmonary embolic events.

Additional evidence supporting this viewpoint can be summarized as follows: (1) a good correlation has been found between tissue parasite burden and inflammatory intensity in experimental models of *T. cruzi* infection [106]; (2) reinfection or continued exposure (due to continuing residence in areas of active transmission) seems to increase the parasite load and disease severity in experimental models and in human cases [107, 108]; (3) interventions that lessen the parasite burden, such as etiological treatment with benznidazole or nifurtimox, attenuate the cardiomyopathy in experimental animals [109, 110]; (4) in children (mostly with the indeterminate form of chronic Chagas disease), trypanocidal treatment has been associated with better host–parasite relation outcomes, but, for obvious reasons, without demonstration of benefit in terms of hard clinical events [111, 112]; (5) A meta-analysis of observational studies with long-term follow-up and of small randomized (in children) studies described positive effects of trypanocidal therapy [113–115]; (6) despite the fact that the BENEFIT trial could not confirm such findings, the BENEFIT investigators acknowledge that the study limitations may have prevented them from obtaining a more definite evidence for a positive response to trypanocidal therapy [116]; (7) they also recognize that these findings do not challenge the guidelines for the etiologic treatment of patients with some forms of Chagas disease, and recommend that alternative therapies with new (better) drugs or combination of drugs should be envisaged [117]; (8) *T. cruzi* genetic material has been observed consistently in cardiac specimens from patients with chronic Chagas cardiomyopathy, but not in cardiac specimens from seropositive patients who died without clinical signs of cardiac disease [118]; (9) *T. cruzi* DNA was detectable by PCR in the peripheral blood of 86% subjects with well-defined CCC [119]; (10) finally, recent guidelines recommend etiologic treatment for various groups of patients with Chagas disease on the basis of current scientific evidence for benefit [120, 121].

As pointed out above, there is marked phenotype and genotype diversity among the various *T. cruzi* strains, comprising TcI to TcVI classes based on their discrete typing units [122]. This may be responsible for the remarkable differences detected in the pathological and clinical manifestations of Chagas disease (e.g., virtual absence of gastrointestinal disease, or discrepant incidences of sudden death) in various geographical regions [54]. *T. cruzi* genetic diversity, possibly coupled with human genomic and phenotypic diversity responsible for the reaction to the infection, may also be responsible for the heterogeneous natural history of Chagas disease among individuals and geographical regions and could also cause the inconsistent response to trypanocidal therapies in experimental and clinical studies [123].

1.4.2.4 Immune-Mediated Tissue Injury

Experimental and clinical investigations have shown immune-mediated cardiac injury by infiltrating mononuclear cells and release of damaging cytokines; however, myocardial damage through this mechanism is mostly triggered by parasite-driven immunopathology [69, 75, 78, 103–105, 124]. Autoimmunity mechanisms, involving polyclonal activation, molecular self-mimicry by parasite antigens or cryptic epitopes shared by the host and parasites have also been described, in experimental models and human cases of Chagas disease, and postulated to contribute to, or aggravate myocardial damage [125–127]. However, this hypothesis remains controversial and difficult to validate, since data supporting the direct involvement of either molecular mimicry or polyclonal activation in the pathogenesis of myocardial lesions ascribed to *T. cruzi* infection are sparse and inconclusive [128, 129]. Thus, although anti-self-responses are encountered in *T. cruzi* infection, the nature of anti-self-antibodies in experimental and human chronic Chagas disease is heterophilic, with a poor correlation with the heart lesions (i.e., there is no direct and definitive evidence that the immune reactions against the mimicked auto-antigens are actually pathogenic) [104, 128].

Another possible mechanism for autoimmune response in the absence of parasites was recently suggested by the observation that mitochondrial DNA from *T. cruzi* is capable of insertion into the genome of a chicken model system in which infection was induced at the egg stage, but parasite persistence was precluded [127].

In summary, despite the mixed evidence from those studies, the role, relative contribution, and clinical relevance of autoimmunity in triggering myocardial degeneration in the chronic phase of Chagas disease remain to be determined [124].

In contrast, following on the evidence provided by early investigators in the field [130, 131], it is now believed that immune pathological responses are directly dependent on parasite persistence and constitute a crucial mechanism for the development of CCC [124]. The immune-mediated pathology of CCC is a rather complex process, probably involving several interactive factors. This complexity starts from the paradox finding that natural or iatrogenic immunosuppressive conditions usually lead to exacerbation of *T. cruzi* parasitemia and may aggravate the inflammatory process [132, 133]. This epitomizes the double-edged sword type of immune response to the parasite, because the inflammatory lesions found in the myocardium of patients and animals chronically infected with the *T. cruzi* exhibit the typical composition of macrophages and a predominant profile of $CD8^+$ over the $CD4^+$ Th1 cells [134]. The pathogenic picture is further compounded by the enhanced expression of genes responsible for an increased release of several pro-inflammatory cytokines and chemokines, especially INF-γ and TGF-α [135], and, simultaneously, a reduced production of Treg cells and their related cytokines IL-10 and IL-17. This shows the occurrence of an immunological imbalance related to upregulation of Th-1 cells, and deficient suppressor activity of regulatory T cells that otherwise would act to control myocardial inflammation [136].

Immunopathology in chronic Chagas cardiomyopathy is also influenced by genetic polymorphisms of the host, thus modulating the expression of immune inhibitory molecules and potentially altering the equilibrium between host and parasite. For instance, alleles, genotypes, and haplotypes associated with enhancement of the regulatory CTLA-4 system expression have been shown to predominate in patients with the indeterminate form of Chagas disease, probably averting the development of cardiomyopathy [137]. In contrast, a genetic and proteomic study in Chagas disease patients who were either asymptomatic or had CCC, documented that polymorphism in the alfa cardiac actin-1 gene (ACTC1) was associated with an increased susceptibility to maintenance of the inflammatory status, possibly by modulating transcription factor binding to ACTC1 promoter regions [138]. These results and others from several investigators confirm previous evidence of familial aggregation of cases with chronic Chagas cardiomyopathy and may explain why only around one-third of infected patients develop the clinically significant complications of the disease, based on a genetic component that confers susceptibility to disease after the infection [139].

Finally, corroborating the observation from the murine models of *T. cruzi* infection, that complete eradication of the parasite remains an elusive target, an interesting hypothesis was developed to explain why the immune system is not usually capable of sterilizing the human host: instead of an inherently deficient immune response, it is possible that the parasite could escape the cytotoxic $CD8^+$ cells killing effects due to his ability to remain unnoticed within myocardial and other harboring structural cells [140].

1.5 Involvement of the Gastrointestinal Tract in Chagas Disease

In the light of today's knowledge, it is surprising that it took so long for digestive form of Chagas disease to be recognized. In fact, in 1913, in Lassance, Carlos Chagas masterly described the swallowing troubles of a patient with Chagas disease. He speculated about the possible association of the symptoms with the disease that he first recognized but, cautiously, preferred to wait for further evidence before drawing a definite conclusion [141]. From then on, until 1934, Carlos Chagas did not turn back to the issue. Still, in the year of 1913, Belisario Penna and Arthur Neiva reported high prevalence of dysphagia ("mal de engasgo") and severe constipation in regions where they encountered human *T. cruzi* infection in epidemic proportions. Again, the possibility of connection between digestive symptoms and Chagas disease was not raised [142]. Even not intending to offer a complete solution for the riddle, a number of features of both the natural history and clinical presentation of Chagas disease, to be presented below in this chapter, might have averted the sight on the chronic digestive form, thereby precluding its early recognition.

The diffusion of knowledge about the gastrointestinal tract (GI) features of Chagas disease is important because it is no longer just a matter of concern for practitioners and health authorities in endemic areas only. Strong rural-urban migratory trends were created by social moves and economic changes in Latin America that pushed thousands of *T. cruzi*-infected persons from endemic areas into wealthier regions where the transmission of the disease by classical vectorial route is absent. As a consequence, the estimated number of *T. cruzi*-infected persons living in South American metropolitan areas (Buenos Aires, Rio de Janeiro, and Sao Paulo) is about one million [121]. Over the last decades, the destinations of migration have been increasingly scattered worldwide, mainly in the North America and Europe, where the immigrants can become ill or transmit the disease by blood or organ donation [50, 143]. Emphasizing the importance of Chagas disease as a serious public health problem, recent data reveal that some 50,000–80,000 Chagas disease patients are living in Spain [144]. Moreover, the study of epidemiologic and clinical aspects of Chagas disease may yield information useful for mounting strategies to face the challenges posed by the risks of other pandemics fostered by globalization.

1.5.1 Digestive Manifestations During the Acute Phase

During the acute phase of Chagas disease, abdominal pain, vomiting, and diarrhea may occur [141] and are ascribed to the systemic inflammatory response in patients with high number of parasites circulating in the blood. Those digestive symptoms have been reported by nearly one-third of patients with acute Chagas disease acquired by oral transmission [145].

Dysphagia, the cardinal manifestation of the digestive form in the chronic phase, may also ensue shortly after *T. cruzi* infection. Four of 115 patients studied during the acute phase presented transient dysphagia [146]. Rassi and Rezende [147] carried out clinical assessment and radiologic study of the esophagus in 94 patients with acute Chagas disease; eight of them (8.5%) reported mild dysphagia. The radiological examination revealed transient esophageal aperistalsis in one of them, and also in 5 of the 86 patients without dysphagia. Barium enema performed in 59 of these patients revealed normal colon in 54, dolichosigmoid in 4, and sigmoid dilation in another one.

1.5.2 Gastrointestinal (GI) Involvement in the Chronic Phase

1.5.2.1 Pathophysiology

Abnormalities of the enteric nervous system, namely degeneration and reduction in number of intrinsic neurons, are considered essential elements in the pathogenesis of gastrointestinal disorders of chronic Chagas disease [76] (Fig. 1.1). These abnormalities are found along the entire length of the gastrointestinal tract, although the most frequent clinical manifestations are due to involvement of esophagus and colon. In addition, reduction in number of interstitial cells of Cajal [148] and enteroglia of myenteric plexuses have been described but their pathophysiologic roles remain to be elucidated [149].

The relevance of damage of the enteric nervous system is reinforced by aberrant motor responses of esophageal smooth muscle to certain stimuli seen in Chagas disease patients, such as the hypersensitivity to cholinergic agents [150] and the paradoxical response to cholecystokinin-like peptides [151]. Also, impairment of certain physiological reflexes, namely, relaxation of the lower esophageal sphincter, gastric accommodation to distension [152] and motor response of the colon to a caloric meal [153] as well as abnormal responses to physiological stimuli such as exaggerated secretion of gastrin in response to a

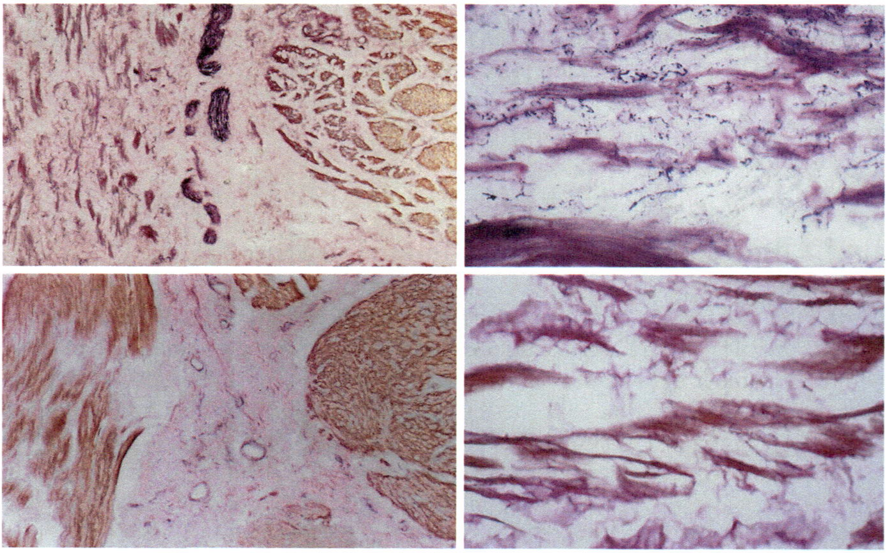

Fig. 1.1 Histochemical identification of NADPH-diaphorase activity in the esophagogastric junction. Dark blue stained neuronal cells and fibers stained in the myenteric plexus (left, upper panel 250×) and circular muscle layer (right, upper panel, 400×) of a patient with gastric cancer; these histological components are absent from the myenteric plexus (left, lower panel, 250×) and circular muscle layer (right, lower panel, 400×) in the material obtained from a patient with Chagas disease. Sections were counter-stained with van Gieson stain

meal [154] are attributable to faulty neural influences on digestive organs, therefore consistent with enteric nervous system damage.

1.5.2.2 Pathogenesis

Exactly how the injury to intrinsic nerves in Chagas disease is caused has not yet been completely clarified. Conventional histopathology analysis shows hypertrophy of muscle layers, mononuclear infiltrates and fibrosis in smooth muscle and intramural ganglia of esophagus, the intensity of inflammation and fibrosis being significantly higher in dilated esophagi than in non-dilated ones [155]. Immuno-histochemical characterization of the inflammatory cells revealed a large proportion of $CD3^+$ lymphocytes, a variable number of macrophages, and a few B lymphocytes [156]. The presence of both *T. cruzi* kinetoplast DNA (kDNA) and cells with cytolytic potential was demonstrated in histologic sections of esophageal tissue from patients with low counts of neurons in intrinsic plexuses; in patients without megaesophagus and normal or near normal neuronal counts, kDNA and inflammatory infiltrates were not found. In addition, NK cells and cytotoxic lymphocytes were detected in the myenteric plexus of patients with megaesophagus [157]. Morphometric studies of inflammatory cells provided evidence for participation of eosinophils, mast cells, and macrophages in the genesis of fibrosis and dilation of the colon [158], and a combination of techniques highlighted the roles of mast cells and eosinophils in the pathogenesis of the megacolon in Chagas disease [159]. In summary, evidence exists that *T. cruzi* persistence in esophageal and colonic muscle is required for megaesophagus and megacolon development, presumably by triggering inflammation and consequent damage to the enteric nervous system; the cascade of this inflammatory response is only partially elucidated. Of note, although several polymorphic sites at immunoregulatory genes may influence the development of Chagas disease variants [160], none of them was found associated specifically with damage to the enteric nervous system and no autoimmune mechanism has been described in the pathogenesis of the digestive form of Chagas disease.

1.5.2.3 Natural History of the Digestive Form in Chagas Disease: Prevalence, Incidence, Spectrum of Severity Disease in Endemic Areas

Capturing the whole spectrum of current trends in Chagas disease morbidity would require recent data that are not yet available. Epidemiological data in areas where human *T. cruzi* infection has been highly prevalent were obtained decades ago in inland Brazil populations with broad-based age pyramids, with most of individuals aged less than 30 years. There is compelling evidence that the ongoing demographic transition in Brazil has been rendering these data obsolete [161]. In fact, in Brazilian metropolitan areas of destination for rural migrants as well as in former endemic areas where the vector transmission of the disease no longer occurs, the age of the Chagas disease population is much older than three decades ago [162–164]. Nevertheless, exploration of available results of cross-sectional and longitudinal studies carried out decades ago in endemic areas may provide useful information on the natural history of the GI involvement in chronic phase of Chagas disease.

Prevalence, Incidence, and Severity Spectrum of Esophageal Involvement

Several studies in endemic areas employing radiologic techniques produced substantial information on the esophageal involvement. The results of several cross-sectional studies are fairly coincident and point to an overall prevalence of esophageal involvement of nearly 10% in chronic Chagas disease individuals, as generally accepted [51, 165] (Table 1.1). That figure may constitute an underestimate, because the prevalence of esophageal disorder among Chagas disease individuals increases sharply with age and reaches 20% at the end of the sixth decade (Table 1.2). Moreover, the annual incidence rate of new radiological evidence of esophageal involvement among individuals with the indeterminate form varies between 0.3% and 1.0% (Table 1.3). Nevertheless, in an area where the *T. cruzi* transmission was virtually suppressed by 1980 [175], the incidence of esophageal involvement among individuals with the indeterminate form was also shown to increase with age, the estimated average annual risk for sexagenarians being around 3% (Table 1.4). Therefore, it is plausible that the current prevalence of digestive form would keep increasing due to the ongoing demographic transition.

Table 1.1 Prevalence of radiologically detected esophageal disorders in *T. cruzi*-infected individuals living in Chagas disease endemic areas

Authors	Municipality	Population sample (*n*)	Prevalence (IC 95%)
Macedo VO (1973) [166]	São Felipe (BA)	840	**8.7%** (7.0–10.8)
Dias JCP et al. (1983) [145]	Bambui (MG)	566	**8.8%** (6.6–11.4)
Castro C et al. (1987) [167]	Mambaí (GO)	1006	**7.1%** (5.6–8.9)
Pereira JB and Coura JR (1986) [168]	Virgem da Lapa (MG)	255	**11.0%** (7.5–15.6)
Coura JR et al. (1996) [169]	Catinga do Piaui (PI)	92	**14.1%** (8.4–22.1)
Penaranda-Casillo R et al. (2006) [170]	Mambaí (GO)	468	**13.8%** (11.0–17.3)

Table 1.2 Prevalences of radiologically detected esophageal disorders in *T. cruzi*-infected individuals from three endemic areas distributed according to the age range

Municipality	Sample size (*n*)	Age range (years)			
		<20	20–39	40–59	≥60
Bambui (MG) [145]	555	3.0% (08–10.4%)	7.5% 5.2–10.7%)	15.0% (9.7–22.2%)	–
Virgem da Lapa (MG) [168]	255	0% (0–12.1%)	9.1% (4.2–8.4%)	10.1% (5.7–17.1%)	22.4% (13.0–35.9%)
Mambai (GO) [167]	1006	1.1% (0.4–2.7%)	6.0% (4.0–8.9%)	15.3% (11.0–21.0%)	21.5% (13.3–33.0%)

Data are percentage of *T. cruzi*-infected subjects, with 95% CI in parenthesis

Table 1.3 Incidence of radiologically detected esophageal disorders in individuals with the indeterminate form of Chagas disease living in endemic areas

Municipality	Population sample (*n*)	Risk period (years)	Number of cases	Incidence	Average annual incidence rate
Bambuí (MG) 1983 [145]	115	27	21	18.3% (12.3–26.3%)	0.6%
Pains-Iguatama 1985 [171]	110	10	3	2.7% (0.9–7.7%)	0.3%
Virgem da Lapa 1985 [172]	124	6	5	4.0% (1.7–9.0%)	0.7%
São Felipe (BA) 1980 [173]	400	10	5	3.9% (2.7–5.5%)	0.4%
Mambaí (GO) 1994 [174]	345	13	45	13.0% (9.9–17.2%)	1.1%
Mambaí (GO) 2006 [170]	445	25	51	11.5% (8.7–14.9%)	0.5%

Table 1.4 Incidence of radiologically detected esophageal disorders according to the age range in individuals with the indeterminate form of Chagas disease living in Mambaí (GO) in 1975–1976 [174]

Age range (years)	Sample size (*n*)	Number of cases	Incidence (IC 95%)	Average annual incidence (rate)
<20	125	10	8.0% (4.4–14.1%)	0.6%
20–39	151	22	14.6% (9.8–21.4%)	1.1%
40–59	61	14	22.9% (10.3–27.1%)	1.8%
>60	8	3	37.5% (13.7–69.4%)	2.8%
Total	347	49		

The esophageal disorder is mild in the large majority of the Chagas disease patients living in endemic areas, and the severity is higher among individuals aged 40 years and older (Table 1.5). The evolution of the established esophageal disorder is slow, if any. Castro et al. [174] reported the same degree of severity of esophageal involvement in 37 of 77 patients studied in 1975–1976 and again in 1988–1989. A similar worsening rate was seen in a 25-year long follow-up study where two patients evolved straight from indeterminate form into group IV, and two presented with esophageal cancer [170].

Evaluation of Esophageal Involvement by High Resolution Manometry (HRM)

HRM is considered the best currently available technique to evaluate esophageal motility [177]. It is plausible that it might widen the spectrum of motor patterns in Chagas disease by increasing the detection of subtle esophageal involvement in individuals who otherwise would be considered in the indeterminate form as well as by refining the

Table 1.5 Distribution of severity of radiologically detected esophageal disorders in *T. cruzi*-infected individuals according to age range and Rezende criteria

			Rezende criteria			
Municipality	Age range (years)	Sample size (*n*)	I	II/III	IV	Total
Virgem da Lapa [176]	40<	94	5.3% (2.3–11.8%)	1.0% (0.2–5.8%)	0% (0–0.04%)	6.4% (2.6–13.9%)
	>40	157	8.9% (5.4–14.4%)	(5.1%) (2.6–9.7%)	0 (0–0.03%)	14.0% (9.2–20.3%)
Mambaí (GO) [170]	40<	137	5.0% (2.3–10.2%)	(3.0%) (0.9–7.8%)	(0.7) (0.1–4.0%)	8.9% (4.8–15.1%)
	>40	300	9.3% (6.4–9.3%)	(5.3%) (3.3–8.5%)	(3.0%) (1.2–4.2%)	17.7% (13.6–22.6%)

characterization of dysmotility. Studies carried out on patients with severe esophageal symptoms in a tertiary Brazilian hospital show typical achalasia patterns in most of them [76, 77]. On the other hand, studies carried out on mostly asymptomatic individuals seropositive for Chagas disease showed only minor, nonspecific abnormalities, namely isolated hypotensive lower esophageal sphincter pressure, ineffective esophageal peristalsis, and fragmented peristalsis [178–180]. These studies suffer from selection bias, by encompassing only the opposite ends of the clinical severity, so that their results do not delineate properly the entire spectrum of esophageal dysmotility in Chagas disease. Results of an ongoing prospective study involving Chagas disease patients recruited from different areas of a large general hospital (blood donors, general gastroenterology clinics, cardiology clinics, surgical unit) depict a wide variety of esophageal phenotypes among which achalasia is the minority (Fig. 1.2).

Prevalence and Severity Spectrum of Colonic Involvement

Data regarding colonic involvement in community-based Chagas disease cases are scanty in autopsy studies, megacolon was found slightly more frequently than megaesophagus [181]. In an endemic area, the prevalence of constipation was significantly higher among *T. cruzi*-infected individuals than that in non-infected subjects (3.35% vs. 0.07%) [182]. In another endemic area gross colonic dilation (megacolon) was found in 14 of 222 subjects (6.3%) and abnormally wide diameter of the sigmoid was found in 13 (5.9%); a subset of patients had an abnormally long sigmoid colon (dolichocolon) that presumably increases the risk for volvulus [183].

Ximenes et al. found that 26% of the patients with megacolon had normal intestinal function and 56% had severe constipation, among patients with normal barium enemas, 61% had normal bowel movements, and 18% had severe constipation [184]. Taken together, these data suggest that constipation is not a sensitive marker of megacolon or sigmoid volvulus, whose presence should be suspected in Chagas disease patients with other digestive symptoms.

1.5.2.4 Digestive Form of Chagas Disease Outside the Known Endemic Areas

Most of the Latin-American immigrants in non-endemic areas are young or middle-aged adults in apparent good health, seeking for work opportunities [178], so that

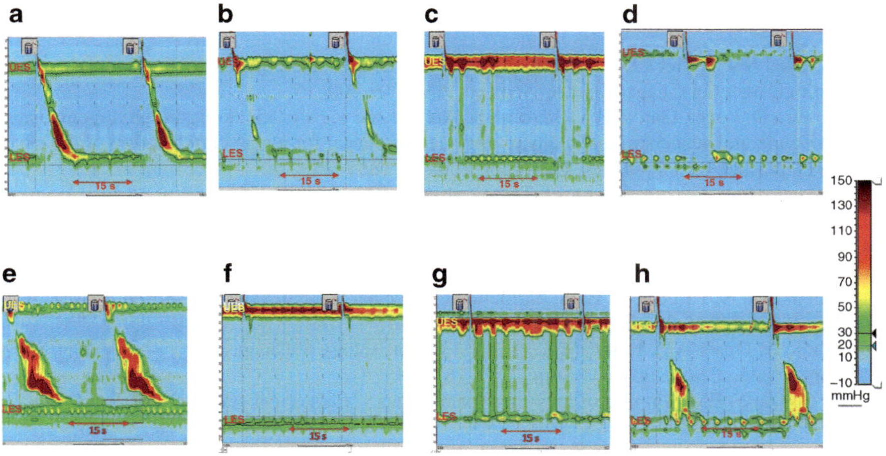

Fig. 1.2 Pressure topography plots of esophageal deglutitive responses from eight seropositive individuals for *T. cruzi* infection demonstrating the spectrum of esophageal dysmotility in Chagas disease. (**a**) Normal peristalsis plus normal LES relaxation (asymptomatic blood donor candidate); (**b**) fragmented peristalsis plus normal LES relaxation (ineffective esophageal motility); (**c**) absent peristalsis plus normal LES relaxation (ineffective esophageal motility); (**d**) absent peristalsis plus abnormally low LES pressure (ineffective esophageal motility); (**e**) normal peristalsis plus impaired LES relaxation (esophagogastric junction output obstruction); (**f**) absent peristalsis plus impaired LES relaxation (achalasia type 1); (**g**) panesophageal isobaric pressurization plus impaired LES relaxation (achalasia type 2); (**h**) rapid progression of peristalsis plus impaired LES relaxation (achalasia type 3)

the majority of *T. cruzi*-infected immigrants are also young or middle-aged adults in apparent good health, seeking for work opportunities [178], who should have either the indeterminate form of the disease or mild symptoms. Results of studies for characterization of esophageal disorders in *T. cruzi*-infected individuals recruited in centers of reference for Chagas disease diagnosis in Spain are consistent with this supposition. In each of four studies [179, 185], the frequencies of both esophageal symptoms and abnormal results of esophageal radiology were low, similarly to results emanating from epidemiological studies in endemic areas in Brazil.

Thus, *T. cruzi*-infected immigrants seldom present medical problems at their arrival, but may evolve to symptomatic forms of the disease thereby representing a burden to the medical services [121]. Although information on effects of environmental, dietary, and lifestyle factors on the natural history of the digestive form of Chagas disease are lacking, it is plausible that the same patterns of disease progression seen in endemic areas occur elsewhere. If so, an increasing prevalence of digestive disorders over the next decades can be expected among Chagas disease immigrants, determining an extra burden to health services in their destination areas. In addition, *T. cruzi*-infected migrants are not usually aware of their chronic infection, thus remaining potential sources of transmission via blood transfusion, organ donation, and mother-to-child vertical transmission [178].

References

1. El-Sayed NM, Myler PJ, Bartholomeu DC, et al. The genome sequence of *Trypanosoma cruzi*, etiologic agent of Chagas disease. Science. 2005;309:409–15.
2. Moreira D, López-García P, Vickerman K. An updated view of kinetoplastid phylogeny using environmental sequences and a closer outgroup: proposal for a new classification of the class Kinetoplastea. Int J Syst Evol Microbiol. 2004;54:1861–75.
3. Ruggiero MA, Gordon DP, Orrell TM, et al. A higher level classification of all living organisms. PLoS One. 2015;10:e0119248.
4. Teixeira ARL, Hecht MM, Guimaro MC, et al. Pathogenesis of Chagas' disease: parasite persistence and autoimmunity. Clin Microbiol Rev. 2011;24:592–630.
5. Schenkman S, Pascoalino BDS, Nardelli SC. Nuclear structure of *Trypanosoma cruzi*. Adv Parasitol. 2011;75:251–83.
6. Araújo A, Jansen AM, Reinhard K, et al. Paleoparasitology of Chagas disease – a review. Mem Inst Oswaldo Cruz. 2009;104(Suppl 1):9–16.
7. De Souza W. Electron microscopy of trypanosomes – a historical review. Mem Inst Oswaldo Cruz. 2008;103:313–25.
8. Gonçalves CS, Ávila AR, de Souza W, et al. Revisiting the *Trypanosoma cruzi* metacyclogenesis: morphological and ultrastructural analyses during cell differentiation. Parasit Vectors. 2018;11(1):83.
9. Ward H, Ajjampur SSR. Appendix 2. Parasitic protozoa. In: Farrar J, Hotez P, Junghanss T, Kang G, Lalloo D, Nicholas J, editors. Manson's tropical diseases. 23rd ed. London: Elsevier Saunders; 2014. p. 1238–49.
10. Messenger LA, Yeo M, Lewis MD, et al. Molecular genotyping of *Trypanosoma cruzi* for lineage assignment and population genetics. Methods Mol Biol. 2015;1201:297–337.
11. Andrade SG. Caracterização de cepas de *Trypanosoma cruzi* isoladas de Recôncavo Baiano. Rev Patol Trop. 1974;3:65–121.
12. Miles MA, Toye PJ, Oswald SC, et al. The identification by isoenzyme patterns of two distinct strain-groups of *Trypanosoma cruzi*, circulating independently in a rural area of Brazil. Trans R Soc Trop Med Hyg. 1977;71:217–25.
13. Miles MA, Souza A, Povoa M, et al. Isozymic heterogeneity of *Trypanosoma cruzi* in the first autochthonous patients with Chagas' disease in Amazonian Brazil. Nature. 1978;272:819–21.
14. Tibayrenc M, Ward P, Moya A, et al. Natural populations of *Trypanosoma cruzi*, the agent of Chagas disease, have a complex multiclonal structure. Proc Natl Acad Sci U S A. 1986;83:115–9.
15. Tibayrenc M, Ayala FJ. Isoenzyme variability in *Trypanosoma cruzi*, the agent of Chagas' disease. Genetical, taxonomical and epidemiological significance. Evolution. 1988;42:277–92.
16. Morel C, Chiari E, Plessman Camargo E, et al. Strains and clones of *Trypanosoma cruzi* can be characterized by pattern of restriction endonuclease products of kinetoplast DNA minicircles. Proc Natl Acad Sci U S A. 1980;77:6810–4.
17. Tibayrenc M. Genetic epidemiology of parasitic protozoa and other infectious agents: the need for an integrated approach. Int J Parasitol. 1998;28:85–104.
18. Guhl F. Epidemiología molecular de *Trypanosoma cruzi*. Rev Esp Salud Pública. 2013;2013:1–8.
19. Anonymous. Recommendations from a satellite meeting. Mem Inst Oswaldo Cruz. 1999;94:429–32.
20. Brisse S, Barnabé C, Tibayrenc M. Identification of six T*rypanosoma cruzi* phylogenetic lineages by random amplified polymorphic DNA and multilocus enzyme electrophoresis. Int J Parasitol. 2000;30:35–44.
21. Brisse S, Verhoef J, Tibayrenc M. Characterisation of large and small subunit rRNA and mini-exon genes further supports the distinction of six *Trypanosoma cruzi* lineages. Int J Parasitol. 2001;31:1218–26.

22. Zingales B, Andrade SG, Briones MRS, et al. A new consensus for *Trypanosoma cruzi* intraspecific nomenclature: second revision meeting recommends TcI to TcVI. Mem Inst Oswaldo Cruz. 2009;104:1051–4.
23. Cura CI, Mejía-Jaramillo AM, Duffy T, et al. *Trypanosoma cruzi* I genotypes in different geographic regions and transmission cycles based on microsatellite motife of the intergenic spacer of spliced leader genes. Int J Parasitol. 2010;40:1599–607.
24. Ramírez JD, Duque MC, Montilla M, et al. Multilocus PCR-RFLP profiling in *Trypanosoma cruzi* I highlights an intraspecific genetic variation pattern. Infect Genet Evol. 2012;12:1743–50.
25. Zingales B, Miles MA, Campbell DA, et al. The revised *Trypanosoma cruzi* subspecific nomenclature: rationale epidemiological relevance and research applications. Infect Genet Evol. 2012;12:240–53.
26. Marcili A, Lima L, Cavazzana M Jr, et al. A new genotype of *Trypanosoma cruzi* associated with bats evidenced by phylogenetic analyses using SSU rDNA cytochrome b and Histone H2B genes and genotyping based on ITS1 rDNA. Parasitology. 2009;136:641–55.
27. Muñoz C, Solari A, Apt W, et al. Caracterización de las Unidades Discretas de Tipificación de *Trypanosoma cruzi* según sus marcadores moleculares. Rev Ibero Latinoam Parasitol. 2013;72:5–21.
28. Tibayrenc M, Ayala FJ. The populations genetics of *Trypanosoma cruzi* revisited in the light of the predominant clonal evolution model. Acta Trop. 2015;151:156–65.
29. Aufderheide AC, Salo W, Madden M, et al. A 9,000-year record of Chagas' disease. Proc Natl Acad Sci U S A. 2004;101(7):2034–9.
30. Fornaciari G, Castagna M, Viacava P, et al. Chagas' disease in a Peruvian Inca mummy. Lancet. 1992;339:128–9.
31. Guhl F, Jaramillo C, Yockteng R, et al. *T. cruzi* DNA in human mummies. Lancet. 1997;349:1370.
32. Rothhammer F, Allison MJ, Nuñez L, et al. Chagas disease in pre-Columbian South America. Am J Phys Anthropol. 1985;68:495–8.
33. Guhl F, Jaramillo C, Vallejo GA, et al. Isolation of *T. cruzi* DNA in 4,000-year-old mummified human tissue from northern Chile. Am J Phys Anthropol. 1999;108:401–7.
34. Madden M, Salo WL, Streitz J, et al. Hybridization screening of very short PCR products for paleoepidemiological studies of Chagas' disease. BioTechniques. 2001;30:102–4.
35. Dias JCP, Schofield CJ. History of Chagas disease as a public health problem in Latin America. In: Teixeira A, Vinaud M, Castro AM, editors. Emerging Chagas disease. Sharjah: Benthan Science Publisher; 2011. p. 1–9.
36. Miles MA. The discovery of Chagas disease: progress and prejudice. Infect Dis Clin N Am. 2004;18:247–60.
37. Steverding D. The history of Chagas disease. Parasit Vectors. 2014;7:317.
38. Reyes López PA. Life and work of Carlos Chagas, on its centennial description of Chagas-Mazza disease. Arch Cardiol Mex. 2009;79(4):237–9.
39. Werner Apt B, Arribada CA, Zulantay AI. Centennial of Chagas disease (1909–2009). Rev Med Chil. 2009;137(5):721–2.
40. Coura JR, Viñas PA, Junqueira AC. Ecoepidemiology, short history and control of Chagas disease in the endemic countries and the new challenge for non-endemic countries. Mem Inst Oswaldo Cruz. 2014;109(7):856–62.
41. Chagas C. Nova tripanozomiaze humana. Estudos sobre a morfolojia e o ciclo evolutivo de *Schizotrypanum cruzi* n. gen., n. sp., ajente etiolojico de nova entidade morbida do homem. Mem Inst Oswaldo Cruz. 1909;1:159–218.
42. Kropf SP, Sá MR. The discovery of Trypanosoma cruzi and Chagas disease (1908–1909): tropical medicine in Brazil. Hist Cienc Saude Manguinhos. 2009;16(suppl 1):13–34.
43. Morel CM. Chagas disease, from discovery to control – and beyond: history, myths and lessons to take home. Mem Inst Oswaldo Cruz. 1999;94(Suppl 1):3–16.
44. Tarleton RL, Reithinger R, Urbina JA, et al. The challenges of Chagas disease—grim outlook or glimmer of hope? PLoS Med. 2007;4(12):e332.

45. Zingales B. *Trypanosoma cruzi* genetic diversity: something new for something known about Chagas disease manifestations, serodiagnosis and drug sensitivity. Acta Trop. 2018;184:38–52.
46. Alarcón de Noya B, Noya González O. An ecological overview on the factors that drives to *Trypanosoma cruzi* oral transmission. Acta Trop. 2015;151:94–102.
47. Kirchhoff LV. Epidemiology of American trypanosomiasis (Chagas disease). Adv Parasitol. 2011;75:1–18.
48. Coura JR, Albajar PV. Chagas disease: a new worldwide challenge. Nature. 2010;465(7301):S6–7.
49. Barreto de Albuquerque J, Silva dos Santos D, . Stein JV et al (2018). Oral versus intragastric inoculation: similar pathways of *Trypanosoma cruzi* experimental infection? From target tissues, parasite evasion, and immune response. Front Immunol 9:1734.
50. Bern C. Chagas' disease. N Engl J Med. 2015;373:456–66.
51. Prata A. Clinical and epidemiological aspects of Chagas disease. Lancet Infect Dis. 2001;1:92–100.
52. Gascón J, Bern C, Pinazo MJ. Chagas disease in Spain, the United States and other non-endemic countries. Acta Trop. 2010;115:22–7.
53. Schmunis GA, Yadon ZE. Chagas disease: a Latin American health problem becoming a world health problem. Acta Trop. 2010;115:14–21.
54. Rassi A Jr, Rassi A, Marin-Neto JA. Chagas disease. Lancet. 2010;375(9723):1388–402.
55. Rassi A Jr, Rassi A, Marcondes de Rezende J. American trypanosomiasis (Chagas disease). Infect Dis Clin N Am. 2012;26(2):275–91.
56. Bern C, Martin DL, Gilman RH. Acute and congenital Chagas disease. Adv Parasitol. 2011;75:19–47.
57. Córdova E, Maiolo E, Corti M, et al. Neurological manifestations of Chagas' disease. Neurol Res. 2010;32:238–44.
58. Cura C, Schijman AG. Relación entre los genotipos de *T. cruzi* y la presentación clínica de la enfermedad de Chagas. Rev Esp Salud Pública. 2013;86:9–16.
59. Álvarez-Hernández DA, Franyuti-Kellya GA, Díaz-López-Silva R, et al. Chagas disease: current perspectives on a forgotten disease. Rev Med Hosp Gen Méx. 2018;81(3):154–64.
60. de la Rosa E, Paglini-Oliva P, Prato LB, et al. Early detection of chronic asymptomatic Chagas infection. Med Sci Monit. 2018;24:4567–71.
61. Alonso-Padilla J, Cortés-Serra N, Pinazo MJ, et al. Strategies to enhance access to diagnosis and treatment for Chagas disease patients in Latin America. Expert Rev Anti Infect Ther. 2019;17(3):145–57.
62. Balouz V, Agüero F, Buscaglia CA. Chagas disease diagnostic applications: present knowledge and future steps. Adv Parasitol. 2017;97:1–45.
63. Alonso-Padilla J, Gállego M, Schijman AG, et al. Molecular diagnostics for Chagas disease: up to date and novel methodologies. Expert Rev Mol Diagn. 2017;17(7):699–710.
64. Schijman AG. Molecular diagnosis of *Trypanosoma cruzi*. Acta Trop. 2018;184:59–66.
65. Bern C. A new epoch in antitrypanosomal treatment for chagas disease. J Am Coll Cardiol. 2017;69(8):948–50.
66. Chatelain E. Chagas disease research and development: is there light at the end of the tunnel? Comput Struct Biotechnol J. 2016;15:98–103.
67. Viotti R, Vigliano C, Lococo B, et al. Long-term cardiac outcomes of treating chronic Chagas disease with benznidazole versus no treatment: a nonrandomized showing greater effectiveness in patients. Ann Intern Med. 2006;144(10):724–34.
68. Okumura M, Brito T, Silva LHP, Silva AC, Netto AC. The pathology of experimental Chagas' disease in mice. I. Digestive tract changes with special reference to necrotizing arteritis. Rev Inst Med Trop São Paulo. 1960;2:17.
69. Teixeira AR, Teixeira ML, Santos-Buch CA. The immunology of experimental Chagas' disease. IV. Production of lesions in rabbits similar to those of chronic Chagas' disease in man. Am J Pathol. 1975;80:163.
70. Andrade ZA, Andrade SG, Correa R, Sadigursky M, Ferrans VJ. Myocardial changes in acute Trypanosoma cruzi infection. Ultrastructural evidence of immune damage and the role of microangiopathy. Am J Pathol. 1994;144(6):1403.

71. Coates BM, Sullivan DP, Makanji MY, Du NY, Olson CL, Muller WA, et al. Endothelial transmigration by Trypanosoma cruzi. PLoS One. 2013;8(12):e81187.
72. Laranja FS, Dias E, Miranda A, Nobrega G. Chagas' disease; a clinical, epidemiologic, and pathologic study. Circulation. 1956;14:1035.
73. Kumar R, Kline IK, Abelmann WH. Experimental Trypanosoma cruzi myocarditis: relative effects upon the right and left ventricles. Am J Pathol. 1969;57:31.
74. Bastos CJ, Aras R, Mota G, Reis F, Dias JP, de Jesus RS, Freire MS, de Araújo EG, Grassi MF. Clinical outcomes of thirteen patients with acute Chagas disease acquired through oral transmission from two urban outbreaks in northeastern Brazil. PLoS Negl Trop Dis. 2010;4(6):e711.
75. Tarleton RL, Koller BH, Latour A, Postan M. Susceptibility of β2-microglobulin-deficient mice to Trypanosoma cruzi infection. Nature. 1992;356(6367):338.
76. Köberle F. Chagas' disease and Chagas' syndromes: the pathology of American trypanosomiasis. Adv Parasitol. 1968;6:63.
77. Meneghelli UG. Chagas' disease: a model of denervation in the study of digestive tract motility. Braz J Med Biol Res. 1985;18:255.
78. Marin-Neto JA, Cunha-Neto E, Maciel BC, Simões MV. Pathogenesis of chronic Chagas heart disease. Circulation. 2007;115:1109.
79. Amorim DS, Manço JC, Gallo L Jr, Marin Neto JA. Chagas' heart disease as an experimental model for studies of cardiac autonomic function in man. Mayo Clin Proc. 1982;57(Suppl):48.
80. Soares Barreto-Filho JA, Consolim-Colombo FM, Ferreira Lopes H, Martins Sobrinho CR, Guerra-Riccio GM, Krieger EM. Dysregulation of peripheral and central chemoreflex responses in Chagas' heart disease patients without heart failure. Circulation. 2001;104:1792.
81. Ribeiro ALP, Moraes RS, Ribeiro JP, Ferlin EL, Torres RM, Oliveira E, et al. Parasympathetic dysautonomia precedes left ventricular systolic dysfunction in Chagas disease. Am Heart J. 2001;141:260.
82. Köberle F. Cardiopathia parasympathicopriva. Munch Med Wochenschr. 1959;101:1308.
83. Marin-Neto JA, Gallo L Jr, Manco JC, Rassi A, Amorim DS. Mechanisms of tachycardia on standing: studies in normal individuals and in chronic Chagas' heart patients. Cardiovasc Res. 1980;14:541.
84. Marin-Neto JA, Rassi A Jr, Maciel BC, Simoes MV, Schmidt A. Chagas heart disease. In: Yusuf S, Cairns JA, Camm AJ, Fallen EL, Gersh BJ, editors. Evidence-based cardiology. 3rd ed. London: BMJ Books; 2010. p. 823–41.
85. Simões MV, Pintya AO, Bromberg-Marin G, Sarabanda AV, Antloga CM, Pazin-Filho A, Maciel BC, Marin-Neto JA. Relation of regional sympathetic denervation and myocardial perfusion disturbance to wall motion impairment in Chagas' cardiomyopathy. Am J Cardiol. 2000;86:975–81.
86. Miranda CM, Figueiredo AB, Maciel BC, Marin-Neto JA, Simões MV. Sustained ventricular tachycardia is associated with regional myocardial sympathetic denervation assessed with 123I-metaiodobenzylguanidine in chronic Chagas cardiomyopathy. J Nucl Med. 2011;52:504.
87. Miranda CH, Gadioli LP, Pintya AO, Figueiredo AB, Maciel BC, Schmidt A, Marin-Neto JA. The severity of ventricular arrhythmia correlates with the extent of myocardial sympathetic denervation, but not with myocardial fibrosis extent in chronic Chagas cardiomyopathy. Am J Nucl Med. 2018;25(1):75–83.
88. Cuba MB, Machado MPR, Farnesi TS, Alves AC, Martins LA, Oliveira LF, et al. Effects of cholinergic stimulation with pyridostigmine bromide on chronic chagasic cardiomyopathic mice. Mediat Inflamm. 2014:475946. 13 pages.
89. Rossi MA. Microvascular changes as a cause of chronic cardiomyopathy in Chagas' disease. Am Heart J. 1990;120:233.
90. Marin-Neto JA, Marzullo P, Marcassa C, Gallo L Jr, Maciel BC, Bellina CR, et al. Myocardial perfusion abnormalities in chronic Chagas' disease as detected by thallium-201 scintigraphy. Am J Cardiol. 1992;69:780.
91. Hiss FC, Lascala TF, Maciel BC, Marin-Neto JA, Simões MV. Changes in myocardial perfusion correlate with deterioration of left ventricular systolic function in chronic Chagas' cardiomyopathy. JACC Cardiovasc Imaging. 2009;2:164.

92. Rossi MA, Tanowitz HB, Malvestio LM, Celes MR, Campos EC, Blefari V, et al. Coronary microvascular disease in chronic Chagas cardiomyopathy including an overview on history, pathology, and other proposed pathogenic mechanisms. PLoS Negl Trop Dis. 2010;4(8):e674. https://doi.org/10.1371/journal.pntd.0000674. Review.
93. Freeman BD, Machado FS, Tanowitz HB, Desruisseaux MS. Endothelin-1 and its role in the pathogenesis of infectious diseases. Life Sci. 2014; https://doi.org/10.1016/j.lfs.2014.04.0210.
94. Torres FW, Acquatella H, Condado JA, Dinsmore R, Palacios IF. Coronary vascular reactivity is abnormal in patients with Chagas' heart disease. Am Heart J. 1995;129:995.
95. Marin-Neto JA, Simões MV, Rassi JR A. Pathogenesis of chronic Chagas cardiomyopathy: the role of coronary microvascular derangements. Rev Soc Bras Med Trop. 2013;46(4). https://doi.org/10.1590/0037-8682-0028-2013.
96. Sambiase NV, Higuchi ML, Benvenuti LA. Narrowed lumen of the right coronary artery in chronic Chagasic patients is associated with ischemic lesions of segmental thinnings of ventricles. Investig Clin. 2010;51(4):531.
97. Lemos de Oliveira LF, Thackeray JT, Marin Neto JA, Dias Romano MM, Vieira de Carvalho EE, Mejia J, Tanaka DM, Kelly da Silva G, Abdalla DR, Malamut C, Bengel FM, de Lourdes Higuchi M, Schmidt A, Cunha-Neto E, Simões MV. Regional myocardial perfusion disturbance in experimental chronic Chagas cardiomyopathy. J Nucl Med. 2018;59(9): 1430–6.
98. Tanaka DM, de Oliveira LFL, Marin-Neto JA, Romano MMD, de Carvalho EEV, de Barros Filho ACL, Ribeiro FFF, Cabeza JM, Lopes CD, Fabricio CG, Kesper N, Moreira HT, Wichert-Ana L, Schmidt A, Higuchi ML, Cunha-Neto E, Simões MV. Prolonged dipyridamole administration reduces myocardial perfusion defects in experimental chronic Chagas cardiomyopathy. J Nucl Cardiol. 2018; https://doi.org/10.1007/s12350-018-1198-7. [Epub ahead of print].
99. Macedo LGR, Lemos DC, Lago IM, Figueiredo GL, Lima Filho MO, Schmidt A, Marin-Neto JA. Desenho de Estudo. Base racional e plano de estudo prospectivo para avaliar o efeito de terapêutica antiplaquetária e vasodilatadora microcirculatória em pacientes com cardiopatia chagásica crônica e distúrbios microvasculares coronários. Rev Bras Cardiol Invasiva. 2012;20(1):82.
100. Rossi MA. The pattern of myocardial fibrosis in chronic Chagas' heart disease. Int J Cardiol. 1991;30:335.
101. Bellotti G, Bocchi EA, de Moraes AV, Higuchi ML, Barbero-Marcial M, Sosa E, et al. In vivo detection of Trypanosoma cruzi antigens in hearts of patients with chronic Chagas' heart disease. Am Heart J. 1996;131:301.
102. Higuchi ML, De Brito T, Reis MM, Barbosa A, Bellotti G, Pereira-Barreto AC, et al. Correlation between T.cruzi parasitism and myocardial inflammatory infiltrate in human chronic chagasic myocarditis: light microscopy and immunohistochemical findings. Cardiovasc Pathol. 1993;2:101.
103. Tarleton RL. Trypanosoma cruzi and Chagas disease: cause and effect. In: Tyler KM, Miles MA, editors. World class parasites: American trypanosomiasis, vol. 7. Dordrecht: Kluwer Academic Publisher; 2003. p. 107.
104. Kierszenbaum F. Mechanisms of pathogenesis in Chagas disease. Acta Parasitol. 2007;52:1.
105. Bonney KM, Engman DM. Chagas heart disease pathogenesis: one mechanism or many? Curr Mol Med. 2008;8:510.
106. Zhang L, Tarleton RL. Parasite persistence correlates with disease severity and localization in chronic Chagas'disease. J Infect Dis. 1999;180:480.
107. Bustamante JM, Rivarola HW, Fernández AR, Enders JE, Fretes R, Palma JA, et al. Trypanosoma cruzi reinfections in mice determine the severity of cardiac damage. Int J Parasitol. 2002;32:889.
108. Storino R, Auger S, Caravello O, Urrutia MI, Sanmartino M, Jörg M. Chagasic cardiopathy in endemic area versus sporadically infected patients. Rev Saude Publica. 2002;36:755.
109. Andrade SG, Stocker-Guerret S, Pimentel AS, Grimaud JA. Reversibility of cardiac fibrosis in mice chronically infected with Trypanosoma cruzi, under specific chemotherapy. Mem Inst Oswaldo Cruz. 1991;86:187.

110. Garcia S, Ramos CO, Senra JF, Vilas-Boas F, Rodrigues MM, Campos-de-Carvalho AC, et al. Treatment with benznidazole during the chronic phase of experimental Chagas' disease decreases cardiac alterations. Antimicrob Agents Chemother. 2005;49:1521.
111. de Andrade AL, Zicker F, de Oliveira RM, Almeida Silva S, Luquetti A, Travassos LR, Almeida IC, de Andrade SS, de Andrade JG, Martelli CM. Randomised trial of efficacy of benznidazole in treatment of early Trypanosoma cruzi infection. Lancet. 1996;348(9039):1407–13.
112. Sosa-Estani S, Segura EL, Ruiz AM, Velazquez E, Porcel BM, Yampotis C. Efficacy of chemotherapy with benznidazole in children in the indeterminate phase of Chagas' disease. Am J Trop Med Hyg. 1998;59(4):526–9.
113. Viotti RJ, Vigliano C, Lococo B, Bertocchi G, Petti M, Alvarez MG, et al. Long-term cardiac outcomes of treating chronic Chagas disease with benznidazole versus no treatment: a non-randomized trial. Ann Intern Med. 2006;144:724.
114. Fabbro DL, Streiger ML, Arias ED, Bizai ML, del Barco M, Amicone NA. Trypanocide treatment among adults with chronic Chagas disease living in Santa Fe city (Argentina), over a mean follow-up of 21 years: parasitological, serological and clinical evolution. Rev Soc Bras Med Trop. 2007;40:1.
115. Pérez-Molina JA, Pérez-Ayala A, Moreno S, Fernández-González MC, Zamora J, López-Velez R. Use of benznidazole to treat chronic Chagas' disease: a systematic review with a meta-analysis. J Antimicrob Chemother. 2009;64(6):1139–47.
116. Rassi A Jr, Marin-Neto JA, Rassi A. Chronic Chagas cardiomyopathy: a review of the main pathogenic mechanisms and the efficacy of aetiological treatment following the BENznidazole Evaluation for Interrupting Trypanosomiasis (BENEFIT) trial. Mem Inst Oswaldo Cruz. 2017;112(3):224–35.
117. Morillo CA, Marin-Neto JA, Avezum A, Sosa-Estani S, Rassi A Jr, Rosas F, Villena E, Quiroz R, Bonilla R, Britto C, Guhl F, Velazquez E, Bonilla L, Meeks B, Rao-Melacini P, Pogue J, Mattos A, Lazdins J, Rassi A, Connolly SJ, Yusuf S, BENEFIT Investigators. Randomized trial of benznidazole for chronic Chagas' cardiomyopathy. N Engl J Med. 2015;373(14):1295–306.
118. Jones EM, Colley DG, Tostes S, Lopes ER, Vnencak-Jones CL, McCurley TL. A Trypanosoma cruzi DNA sequence amplified from inflammatory lesions in human chagasic cardiomyopathy. Trans Assoc Am Phys. 1992;105:182.
119. Salomone OA, Juri D, Omelianiuk MO, Sembaj A, Aguerri AM, Carriazo C, et al. Prevalence of circulating Trypanosoma cruzi detected by polymerase chain reaction in patients with Chagas' cardiomyopathy. Am J Cardiol. 2000;85:1274.
120. Andrade JP, Marin Neto JA, Paola AA, Vilas-Boas F, Oliveira GM, Bacal F, Bocchi EA, Almeida DR, Fragata Filho AA, Moreira Mda C, Xavier SS, Oliveira Junior WA, Dias JC. I. Latin American Guidelines for the diagnosis and treatment of Chagas' heart disease: executive summary. Arq Bras Cardiol. 2011;96(6):434–42.
121. Dias JC, Ramos AN Jr, Gontijo ED, Luquetti A, Shikanai-Yasuda MA, Coura JR, Torres RM, Melo JR, Almeida EA, Oliveira W Jr, Silveira AC, Rezende JM, Pinto FS, Ferreira AW, Rassi A, Filho AAF, Sousa AS, Correia D, Jansen AM, Andrade GM, Britto CF, Pinto AY, Rassi A Jr, Campos DE, Abad-Franch F, Santos SE, Chiari E, Hasslocher-Moreno AM, Moreira EF, Marques DS, Silva EL, Marin-Neto JA, Galvão LM, Xavier SS, Valente SA, Carvalho NB, Cardoso AV, Silva RA, Costa VM, Vivaldini SM, Oliveira SM, Valente VD, Lima MM, Alves RV. 2 nd Brazilian Consensus on Chagas Disease, 2015. Rev Soc Bras Med Trop. 2016;49(Suppl 1):3–60. https://doi.org/10.1590/0037-8682-0505-2016. Erratum in: Rev Soc Bras Med Trop. 2017 Jan–Feb;50(1):149.
122. Zingales B, Miles MA, Campbell DA, Tibayrenc M, Macedo AM, Teixeira MMG, et al. The revised *Trypanosoma cruzi* subspecific nomenclature: rationale, epidemiological relevance and research applications. Infect Genet Evol. 2012;12:240.
123. Zingales B, Miles MA, Moraes CB, Luquetti A, Guhl F, Schijman AG, et al. Drug discovery for Chagas disease should consider Trypanosoma cruzi strain diversity. Mem Inst Oswaldo Cruz. 2014;109(6):828.
124. Marin-Neto JA, Rassi Jr A, Maciel BC. Pathology and pathogenesis of Chagas disease. In: UpToDate; 2014.

125. Minoprio P. Parasite polyclonal activators: new targets for vaccination approaches? Int J Parasitol. 2001;31:588.
126. Cunha-Neto E, Bilate AM, Hyland KV, Fonseca SG, Kalil J, Engman DM. Induction of cardiac autoimmunity in Chagas heart disease: a case for molecular mimicry. Autoimmunity. 2006;39(1):41. Review.
127. Teixeira AR, Gomes C, Nitz N, Sousa AO, Alves RM, Guimaro MC, et al. Trypanosoma cruzi in the chicken model: Chagas-like heart disease in the absence of parasitism. PLoS Negl Trop Dis. 2011;5(3):e1000. https://doi.org/10.1371/journal.pntd.0001000.
128. Tarleton RL. Chagas disease: a role for autoimmunity? Trends Parasitol. 2003;19:447.
129. Tarleton RL, Zhang L. Chagas disease etiology: autoimmunity or parasite persistence? Parasitol Today. 1999;15:94–9.
130. Tarleton RL, Zhang L, Downs MO. "Autoimmune rejection" of neonatal heart transplants in experimental Chagas disease is a parasite-specific response to infected host tissue. Proc Natl Acad Sci U S A. 1997;94:3932.
131. Soares MB, Pontes-De-Carvalho L, Ribeiro-Dos-Santos R. The pathogenesis of Chagas' disease: when autoimmune and parasite-specific immune responses meet. An Acad Bras Cienc. 2001;73:547.
132. Rassi A, Amato Neto V, de Siqueira AF, Doles J, Leite MS, Silva OQ, et al. The influence of corticoids, in chronic Chagas disease, administered in virtue of associated disorders. Rev Soc Bras Med Trop. 1997;30:93.
133. Sartori AM, Ibrahim KY, Nunes Westphalen EV, Braz LM, Oliveira OC Jr, Gakiya E, et al. Manifestations of Chagas disease (American trypanosomiasis) in patients with HIV/AIDS. Ann Trop Med Parasitol. 2007;101:31.
134. Dutra WO, Gollob KJ. Current concepts in immunoregulation and pathology of human Chagas disease. Curr Opin Infect Dis. 2008;21(3):287.
135. Ferreira LRP, Frade AF, Baron MA, Navarro IC, Kalil J, Chevillard C, et al. Interferon-γ and other inflammatory mediators in cardiomyocyte signaling during Chagas disease cardiomyopathy. World J Cardiol. 2014;6(8):782.
136. Nogueira LG, Santos RH, Fiorelli AI, Mairena EC, Benvenuti LA, Bocchi EA, et al. Myocardial gene expression of T-bet, GATA-3, Ror-γt, FoxP3, and hallmark cytokines in chronic Chagas disease cardiomyopathy: an essentially unopposed TH1-type response. Mediat Inflamm. 2014;2014:914326, . 9 pages. https://doi.org/10.1155/2014/914326.
137. Dias FC, Medina Tda S, Mendes-Junior CT, Dantas RO, Pissetti CW, Rodrigues Junior V, et al. Polymorphic sites at the immunoregulatory CTLA-4 gene are associated with chronic Chagas disease and its clinical manifestations. PLoS One. 2013;8(10):e78367. https://doi.org/10.1371/journal.pone.0078367. eCollection 2013.
138. Frade AF, Teixeira PC, Ianni BM, Pissetti CW, Saba B, Wang LH, et al. Polymorphism in the alpha cardiac muscle actin 1 gene is associated to susceptibility to chronic inflammatory cardiomyopathy. PLoS One. 2013;8(12):e83446. https://doi.org/10.1371/journal.pone.0083446. eCollection 2013.
139. Cunha-Neto E, Chevillard C. Chagas disesase cardiomyopathy: immunopathology and Genetics. Mediat Inflamm. 2014;2014:683230.
140. Álvarez JM, Fonseca R, Borges da Silva H, Marinho CR, Bortoluci KR, Sardinha LR, et al. Chagas disease: still many unsolved issues. Mediat Inflamm. 2014;2014:912965, . 9 pages. https://doi.org/10.1155/2014/912965.
141. Chagas C. Tripanosomiase americana. Forma aguda da moléstia. Mem Inst Oswaldo Cruz. 1916;8:37–60.
142. Neiva A, Penna B. Viagem científica pelo norte da Bahia, sudeste do Piauí e de norte a sul de Goias. Mem Inst Oswaldo Cruz. 1916;8:74–224.
143. Schmunis GA, Yadon ZE. Chagas disease: a Latin américa health problem becoming a world health problem. Acta Trop. 2010;115:14–21.
144. Perez-Ayala A, Perez-Molina JA, Norman F, Maillo-Merige B, Faro MV, Lopez-Velez R. Gastrointestinal Chagas disease in migrants to Spain. Prevalence and methods for early diagnosis. Ann Trop Med Parasitol. 2011;105:23–9.

145. Dias JCP. Doença de Chagas em Bambuí, Minas Gerais, Brasil. Estudo clínico-epidemiológico a partir da fase aguda, entre 1940 e 1982. Tese de Doutoramento, Universidade Federal de Minas Gerais, Belo Horizonte; 1982.
146. Dias JCP, Camacho LAB, Silva JC, Magalhães JS, Krieger H. Esofagopatia chagásica na área endêmica de Bambui, MG, Brasil. Rev Soc Bras Med Trop. 1983;16:46–57.
147. Rassi A, Rezende JM. Estudo clínico-radiológico do esôfago e dos cólons na fase aguda da doença de Chagas com relato de tres casos de remissão espontânea da aperistalse do esôfago do grupo I. Rev Soc Bras Med Trop. 2011;44(1):70–5.
148. Hagger R, Finlayson C, Kahn F, De Oliveira R, Chimelli L, Kumar D. A deficiency of interatitial cells of Cajal in Chagasic megacolon. J Auton Nerv Syst. 2000;80:108–11.
149. Iantorno G, Bassotti G, Kogan Z, Lumi CM, Cabanne AM, Fisogni S, et al. The enteric nervous system in chagasic and idiopathic megacolon. Am J Surg Pathol. 2007;31:460–8.
150. Godoy RA, Vieira CB. Effects of cholinergic drugs on the esophagus of patients with Chagas' disease. Acta Physiol Latinoam. 1961;11:107.
151. Meneghelli UG, Dantas RO, Godoy RA. Effect of caerulein on the lower esophageal sphincter in chagasic esophagopathy. Book of abstracts of the 4th Congress of Organization d'Etudes Statistiques pour les maladies de 'Oesophage (OESO), 1993, Paris, p 180.
152. Oliveira RB, Troncon LEA, Meneghelli UG, Padovan W, Dantas RO, Godoy RA. Impaired gastric accommodation to distension and rapid gastric emptying in patients with Chagas' disease. Dig Dis Sci. 1980;25:790–4.
153. Meneghelli UG. Motilidade do sigmoide e do reto de portadores da moléstia de Chagas em condições basais e sob a ação da pentagastrina. Tese: Faculdade de Medicina de Ribeirão Preto da Universidade de São Paulo; 1977.
154. Troncon LEA, Oliveira RB, Meneghelli UG, Dantas RO, Godoy RA. Fasting and food-stimulated plasma gastrin levels in chronic Chagas disease. Digestion. 1984;29:171–6.
155. Adad SJ, Andrade DCS, Lopes ER, Chapadeiro E. Contribuição ao estudo da anatomia patológica do megaesôfago chagásico. Rev Inst Med Trop São Paulo. 1991;33:443–50.
156. D'Avila Reis D, Lemos EM, Silva GC, Adad SJ, McCurley T, et al. Phenotypic characterization of the inflammatory cells in chagasic mega-esophagus. Trans R Soc Trop Med Hyg. 2001;95:177–8.
157. Da Silveira ABM, Arantes RME, Vago AR, Lemos EM, Adad SJ, et al. Comparative of the presence of Trypanosoma cruzi kDNA, inflammation and denervation in chagasic patients with and without megaesophagus. Parasitology. 2005;131:627–34.
158. Da Silveira ABM, Adad SJ, Correa-Oliveira R, Furness JB, Dávila RD. Morphometric study of eosinophils, mast cells, macrophages, and fibrosis in the colono of chronic chagasic patients with and without megacolon. Parasitology. 2007;134:789–96.
159. Martins PR, Nascimento RD, Lopes JG, Santos MM, Oliveira CA, et al. Mast cell in the colon of *Trypanosoma cruzi*-infected patients: are they involved in the recruitment, survival and/or activation of eosinophils? Parasitol Res. 2015;114:1847–56.
160. Dias FC, Medina TDS, Mendes-Junior CT, Dantas RO, Pissetti CW, Rodrigues Junior V, Dellalibera-Joviliano R, Marin-Neto JÁ, et al. Polymorphic sites at the immunoregulatory CTLA-4 gene are associated with chronic Chagas disease and its clinical manifestations. PLoS One. 2013;8:e78367.
161. http://siteresources.worldbank.org/BRAZILINPOREXTN/Resources/3817166-1302102548192/Brazil_Aging_Full_Eng_final.pdf.
162. Meneguelli UG, Ejima FH, Rosa e Silva L. Evidências do declínio da ocorrência do megaesôfago e do megacólon: Estudo epidemiológico no Hospital das Clínicas de Ribeirão Preto. Medicina Ribeirão Preto. 1991;24:218–24.
163. Kamiji MM, Oliveira RB. O perfil dos portadores de doença de Chagas com ênfase na forma digestiva, em hospitsl terciário de Ribeirão Preto, SP. Ver Soc Bras Med Trop. 2005;38:305–9.
164. Souza DHS, Vaz MGM, Fonseca CR, Luquetti A, Rezende-Filho J, Oliveira EC. Current epidemiological profile of Chagasic megaesophagus in Central Brazil. Rev Soc Bras Med Trop. 2013;46:316–21.
165. Rassi A Jr, Rassi A, Marin-Neto JA. Chagas disease. (Review). Lancet. 2010;375:1388–402.

166. Macedo VO. Influencia da exposição à reinfecção na evolução da doença de Chagas. Rev Patol Trop. 1976;5:33–116.
167. Castro C, Rezende JM, Camargo C, Prata A, Macêdo V. Prevalência de esofagopatia chagásica no Município de Mambaí, Goiás—Brasil. Rev da Soc Bras Med Trop. 1987;20:13–7.
168. Pereira JB, Coura JR. Morbidade da doença de Chagas: estudo seccional em uma região endêmica, Virgem da Lapa, MG. Rev Soc Bras Med Trop. 1986;19:138–48.
169. Coura JR, Borges-Pereira J, Alves Filho FI, Castro JAF, Cunha RV, et al. Morbidade da doença de Chagas em áreas do sertão da Paraíba e da catinga do Piauí. Rev Soc Bras Med Trop. 1996;29:197–205.
170. Penaranda-Carrillo R, Castro C, Rezende J, Prata A, Macedo A. Radiographic study of the esophagus of chagasic patients in 25 years of the Mambaí Project. Rev Soc Bras Med Trop. 2006;39:152–5.
171. Coura JR, Abreu LL, Pereira JB, Willcox HP. Morbidade da Doença de Chagas: IV. Longitudinal study of 10 years in Pains and Iguatama, Minas Gerais, Brasil. Mem Inst O Cruz. 1985;80:73–80.
172. Borges Pereira J, Willcox HP, Coura JR. Morbidade da doença de Chagas. III. Estudo longitudinal de seis anos, em Virgem da Lapa, MG, Brasil. Mem Inst O Cruz. 1985;80:63–71.
173. Macedo VO, Silveira CA. Perspectiva de terapêutica específica na doença de Chagas—Experiência na forma indeterminada. Rev Soc Bras Med Trop. 1987;20(supl):24–6.
174. Castro C, Macedo V, Rezende JM, Prata A. Estudo radiológico longitudinal do esôfago em área endêmica de doença de Chagas, em um período de 13 anos. Rev Soc Bras Med Trop. 1994;27:227–33.
175. Penaranda-Carrillo R, Moreira E, Silveira A, Leite J, Vinhaes M, et al. Evaluation of the impact of vector control programs through serological testings in Mambaí, Buritinópolis, Goias State. Rev Soc Bras Med Trop. 2002;35:331–8.
176. Pereira JB, Coura JR. Morbidade da Doença de Chagas. Estudo Seccional em uma área Endêmica, Virgem da Lapa, MG. Rev Soc Bras Med Trop. 1986;19:139–48.
177. Kahrilas PJ, Bredenoord AJ, Carlson DA, Pandolfino JE. Advances in management of esophageal motility disorders. Clin Gastroenterol Hepatol. 2018;16:1692–700.
178. Basile L, Jansa JM, Carlier Y, Salamanca DD, Angheben A, et al. Chagas'disease in European countries: the challenge of a surveillance system. Euro Surveill. 2011;16(37):19968.
179. Perez-Ayala A, Perez-Molina JA, Norman F, Maillo-Monge B, Faro MV, Lopez-Velez R. Gastrointestinal Chagas disease in migrants to Spain. Prevalence and methods for early diagnosis. Ann Tro Med Parasitol. 2011;105:23–9.
180. Remes-Troche JM, Torres-Aguilera M, Antonio-Cruz KA, Vazquez-Jimenez G, De-La-Cruz-Patiño E. Esophageal motor disorders in subjects with incidentally discovered Chagas disease: a study using high-resolution manometry and the Chicago classification. Dis Esophagus. 2014;27:24–529.
181. Koberle F, Britto-Costa R, Mello de Oliveira JA, et al. Patologia da moléstia de Chagas. Medicina Ribeirão Preto. 1972;5:5–45.
182. Castro C, Hernandez EB, Rezende J, Prata A. Estudo radiológico do megacólon em área endêmica de doença de Chagas. Rev Soc Bras Med Trop. 2010;43:562–6.
183. Castro C, Hernandez EB, Rezende J, Prata A. Occurrence of dolichocolon without megacolon in chronic Chagas disease patients. Rev Soc Bras Med Trop. 2012;45:353–6.
184. Ximenes CA, Rezende JM, Moreira H, et al. Técnica simplificada para o diagnóstico radiológico do megacólon chagásico. Rev Soc Bras Med Trop. 1984;17(Suppl):23.
185. Sanches-Montalvá A, Moris M, Mego M, Salvador F, Accarino A, Ramírez K, et al. High resolution esophageal manometry in patients with Chagas disease: a cross-sectional evaluation. PLoS Negl Trop Dis. 2016;10:e0004416.

Chagas Disease Epidemiology: From Latin America to the World

2

Belkisyolé Alarcón de Noya and Yves Jackson

2.1 Introduction

American Trypanosomiasis evokes terms such as zoonosis, ancient, rural, poverty, ranches, palm roofs, chinches, chipos, vinchucas, Latin America, heart disease, lack of effective treatment, pacemaker, etc. It is the "traditional" epidemiology of the infection of man by the parasite *Trypanosoma cruzi* acquired through contact with infected hematophagous triatomines. After sucking human blood, they deposit feces containing very infective metacyclic trypomastigotes that invade skin cells liberating sanguineous trypomastigotes. The latter disseminate and deposit into peripheral tissues such as myocardium, fat, and digestive tract mucosa as amastigotes. Diverse immunopathological mechanisms lead to chronic organ damages.

T. cruzi infection is a consequence of different mechanisms of transmission. Two vectorial-related mechanisms are known, the transcutaneous and the oral routes, the latter following contamination of food with feces of the infected vector or with the incorporation of the vector itself into food. In addition, man-to-man infection occurs through trans-placental transmission, or by contaminated blood product, organ or tissue transfusion or transplant. Less frequently, the accidental form by manipulation of *Didelphis* sp. and infection in the laboratory due to inappropriate handling of infectious biological material [1].

The epidemiology of CD is rapidly evolving and changes have occurred both in and out of traditional endemic areas, i.e., in rural parts of North, Central, and South America. Since the mid-twentieth century, vast flux of migrations pushed millions of

B. Alarcón de Noya (✉)
Instituto de Medicina Tropical, Facultad de Medicina, Universidad Central de Venezuela, Caracas, Venezuela

NHEPACHA Network, Barcelona, Spain

Y. Jackson
Geneva University Hospital and University of Geneva, Geneva, Switzerland
e-mail: yves.jackson@hcuge.ch

M.-J. Pinazo Delgado, J. Gascón (eds.), *Chagas Disease*,
https://doi.org/10.1007/978-3-030-44054-1_2

people from rural to urban regions in the context of socioeconomic gradients, political factors, and environmental changes [2]. Following their human hosts, the traditional rural and sylvatic vectors adapted to new environments and new cycles of transmission emerged in so far non-endemic areas of Latin America [3]. Consequently, peri-urban vectorial transmission of *T. cruzi* has been recorded in Argentina [4, 5], Bolivia [6], Brazil [7], Colombia [8], Peru [9], and Venezuela [10]. One consequence was the advent of urban outbreaks following transmission through food-related oral route [11].

Although the transmission mechanisms of the CD are diverse and versatile, Control Programs in Latin America have been mainly based on vector elimination through chemical agents spraying and more recently blood donation screening along with enhanced detection of mother-to-child transmission with early child treatment. Vectorial transmission by the main vector *Triatoma infestans* has decreased in recent decades and even potentially eradicated in some countries such as Chile, Uruguay, Brazil, eastern Paraguay, and in some regions of Argentina [12, 13]. However, vectorial transmission remains very active in Bolivia, Peru, Ecuador, Colombia, Venezuela, the Guyana, large parts of the Central American countries, and Mexico where Vector Control Programs have not been fully consolidated [13]. In absence of sustained monitoring and capacity to rapidly respond, it is likely that currently controlled areas could be re-infested [14].

In addition to the abovementioned changes, the last decades witnessed dramatic changes in transnational human mobility due to a set of factors, including globalization and the development of mass transportation systems. Indeed, large numbers of people moved out of CD endemic areas for political, socio-economical, education-related, or other reasons and settled in countries worldwide previously unaffected and unprepared for tackling this new global health issue.

2.2 Distribution and Burden of the Disease in Non-endemic Countries

The exact geographical distribution of CD and its epidemiology in America have been difficult to determine given the paucity of public health attention, and the limitations of the existing surveillance systems. Despite progress, epidemiological data remain mainly based on assumptions and models with hardly no real-life data at country levels. WHO estimated that 17.4 million persons were infected in 18 countries in 1985 with 100 million (25% of the total population) at risk of infection [15]. Thanks to improvement in living conditions, multi-pronged control programs discussed above and improved access to treatment, the prevalence has been reduced to 8–10 million in 2005 and 6–7 million in 2010 with more than 100 million persons persistently at risk of infection [16, 17]. Argentina, Brazil, Mexico, and Bolivia host the largest numbers of persons infected. The latest estimates suggest that 40,000 new infections and 15,000 deaths of CD occur every year.

In recent times, millions of people at risk have moved across international borders. This trend has accelerated since 1990 and has contributed to disseminate the infection to vector-free, subsequently called non-endemic, regions such as North

America, Japan, Australia, and Western Europe [18]. As for endemic regions, the lack of surveillance systems decreases the ability to provide a comprehensive understanding of the global epidemiological situation. Most estimates rely on epidemiological models based on the number of migrants from each country at risk multiplied by the average infection rates in those countries [17, 19–21].

In Europe, Spain has received the largest number of migrants from endemic areas, but Italy, France, Portugal, Switzerland, and the United Kingdom also host large communities. Initially, most cases were identified in adult migrants suffering severe cardiac damages or reactivation following immunosuppression [22]. In the late 1990s and the early next decade, an increasing number of *T. cruzi* transmission through congenital route and blood transfusion were reported in Europe and the USA [23–25]. This supported the idea of this health issue spreading outside Latin America and to the framing of the concept of non-endemic countries requiring specific responses [15]. Yet, in absence of epidemiological data outside Latin America, it took years before the first public health responses started to be implemented in non-endemic countries with Spain leading the way, soon followed by neighboring countries. To date, Europe has been more responsive to this emerging health problems compared to the USA, Japan, or Australia. The latest estimates pointed to 68,000–123,000 persons infected in Europe [19] and 200,000–300,000 in the USA [26] while Japan, Canada, Australia, and New Zealand were likely to host only a limited number of cases [21].

Given its chronicity and the potentially severe cardiac consequences, CD entails an economic burden of $7 billion per year, similar to or worse than other well-known diseases like rotavirus or cervical cancer [27].

2.3 Factors Affecting the Epidemiology of Chagas Disease in Europe

The origin and geographical clustering of sub-groups of migrants in Europe has had a strong influence over the distribution of cases. Migrant communities tend to cluster geographically according to origin for different reasons, including better support for new arrivants. In addition, Latin American communities have mainly settled in urban areas because of enhanced job activities opportunities. Indeed, migrants from Bolivia and Paraguay show the highest prevalence of infection [28]. In almost all seroprevalence studies, migrants coming from highly endemic areas of Bolivia account for 80% or more of cases.

Spain and Portugal for historical, cultural, political, and economic reasons have been the favored places of entry for the majority of migrants at the turn of the century. Many have acquired European Passport facilitating their integration into the country of destination. Yet, large numbers of them moved out of these countries following the 2008 financial crisis and either returned home or settled in other European countries, highlighting the dynamic, cyclical, and multi-step contemporary migratory patterns. Later on, the rise of restrictive immigration policies and the anti-immigrants rhetoric prevailing in many European countries have also influenced the mobility and distribution of communities. A key aspect pertains to the vulnerability of labor migrants in Europe who frequently lacked access to social and financial security, usually employed in 3-D (dirty, degrading, and dangerous) and poorly paid jobs. In terms of access to care, migrant's ability to benefit from the public or private health sector widely differs from country to country and across time. Spain for example had initially rather liberal policies until a Royal Decree in 2012 severely limited access to comprehensive health services. In Switzerland, most migrants at risk have no residency permit (undocumented) and the monthly income they generate is not sufficient to cover the cost of the mandatory private health insurance which regulates access to care. In some regions, consulting at a public hospital may represent a consequent danger as health authorities have to denounce irregular migrants to the immigration departments. In many settings, language barriers, the need for out-of-pocket direct payment, geographical distances, and lack of knowledge about how to navigate the healthcare system have contributed restricting the access to medical care.

European host countries were largely unprepared for addressing the new public health and clinical challenges pertaining to the emergence of CD at the turn of the twentieth century. Most countries lacked policy regarding blood donor screening and congenital transmission screening. Still in 2018, most countries hosting substantial numbers of migrants at risk have limited diagnostic capacities and lack scheme facilitating access to anti-parasitic drugs outside the main tertiary health centers. Moreover, health professional's awareness of CD and knowledge about the optimal preventive, diagnostic, and therapeutic strategies remain very limited in Europe. As a consequence, services to communities at risk are mainly delivered by tertiary health centers in the main cities in Europe, complemented with non-governmental organizations in other parts which is largely insufficient to cover the global needs. Of note, specific programs have flourished in a few cities in Spain, Italy, Germany, and Switzerland which usually combine both social and medical strategies to improve the access to diagnosis and treatment.

Cultural representations about CD, which both take roots in the culture of origin and evolve with the new experiences overseas, impact on the health seeking behavior of people at risk [29]. Frequently associated with low social position, rurality, and lack of education which might lead to stigmatization of affected people and perceived with fatalism as an incurable infection, it tends to rank low in peoples' health priorities, especially in the context of socioeconomic vulnerability. All these factors combine to explain the low number of cases which have been diagnosed and received medical attention, including treatment, in Europe so far. Indeed, experts estimate that less than 10% of people affected by CD in non-endemic regions have been identified to date, which contributes to fuel the ongoing vertical transmission and highlights the need to implement multi-pronged strategies to curb *T. cruzi* transmission and its toll on human health.

2.4 Risk of Vectorial Transmission in Non-endemic Areas

The most important route of *T. cruzi* transmission worldwide remains vectorial despite the increasing relative importance of congenital and transfusion/transplant routes in areas where vectorial control programs have been deployed. The area of endemicity of CD in America relates to the spatial distribution of hematophagous triatomines of the genera *Triatoma*, *Rhodnius*, and *Panstrongylus*. The disease is maintained in nature as an enzootia. Recently, the rapid encroachment of urban development into wild land, the development of intensive agricultural practices, and the adaptation of vectors to peri-domiciliary and domiciliary environments have increased human contacts with vectors [30]. *Triatoma infestans* is the main vector in South America, probably originated in Bolivia. Its distribution has extended from northeastern Brazil to southern Argentina with some areas under vector control. The same dispersion has occurred with *Rhodnius prolixus*: originating from Colombia and Venezuela, it extended to The Andean region, the Guyana, and Central America [30].

The presence of hematophagous triatomines infected with *T. cruzi* (notably *T. sanguisuga*) has been reported in various Southern States of the United States of America. Zoonotic transmission has led to sporadic autochthonous human cases. Many indigenous triatomine species are susceptible to *T. cruzi* infection, thus making a source of potential vectors to human [31].

In the Old World, eight species of *Triatoma* have been identified as having a potential of transmission to humans. *Triatoma rubrofasciata* has the largest area of distribution worldwide. It spans across North (Mexico, Florida) and South America (Argentina, Brazil, Cuba and most Caribbean islands, French Guiana, Suriname, and Venezuela), Africa (Angola, Democratic Republic of Congo, Guinea-Conakry, Sierra Leone, South Africa, Tanzania, Madagascar, Mauritius, Rodriguez Islands, Seychelles), The Middle East (Saudi Arabia), South Asia and the Western Pacific Region (India-Tamil Nadu, China, Indonesia, Malaysia, Sri Lanka, Singapore, Japan, Philippines, Taiwan, Thailand, Vietnam, Andaman Islands, Tonga, Burma-Myanmar, Cambodia, Carolina Islands, Comoros Islands, Hawaii) [32, 33]. High density of this vector has been found in some urban areas causing frequent stings to humans. It is likely that cargo ship represents the main mode of transportation from the New World to other continents [33].

In recent years, evidence have shown that bedbugs (*Cimex* spp.), that have widely disseminated across the world, may be susceptible to carry *T. cruzi* and to transmit it to mice in experimental conditions [34]. *T. cruzi* is able to survive inside *Cimex* organisms and throughout the insect's molting process guaranteeing the transstadial persistence of *T. cruzi* [35]. Of concern is the fact that *Cimex* has developed resistance to insecticides such as pyrethroid [36].

2.5 Public Health Responses of Non-endemic Countries

The Chagas Disease Control Programs of the different Latin American countries have been supported by the Pan American Health Organization (PAHO) since 1991 and based on the regions diversity four different control initiatives were designated, namely the Southern Cone Initiative, the Initiative of the Andean Region, the Amazonian and the Initiative of Central America [37]. However, since 1981, an increasing number of sporadic cases of CD has occurred in other latitudes which led to the proposal of a WHO-led "Non-endemic Countries Initiative" [37]. Epidemiological surveillance activities have been progressively implemented in blood banks in the United Kingdom (1999), Spain and Italy (2005), the USA (2007), France (2009), and Switzerland (2013) [38, 39]. Of concern, several countries hosting large populations at risk still do not screen blood donors.

Some additional specific health policies targeting CD transmission exist at national or local level in Europe. National guidelines for solid organ transplant in Italy, Spain, and the United Kingdom specifically mention the need to screen for CD in donors and recipients at risk [40]. Three autonomous communities in Spain, one region in Italy (Tuscany), and some health institutions in neighboring countries have health policy for the screening of congenital transmission in pregnant women at risk and their babies [40]. In most of the European countries that do not have specific programs for CD, the rules and recommendations of the Council of Europe are followed [41]. Extra-European countries such as Australia and Japan have very limited policies, whereas the USA which hosts the largest number of infected people outside the endemic area has no other policy than blood donor screening [42].

We present below three examples of non-endemic countries with different risks of *T. cruzi* transmission to highlight the challenges pertaining to the control of the CD.

The responses in the United States of America are characterized by large heterogeneity across States and their limited scope. In 2007, blood donors screening was introduced in the USA. After 10 years, 2300 infected donors have been identified [43], mainly in Arizona, Massachusetts, Tennessee, and Texas [39]. According to the Center for Diseases Control (CDC), only seven states acted to control CD blood-donation transmission. During the 2008–2013 periods, Massachusetts complemented blood donors screening by a program of medical care to affected people but surveillance was discontinued in 2014. In 2008, there were reports of triatomines naturally infected with *T. cruzi* in Arizona, where active case detection and surveillance was implemented. Mississippi initiated actions in 2010 to determine if the cases detected in blood banks were caused by local autochthonous transmission or imported. Other States such as Tennessee, Arkansas, Louisiana, and Texas initiated active search for triatomines and naturally infected reservoirs along with medical attention to the infected persons [43]. The USA has an estimate of up to 315 congenital cases per year, but since there is no systematic search for congenital infection, it is likely that most cases remain undetected [43]. Overall, the response is largely insufficient to cover the transmission risk and to provide affected people with adequate medical care, especially in regions where most people at risk live such as California. One concern is that a significant proportion of people at risk has no health insurance and sometimes lacks residency status which hampers access to medical care. Moreover, access to diagnostic procedures and anti-parasitic treatment is severely restricted. A positive note is the registration of benznidazole for the treatment of children.

Spain hosts the largest community of Latin Americans in Europe [44, 45]. In 2011, it was estimated that 2,090,695 Latin Americans were living in Spain in the early twenty-first century, of which 47,738 and 67,423 could be infected with *T. cruzi* [19, 28, 42]. Spain implemented blood donors screening in 2005 and subsequently put into practice policies to identify and prevent transmission by organ and tissue donation and from mother to child. Yet, responses widely differ from region to region and do currently not allow for a full coverage of risk nationwide [28]. In different regions, community-based interventions have been deployed to increase the participation to screening programs. Like in the USA, access to preventive and curative care for the most vulnerable migrants has been threatened by changes in laws.

China has yet to identify the first CD case on its territory but it hosts populations of vectors with transmission potential. The National Institute of Parasitic Diseases at the Chinese Center for Disease Control and Prevention started research in southern China in 2016 based on the concern of the expansion of CD to non-endemic countries and the increasing mobility of potential infected Latin American migrants. A resident of Foshan city in Guangdong Province has found five adults and a nymph which were morphologically and molecularly identified as *T. rubrofasciata* [32]. This vector has been found in other latitudes naturally infected with *Trypanosoma cruzi* and *T. conorhini* [33], hence the growing concern about the possibility of establishment of infection in this very populated region of China. Epidemiological surveillance is based on vector surveillance since there are no health policies related to screening in blood banks, transplants, and congenital infection. The Center for Disease Prevention and Control in Guangdong province has appealed on social media for the public to look out for South American "kissing bugs," even offering free health consultations to the inhabitants who catch triatomines [46].

2.6 Final Remarks

Although the hematophagous triatomine vectors of the American Trypanosomiasis have disseminated to the world since ancient times and that an increasing number of countries worldwide report CD cases, the importance of this disease as a global public health issue has yet to be emerged, notably in Western non-endemic countries ill prepared to face such challenges.

As a testimony of CD inclusion in the list of neglected tropical diseases by the WHO, the public health and clinical challenges remain largely unmet by health authorities and health professionals, even more so in countries where migrant populations at risk suffer inequity in accessing to their social rights such as access to care. In this regard, tackling CD in non-endemic countries can be seen as an indicator of health equity.

References

1. Colmenares C, de Noya A, Noya O. Mechanisms of infection in Chagas disease. In: Alarcón de Noya B, Noya O, Robertson LJ, editors. *Trypanosoma cruzi* as a foodborne pathogen, Springer briefs in food, health and nutrition. New York: Springer; 2015. p. 21–32. ISBN: 978-3-319-23409-0.
2. Coura J, Albajar-Viñas P. Chagas disease: a new worldwide challenge. Nature. 2010;465:S6–7.
3. Reyes-Lugo M. *Panstrongylus geniculatus* Latreille 1811 (Hemiptera: Reduviidae: Triatominae), vector de la enfermedad de Chagas en el ambiente domiciliario del centro-norte de Venezuela. Rev Biomed. 2009;20:180–205.
4. Carrizo Paez R, Pickenhayn J, Carrizo Paez M. Chagas urbano en San Juan. Diagnóstico, revisión y propuesta para un sistema integrado de ataque. Rev Argent Cardiol. 2008;76:480–7.
5. Chuit R. La Enfermedad de Chagas en el Siglo XXI – Argentina. Primer Congreso Virtual de Cardiologia; 2000. p. 5.
6. Medrano-Mercado N, Ugarte-Fernandez R, Butron V, Uber-Busek S, Guerra HL, Araújo-Jorge TC, et al. Urban transmission of Chagas disease in Cochabamba, Bolivia. Mem Inst Oswaldo Cruz. 2008;103:423–30.
7. Conde Sangenis LH, Magalhães Saraiva R, Georg I, Castro L, dos Santos Lima V, Roque AL, et al. Autochthonous transmission of Chagas disease in Rio de Janeiro State, Brazil: a clinical and eco-epidemiological study. BMC Infect Dis. 2015;15:4–16.
8. Reyes M, Torres A, Esteban L, Flórez M, Angulo VM. Riesgo de transmisión de la enfermedad de Chagas por intrusión de triatominos y mamíferos silvestres en Bucaramanga, Santander, Colombia. Biomedica. 2017;37:68–78.
9. Levy MZ, Bowman NM, Kawai V, Waller LA, Cornejo del Carpio JG, Cordova Benzaquen E, et al. Periurban *Trypanosoma cruzi*-infected *Triatoma infestans*, Arequipa, Peru. Emerg Infect Dis. 2006;12:1345–52.
10. Urdaneta-Morales S. Chagas' disease: an emergent urban zoonosis. The Caracas valley (Venezuela) as an epidemiological model. Front Public Health. 2014;2:265–77.
11. Alarcón de Noya B, Díaz-Bello Z, Colmenares C, Ruiz-Guevara R, Mauriello L, Muñoz-Calderón A, et al. Update on oral Chagas disease outbreaks in Venezuela: epidemiological, clinical and diagnostic approaches. Mem Inst Oswaldo Cruz. 2015;110(3):377–86.
12. Abad-Franch F, Diotaiuti L, Gurgel-Goncalves R, Gurtler RE. Certifying the interruption of Chagas disease transmission by native vectors: cui bono? Mem Inst Oswaldo Cruz. 2013;108(2):251–4.
13. Guhl F. Enfermedad de Chagas: Realidad y perspectivas. Rev Biomed. 2009;20:228–34.
14. Tarleton R, Gürtler R, Urbina J, Ramsey J, Viotti R. Chagas disease and the London declaration on neglected tropical diseases. PLoS Negl Trop Dis. 2014;8(10):e3219. https://doi.org/10.1371/journal.pntd.0003219.
15. World Health Organization. Control of Chagas disease. World Health Organ Tech Rep Ser. 2002/07/03 ed. Brasilia (Brazil): WHO; 2002. p. i–vi, 1–109, back cover.
16. PAHO. Estimacion cuantitativa de la enfermedad de Chagas en las Americas. Montevideo, Uruguay: PAHO; 2006.
17. World Health Organization. Working to overcome the global impact of neglected tropical diseases: first WHO report on neglected tropical diseases. Geneva: WHO; 2010.
18. Schmunis GA, Yadon ZE. Chagas disease: a Latin American health problem becoming a world health problem. Acta Trop. 2009;115(1–2):14–21. [Published Online First: 2009/11/26].
19. Basile L, Jansa JM, Carlier Y, Salamanca DD, Angheben A, Bartoloni A, et al. Chagas disease in European countries: the challenge of a surveillance system. Euro Surveill. 2011;16(37):19968.
20. Bern C, Kjos S, Yabsley MJ, Montgomery SP. *Trypanosoma cruzi* and Chagas' disease in the United States. Clin Microbiol Rev. 2011;24(4):655–81. https://doi.org/10.1128/CMR.00005-11. [Published Online First: 2011/10/07].

21. Jackson Y, Pinto A, Pett S. Chagas disease in Australia and New Zealand: risks and needs for public health interventions. Tropical Med Int Health. 2014;19(2):212–8. https://doi.org/10.1111/tmi.12235.
22. Villalba R, Fornes G, Alvarez MA, Roman J, Rubio V, Fernandez M, et al. Acute Chagas' disease in a recipient of a bone marrow transplant in Spain: case report. Clin Infect Dis. 1992;14(2):594–5. [Published Online First: 1992/02/01].
23. Flores-Chavez M, Fernandez B, Puente S, Torres P, Rodríguez M, Monedero C, et al. Transfusional Chagas disease: parasitological and serological monitoring of an infected recipient and blood donor. Clin Infect Dis. 2008;46(5):e44–7. https://doi.org/10.1086/527448. [Published Online First: 2008/02/09].
24. Leiby DA, Read EJ, Lenes BA, Yund AJ, Stumpf RJ, Kirchhoff LV, et al. Seroepidemiology of *Trypanosoma cruzi*, etiologic agent of Chagas' disease, in US blood donors. J Infect Dis. 1997;176(4):1047–52. [Published Online First: 1997/10/23].
25. Muñoz J, Coll O, Juncosa T, Vergés M, del Pino M, Fumado V, et al. Prevalence and vertical transmission of *Trypanosoma cruzi* infection among pregnant Latin American women attending 2 maternity clinics in Barcelona, Spain. Clin Infect Dis. 2009;48(12):1736–40. https://doi.org/10.1086/599223.
26. Manne-Goehler J, Umeh CA, Montgomery SP, Wirtz VJ. Estimating the burden of Chagas Disease in the United States. PLoS Negl Trop Dis. 2016;10(11):e0005033. https://doi.org/10.1371/journal.pntd.0005033. [Published Online First: 2016/11/08].
27. Lee BY, Bacon KM, Wateska AR, Bottazzi ME, Dumonteil E, Hotez PJ. Global economic burden of Chagas disease: a computational simulation model. Lancet Infect Dis. 2013;13(4):342–8. https://doi.org/10.1016/S1473-3099(13)70002-1.
28. Requena-Méndez A, Aldasoro E, de Lazzari E, Sicuri E, Brown M, Moore DA, et al. Prevalence of Chagas disease in Latin-American migrants living in Europe: a systematic review and meta-analysis. PLoS Negl Trop Dis. 2015;9(2):e0003540. https://doi.org/10.1371/journal.pntd.0003540.
29. Ventura-Garcia L, Roura M, Pell C, Posada E, Gascón J, Aldasoro E, et al. Socio-cultural aspects of Chagas disease: a systematic review of qualitative research. PLoS Negl Trop Dis. 2013;7(9):e2410. https://doi.org/10.1371/journal.pntd.0002410.
30. Coura J. The main sceneries of Chagas disease transmission. The vectors, blood and oral transmissions. A comprehensive review. Mem Inst Oswaldo Cruz. 2015;110(3):277–82.
31. Schofield C, Grijalva M, Diotaluti L. Distribución de los vectores de la Enfermedad de Chagas en países no endémicos: la posibilidad de transmisión vectorial fuera de América Latina. Enf Emerg. 2009;11(Suppl 1):20–7.
32. Liu Q, Guo Y, Zhang Y, Zhou Z, Zhang L, Zhu D, et al. First records of *Triatoma rubrofasciata* (De Geer, 1773) (Hemiptera, Reduviidae) in Foshan, Guangdong Province, Southern China. Infect Dis Poverty. 2017;6:129–35.
33. Dujardin JP, Lam TX, Khoa PT, Schofield CJ. The rising importance of *Triatoma rubrofasciata*. Mem Inst Oswaldo Cruz. 2015;110(3):319–23.
34. Salazar R, Castillo-Neyra R, Tustin AW, Borrini-Mayorí K, Náquira C, Levy MZ. Bed bugs (*Cimex lectularius*) as vectors of *Trypanosoma cruzi*. Am J Trop Hyg. 2015;92:331–5.
35. Blakely BN, Hanson SF, Romero A. Survival and trans stadial persistence of *Trypanosoma cruzi* in the bed bug (Hemiptera: Cimicidae). J Med Entomol. 2018;55(3):742–6.
36. Davies TG, Field LM, Williamson MS. The re-emergence of the bed bug as a nuisance pest: implications of resistance to the pyrethroid insecticides. Med Vet Entomol. 2012;26:241–54.
37. WHO. Control and prevention of Chagas disease in Europe. Report of a WHO Informal Consultation. Geneva: Switzerland; 2009. p. 69.
38. Liu Q, Zhou XN. Preventing the transmission of American trypanosomiasis and its spread into non-endemic countries. Infect Dis Poverty. 2015;4:60–71.
39. Montgomery SP, Starr MC, Cantey PT, Edwards MS, Meymandi SK. Neglected parasitic infections in the United States: Chagas disease. Am J Trop Med Hyg. 2014;90(5):814–8.

40. Requena-Méndez A, Albajar-Viñas P, Angheben A, Chiodini P, Gascón J, Muñoz J, et al. Health policies to control Chagas disease transmission in European countries. PLoS Negl Trop Dis. 2014;8(10):e3245. https://doi.org/10.1371/journal.pntd.0003245.
41. Council of Europe Guideline "Safety and Quality assurance for the transplantation of organ, tissues and cells". 4th ed.; 2010 and Council of Europe. Guide to the preparation, use and quality assurance of blood components. 16th ed.; 2010. http://tots.edqm.eu/entry.htm.
42. Pinazo MJ, Gascón J. Chagas disease: from Latin America to the world. Rep Parasitol. 2015;4:1–8.
43. Bennett C, Straily A, Haselow D, Weinstein S, Taffner R, Yaglom H, et al. DVM Chagas disease surveillance activities-seven states, 2017. Morb Mortal Wkly Rep. 2018;67(26):738–41.
44. Schmunis GA. Epidemiology of Chagas disease in non-endemic countries: the role of international migration. Mem Inst Oswaldo Cruz. 2007;102(Suppl 1):75–85.
45. Guerri-Guttenberg RA, Grana D, Ambrosio G, Milei J. Chagas cardiomyopathy: Europe is not spared! Eur Heart J. 2008;29(21):2587–91.
46. South China morning post. Reward offered for 'kissing bugs' after feared Chinese outbreak of fatal tropical disease Chagas; 6 July 2018.

Challenges in Chagas Disease Control Through Transmission Routes

3

Mirko Rojas, Oscar Noya, Sergio Sosa-Estani,
and Joaquim Gascón

3.1 Perspectives on Vector Control and Entomological Surveillance with Community Participation of Triatominae

Of the communicable diseases, Chagas has the greatest prevalence in Latin America. This disease results from infection by the parasite *Trypanosoma cruzi* and is transmitted by insects of the sub-family Triatominae (Hemiptera: Reduviidae) which are widely distributed in rural and peri-urban communities of the American continent [1, 2].

M. Rojas
Fundación Ciencia y Estudios Aplicados para el Desarrollo en Salud y Medio Ambiente (CEADES), Cochabamba, Bolivia

Fundación Salud Naturaleza Integral (SANIT), Cochabamba, Bolivia

O. Noya
Sección de Biohelmintiasis, Instituto de Medicina Tropical de la Facultad de Medicina, Universidad Central de Venezuela, Caracas, Venezuela

Centro para Estudios Sobre Malaria Instituto de Altos Estudios "Dr. Arnoldo Gabaldón" Ministerio del Poder Popular para la Salud, Caracas, Venezuela

Red NHEPACHA, Barcelona, Spain

S. Sosa-Estani
Red NHEPACHA, Barcelona, Spain

Drugs for Neglected Disease initiative (DNDi) Latin America, Rio de Janeiro, Brazil

Epidemiology and Public Health Research Center, CONICET, Buenos Aires, Argentina
e-mail: ssosa@dndi.org

J. Gascón (✉)
Red NHEPACHA, Barcelona, Spain

Instituto de Salud Global (ISGlobal), Hospital Clinic Barcelona, Barcelona, Spain
e-mail: quim.gascon@isglobal.org

M.-J. Pinazo Delgado, J. Gascón (eds.), *Chagas Disease*,
https://doi.org/10.1007/978-3-030-44054-1_3

Infection by *T. cruzi*, originally an enzootic disease, has become a public health issue due to the settling of vector insects in the home. The anthropic action of progressive occupation and settling in wild open spaces is likely to have caused a disequilibrium in ecological niches, increasing the distance between wildlife reservoirs, which are the natural source of blood for Triatominae. As a result, the primary feeding sources represented by different species of wild mammals have either been diminished or depleted. Instead, humans became main food source and consequently became recipients of the parasite *T. cruzi*, etiologic agent of Chagas disease [1, 3, 4].

The process of Triatominae settling in the home was consolidated over time due to the favorable survival conditions offered by the ease of penetrating human dwellings; the microclimate within them; the food supply; economic, social, and cultural conditions; and some Triatominae characteristics, such as hematophagy. The vector's home settling process was facilitated, for example, by the housing material used—adobe or clay pugged with water—and the domestication of animals, such as the domestic breeding of the wild guinea pig and/or guinea pig (*Cavia porcellus*). There is also evidence of several words that describe the domestic vector in the languages of ancestral Andean cultures. For example, "Vinchuca" is the name of the insect in Quechua and means "to fall down," reflecting the habit of the insect detaching from the ceiling at the night and landing on beds to feed on the blood of people sleeping [3, 5–7].

In early prevention of Chagas disease, it is important to distinguish the wild (enzootic) and domestic cycles of transmission of the *T. cruzi* parasite, to identify the Triatominae species involved in each cycle, to determine the ability to adapt and colonize dwellings, vector capacity, and the degree of anthropophilicity of the Triatominae. The risk of people becoming infected is very different from one cycle to another, as is the possibility of carrying out effective vector control interventions [1].

Vector control is carried out by chemical and physical means. The pioneers who implemented these strategies were Venezuela, Brazil, and Argentina at the end of 1950s. Chemical control consists of the systematic and regular use of insecticides of residual action in houses infested by Triatominae. Physical control, on the other hand, is carried out through repairs or reconstruction of infested dwellings or those that may harbor the insect vector. Chemical and physical interventions may have specific indications and, in other cases, may be complementary and not mutually exclusive [1, 8, 9].

The dwelling-adapted species can re-infest from other artificial ecotopes in the area or be reintroduced from geographically distant foci. Wild Triatominae can invade and colonize housing more frequently from the nearest wild environment than autochthonous species. This implies that introduced and/or fully domiciled species are susceptible to a high degree of control within the home through chemical treatment with insecticides, as well as preventing the formation of new colonies. In the case of wild Triatominae, the likelihood of invasion can only be reduced by physically protecting the dwellings and, in the case of colonization of the house, through chemical treatment [6, 7, 10–12].

The effectiveness of the response and the stability of vector control depend on two basic conditions: temporal continuity and spatial contiguity, that is to say that domiciliary chemical treatment must be continuous for as long as necessary and include intervening, contiguous areas. If these requirements are met, together with the absence of permanent vector colonies inside the dwelling, it is possible to achieve interruption of vector transmission, except in exceptional circumstances that cause the infestation to persist [2, 13].

Chagas control programs have historically had a vertical structure and have been characterized, as other grand health campaigns, by (a) actions based on clearly established goals, to be met in a given time; (b) high specificity; (c) normative rigidity and, consequently, a standardized technical and technological treatment; and (d) mobilization of a large amount of resources in a short time frame. Management of the programs was, in turn, centralized and vertical, as were the operations, which were defined and carried out by central levels of national government. These vector control actions were implemented in the 1970s and 1980s. In the 1990s, with the creation of the "Cono Sur" Sub-regional Initiatives (1991), and of the Andean and Central American Countries Initiative (1997) through the technical cooperation of PAHO/WHO, almost all of the 18 countries considered endemic were reached [13–16].

Advances made in the control of Chagas in recent decades shows that existing technical knowledge, where it has been applied properly, has been sufficient to significantly reduce the burden of the disease in Latin America. The consolidation of established processes is limited by the high operational cost and decentralization of the vector control programs. This is due to a large extension of transmission zones or areas at risk of transmission being restored, after having previously interrupted vector transmission [17, 18].

There are additional complexities in vector control: institutional political changes at different governmental levels, a lack of commitment to public health policies, failures in the management of vector control, control failures attributable to the environment, the presence of wild populations with a high propensity to infest dwellings, resistance of Triatominae to pyrethroid insecticides, and the overlapping of various vectors transmitting diseases [6, 19–21].

Given these circumstances, it became clear that there was a need for a "paradigm shift" from the old vertical model of action, to a horizontal model of prevention and control of transmitting disease vectors, with a strong empowerment of government at regional and municipal levels, but always supervised by specialized personnel in all disciplines related to vector control (entomologists, insecticide experts, engineers, etc.) that centrally monitor the local programs. Ideally, each region should have an entomology laboratory.

However, many countries are not ready to face the challenges that lie ahead. Given the significant influence of social and environmental factors on the transmission of pathogens by vectors, it is essential that vector control and monitoring and evaluation systems are flexible, so that they can support approaches adapted to local circumstances. The realignment of domestic programs to optimize the implementation of interventions against multiple vectors and diseases should ensure that available resources are used in such a way that they have maximum impact [22, 23].

The need for a proper focus and "comprehensive management of vector control" to address the impact of vector-borne diseases has never been so pressing. The risk of transmission of etiological agents by insect vectors is changing rapidly due to unplanned urbanization, increased movements of people and goods, alterations in the use of natural assets and services (such as deforestation), climate change, and biological factors such as resistance of vectors to insecticides [22–24].

According to the PAHO/WHO technical recommendations, health systems must be prepared to detect changes and respond to them quickly and effectively. This response not only requires good management of entomological and epidemiological information at different governmental levels, but also requires proven effective control interventions, with well-trained personnel that can create sustainable and comprehensive systems of implementation and intervention. Thus, the participation of local authorities and communities within the framework of a wide comprehensive and cross-sectoral collaboration will be key to improving the execution of vector control, by adapting the interventions to each specific context. Additionally, the participation and collaboration of local communities will be necessary to create sustainable control programs that adequately respond to technical, operational, and economic challenges [22, 24, 25].

Communities play a fundamental role in the success and sustainability of vector control and surveillance. Vector control requires coordination between the many interested parties and depends upon the use of the knowledge and skills of the communities. The entomological surveillance system must be led by the inhabitants, but it must also be a priority health policy, so that it is stable even in the face of institutional and political instability [6, 22, 26, 27].

Chagas control programs should seek to harmonize and highlight these relationships and carry out routine joint actions that are of interest for the entomological surveillance of the disease. Integration could be structured around the axis of community participation, especially in relation to environmental management in homes and peri-domestic areas [21, 22, 25–27].

Directed meetings on organization, participation and, in particular, motivating the community in vector control-surveillance could help generate spaces for discussion and exchange of ideas. Systematic entomological surveillance activities can be complemented with preliminary studies, random checks and focused investigations according to the context. This requires operational research of a multidisciplinary nature on an appropriate spatial and temporal scale [12, 22, 26–28].

Surveillance activities must be strategically and deliberately planned in order to obtain information that allows the stratifying of risk in communities, detecting increases in the risk of transmission, and identifying specific threats for the efficacy of vector control (such as resistance of vectors to insecticides). The essential data needed may differ from one area to another, and through time, so adaptive capacity is required to ensure that appropriate data are collected and unnecessary activities, which are of no use as a basis for planning or implementing the actions on vector control, are avoided [6, 22, 25–27].

The detection of residual foci of infestation or re-infestation of treated dwellings is undoubtedly one of the main challenges of the entomological surveillance of

Chagas disease. The development, application, and evaluation of tactics aimed at increasing the probability of detection of these events is a priority objective, especially when the density of vector populations is low. The sensitivity of vector detection methods is generally unsatisfactory, but awareness of the presence of vectors by the inhabitants (community participation) is clearly superior, including direct searches by trained institutional personnel or "active searches" and the so-called "sensor devices" [6, 27].

Chagas is a zoonosis that cannot be eradicated: re-infestation by native species makes it essential to maintain entomological surveillance systems in the medium and long term. The denunciation of foci of infestation by the inhabitants is the simplest and most direct form of community participation in entomological surveillance; however, these strategies must be enriched, extending and deepening the dialogue between technicians, government, and communities.

3.2 Control of Orally Transmitted Chagas Disease

Oral transmission is the usual form of transmission of *T. cruzi* in wildlife and domestic animals, as fur and thick skin creates a barrier for cutaneous penetration. However, in humans, due to thin skin with little hair, contamination occurs by the cutaneous route [29].

Recently, due to epidemiological and clinical awareness on the part of the medical community, reporting of oral outbreaks in five countries in the Americas has increased [30]. Higher mortality has generated greater interest in Chagas disease. The number of cases of Chagas disease is increasing due to the rapid and marked eco-epidemiological changes that are occurring in the region (deforestation of peri-urban areas, especially in poor neighborhoods, and elimination of wildlife, which is the usual food source for triatomines, etc.) [31, 32].

As a consequence, previously wild vectors have been moved into human dwellings, as is the case for *Panstrongylus geniculatus*, the species with the widest distribution in the Americas [29, 33]. In addition, due to its close interaction with infected domestic and wild mammals, it has a high prevalence of infection with *T. cruzi*. It has an important role as a vector by the oral route, but a limited role in transmission by the cutaneous route, due to its delayed defecation reflex [33]. By analogy, any triatomine with a late defecation reflex could be a potential vector by the oral route. The potential role of other vectors susceptible to infection by *T. cruzi* is a further concern, and is the case for bedbugs (*Cimex lectularius*) [34].

When oral transmission occurs simultaneously in a human grouping (families, schools, nursing homes, etc.), it is easier to attribute the outbreak to oral contamination. On the other hand, early diagnosis of this clinical route is very difficult when there are isolated cases or people who consume contaminated food outside the usual environment, e.g., in the street, making it difficult to identify the vector and thus delaying the epidemiological link, especially when people move away from the place of exposure and are unaware of the possible source of infection. The vast majority of cases involve fruit juices prepared by artisans.

In the Brazilian Amazon, food is frequently consumed away from the place where the fruit is collected and the juice prepared and infection by this route has been called “remote transmission” [35]. Transmission can, therefore, occur in an area where there are no triatomines since the food was prepared in an endemic area away from the site of consumption.

It has been experimentally demonstrated that metacyclic *T. cruzi* trypomastigotes placed on fruit, fruit juices, and milk remain viable and are infectious to laboratory mice for up to 10 h, in coconut water they are maintained for up to 18 h [36], and in sugarcane juice for up to 24 h [37]. Likewise, infectivity in experimental animals has been demonstrated for contaminated fruit juices that are prepared at −20 °C for 26 h for sale as ice cream [38]. This reveals the high resistance of trypomastigotes to extreme conditions.

The lack of knowledge of this mechanism of infection by health professionals has delayed diagnosis of this form of infection. Likewise, health institutions and the academic staff of training centers themselves still do not recognize this clinical form and therefore do not share information about it. These factors hinder education of the population, who are not warned about the need to avoid the consumption of certain foods, especially artisan made juices or fruit ice cream, whose processing does not follow the sanitary norms of food preparation.

Although vector control with residual insecticides is still the main control measure associated with the improvement of housing (mosquito nets in windows and doors, plastering of walls, construction of roofs, and concrete flooring), based on the previous considerations, the following specific proposals have been suggested for the control of oral infection [39]:

1. Supervision of food preparation in canteens of schools, nursing homes, barracks, factories, etc., protecting the kitchens with mosquito netting in windows and doors, which prevent the invasion and domiciliation of triatomines. This measure needs to be accompanied by education of the food handlers. This is equally important in family dwellings, especially in the popular neighborhoods that are usually close to the edges of the forests and jungles that surround the urban centers.
2. Keep foods covered or enclosed in containers to avoid contamination with triatomine feces and to make contamination with debris from marsupials containing metacyclic trypomastigotes in their anal glands less likely.
3. Comply with the rules of food preparation, particularly the pasteurization of fruit juices for consumption as beverages or as artisanal ice creams.
4. Avoid eating food, especially fruit juices, of dubious preparation outside the home, especially when it comes to street vendors without any sanitary control.
5. Avoid the consumption of raw meats of wild and domestic animals since blood-borne trypomastigotes can also infect by the oral route [40, 41].
6. Reject the consumption of raw blood of wild and domestic animals during religious rituals or as therapy, which occurs in some rural communities.

Most of these recommendations should be known by the general population through educational programs at school level, at university and in the mass media.

3.3 Control of Transmission by Transfusion/ Organ Transplant

Transmission through blood is not determined by environmental factors, as in the case of vector and oral transmission, but almost exclusively by sociodemographic determinants associated with the migration of people.

The risk of transmission by blood or organ transplant is present whenever there are infected persons, the main factor to consider when planning public health interventions is the number of infected people in a locality, whether or not the area suffers from triatomine infestation and vector transmission. Thus, when there is intense immigration of people from an area with vector transmission into areas without vectors, there may be a risk of transmission by blood that requires the implementation of measures to control transmission through these routes, such as screening of blood and organ donors [42].

These strategies, started in the 1970s, were not adequately implemented in Latin America [43], but this improved during the last decade and more recently these programs have been implemented in non-endemic countries (see below). The risk of transmission by blood transfusion has been drastically reduced, and in 2015 all Latin-American countries achieved universal blood screening for Chagas disease of blood donors.

A very important issue to guarantee the success of Chagas disease control programs is to implement evaluation programs and to monitor their efficacy. These aspects were recently reviewed by a PAHO/WHO group of experts and, in addition, subjects involving sustainability, coverage, financing, and the fact that the quality of the program must be taken into account in each evaluation were discussed.

3.4 Control of Congenital Transmission

Congenital infection with *T. cruzi* is a global problem, occurring on average in 5% of children born from chronically infected mothers in endemic areas, with variations depending on the region [44]. Considering that etiological treatment of the child is always effective if performed before one year of age, diagnosis of infection in pregnant women and their newborns has to become the standard of care and be integrated into the surveillance programs for syphilis and HIV, as was recently formalized through the EMTI-Plus strategy [45].

The role of parasite genotypes and host genetic factors in parasite transmission and development of infection in fetuses/neonates must be further investigated in order to better estimate the risk factors and impact on health of congenital infection with *T. cruzi* [46].

Strategies for control of congenital transmission began to be implemented in the 1980s. The infection can mainly be controlled by early diagnosis and timely treatment of children with congenital infection [47] and can be prevented by treatment with trypanocides in women of gestational age [48].

It is estimated that between 8000 and 15,000 *T. cruzi* infected babies are born every year to infected mothers in Chagas disease endemic countries. Currently, poor access to and performance of the current diagnostic algorithm, based on microscopy at birth and serology at 8–12 months after delivery, is one of the barriers to congenital Chagas disease (CCD) control [49]. Detection of parasite DNA using molecular diagnostic tools could be an alternative or complement to current diagnostic methods, but its implementation in endemic regions remains limited. Prompt diagnosis and treatment of CCD cases would have a positive clinical and epidemiological impact [50].

The number of babies that would need to be tested per year varies from 157,972 to 214,074, depending on the estimate of the prevalence of *T. cruzi* infection. Argentina, Brazil, Bolivia, and Mexico are the countries where most tests should be conducted. Taking 5% as the average *T. cruzi* transmission rate from infected mothers to newborns, the annual prevalence of CCD cases in Latin America in 2010 ranged from 7899 to 10,704. These numbers are similar to those estimated by WHO in previous reports: 15,000 in 2006 and 8700 in 2010. These estimates can be used as a reference, but some limitations in the calculations should be considered: for example, the performance of serological tests and the congenital transmission rate vary between countries. Accurate and up-to-date data are needed in endemic countries [50].

Timely implementation of more efficient diagnostic algorithms will avoid a significant number of children being at risk of developing Chagas disease and transmitting *T. cruzi* parasites to the next generation.

3.5 Control of Transmission in Areas in Non-endemic Areas

Although in non-endemic areas there is neither vector nor oral transmission, *T. cruzi* can be transmitted through blood transfusion, organ transplantation and from mothers to children during pregnancy and childbirth.

3.5.1 Blood Transfusion and Transplants

In several non-endemic countries, there have been cases of Chagas disease due to transmission by organ transplantation or through blood transfusion [51–54]. In a study from the USA, of 31 patients with orthotopic heart transplantation, 19 (61%) developed evidence of reactivation of Chagas disease. All subjects with evidence of reactivation were alive at follow-up (median: 60 weeks) [55]. Also in the USA, in another study, transmission was confirmed in five recipients from *T. cruzi*+ donors, with four deaths, one of which was directly related to Chagas disease [56].

In a Canadian study in 2011, it was estimated that the risk of acquiring *T. cruzi* infection through blood transfusion and blood products was 2%. The expected number of infected cases was calculated to be between 0 and 20; 13 cases were confirmed [57].

In a study done in blood donors in the Netherlands, with some risk factors for *T. cruzi* infection, no positive cases were found. Most donors were originally from Suriname and Brazil, or were children of women from these countries. The authors concluded that in the Netherlands, interventions to mitigate the risk of *T. cruzi* transmission by transfusion would be highly cost-ineffective and are thus not required [58].

Two studies performed in the North of Italy (Veneto/Tuscany) and Rome show diverging results of 0% and 3.9% *T. cruzi* prevalence, respectively [59, 60].

A study in Spain showed a *T. cruzi* infection prevalence of 0.62% among blood donors coming from several countries, with 11 donors confirmed positive with a high rate (10.2%) in Bolivian donors. One of the positive donors was a long-term Spanish resident in an endemic country [61]. In the United Kingdom, 15,536 donations have been collected and screened since 2005, of which 15,499 (99.8%) were *T. cruzi* antibody negative and released to inventory [62].

Since the prevalence of Chagas disease in endemic areas and the diversity and intensity of migration from Latin-American countries to non-endemic countries are heterogeneous, the preventive policies for the transmission of *T. cruzi* adopted by the recipient countries are different [63].

In non-endemic countries, two basic strategies are used to avoid *T. cruzi* transmission:

- Deferral of donors who know they have the disease or who are at risk of being carriers of *T. cruzi*. These people are mainly detected through questionnaires that include questions about personal issues and a history of transfusions in endemic countries. But because a high percentage of infected patients are asymptomatic, or have symptoms that can be confused with other diseases, many patients do not know if they have the infection and they also do not know the risk factors for acquiring *T. cruzi* infection. Several studies have shown that this strategy is not completely effective and that if it is strictly applied, many potential donors are lost [64].
- Screening Programs. A selective screening program has been adopted in many non-endemic countries with a high number of migrants coming from the Latin-American region (i.e., Spain, France, or the USA). Donation is accepted if a specific serological test is negative. These programs are useful for minimizing the risk of transmission, and also minimizing the cumulative loss of donors and donations that would ensue if only permanent deferral of at-risk donors is adopted. Moreover, it is important to include in the program donors at risk of *T. cruzi* infection, even they were not born in an endemic country. A pre-donation interview is key to identifying at-risk donors if the program is not routinely performed for all donors.

For organ and tissue donors, a positive serology does not totally exclude them from transplant programs. Specific guidelines have been published that include the criteria for the acceptance of organs from patients with positive serology for *T. cruzi*, as well as the strategies for good follow-up of patients who received organs from *T. cruzi* positive donors [65].

3.5.2 Congenital Transmission

Due to the chronicity of Chagas disease and the involvement of immigrant women of childbearing age, congenital transmission of *T. cruzi* is expected in non-endemic countries. In fact, with the control of blood banks, congenital transmission is the primary cause of *T. cruzi* transmission in non-endemic countries.

Several cases of congenital transmission have been detected in Europe [66, 67, 68]. In a study done in two maternity centers in Barcelona (Spain), 3.4% of women of Latin-American origin were *T. cruzi* positive, with a transmission rate of 7.3%; Bolivian women were the most affected [69]. Later, similar results were obtained in other areas of Spain [70, 71]. In Italy, a study in pregnant Bolivian women showed an 8.7% prevalence of *T. cruzi* infection, with a transmission rate of 4% [72].

In general, non-endemic countries have no legislation requiring screening of pregnant women coming from Chagas disease endemic areas and monitoring their offspring [73]. In the UK, serological testing should be offered to pregnant women belonging to at-risk groups, and positive cases should be referred to a specialist center. In only three autonomous communities in Spain (Catalonia, Galicia, and Valencia) and one region in Italy (Tuscany) is there a specific program for the control of congenital transmission of *T. cruzi* [73].

It is important to control vertical transmission since early treatment of infected babies has a very high rate of cure, near to 100%, meaning that long-term, heart/digestive complications of the disease can be avoided.

Other benefits derived from the implementation of these protocols can be realized: in addition to detecting infection in babies, older siblings of the family can also be tested and treated, avoiding complications in adulthood [73]. Also, mothers with *T. cruzi* infection can be treated after pregnancy to reduce the possibility of *T. cruzi* transmission in subsequent pregnancies [48, 74].

From a public health perspective, several studies show that these control programs are also beneficial from an economic point of view [75, 76].

3.6 Taking Care of Infected People as a Method of Control of Chagas Disease

Chagas disease is one of the most neglected public health problems in America, where <1% of the estimated six million cases (including >300,000 in the United States) have been diagnosed and treated. Chagas disease is also a significant public health challenge in Europe, where <10% of the estimated 68,000–122,000 cases have been diagnosed or treated. Globally, Chagas disease creates an annual burden exceeding 800,000 disability adjusted life years and $600,000,000 in healthcare costs [77].

The medical care all over have different systems with subsystems and it must to have into account to resolve access to diagnosis and treatment, coverage, referential system, medical expertise, among other considerations. This is a new goal for an integrated Chagas disease control programs, including the care of people infected as formal objective and activities.

There is absolute consensus that the diagnostic, and treatment early as possible should be offered within primary healthcare and refer to a secondary or higher level only when is needed. The intervention treating Chagas cases in the acute and indeterminate stages reduces transmission and provides economic and health benefits. This supports the need for improved diagnostics and access to safe and effective treatment [78].

3.6.1 The Role of Etiological Treatment for *T. cruzi* Infection on Several Levels of Prevention in Public Health

The etiological treatment allows for actions on several levels of public health prevention [42, 79].

Previous studies provide evidences useful for applying healthcare strategies in control programs in several countries. When a high number of people can get diagnosis, treatment, and cure, it generates a new scenario for the future reduction of the burden of the disease.

- *Primary prevention level*—If the goal is to prevent the occurrence of new infections, etiological treatment could have an indirect effect when applied to children and young people. Curing children and women in reproductive age would prevent future events of congenital *T. cruzi* transmission [48, 74, 80–82]. In addition, blood and organ donors would increase if infected people are treated. The effectiveness of the etiological treatment with respect to these primary prevention indications would be cost-effective as was described [78]. Another strategy would be to obtain a feasible treatment for pregnant women, such as that used for human immunodeficiency virus (HIV) infection, to prevent congenital transmission during pregnancy. However, safe drugs are necessary for this strategy.

  Etiological treatment in cases of accidents with material contaminated with parasites or with blood samples of patients infected with *T. cruzi* could also be considered as an indication for primary prevention. In fact, the treatment is not strictly prophylactic, as it is not possible to prevent infections, but the infection can be aborted immediately after the accident with a timely treatment and an appropriate concentration of specific drugs [83].
- *Secondary prevention level*—If prevention activities cannot prevent infection in children, curing infected children is still possible by prescribing etiological treatment. In this regard, etiological treatment should be indicated when damages

from cardiac or digestive disease are not strongly present in children. This is the best opportunity to get seronegativation [84–89] and avoid disease, thus preserving social, mental, and physical health into adulthood [90, 91].

National control programs have been implemented in more and more Latin-American countries since 1994. They consist of the screening of children populations as a regular strategy providing early diagnosis and treatment [92]. The positive effect of curing infected children detected by serological screening must be assessed by considering the number of infected people, the transmission, evolution and burden of Chagas disease, so that it becomes possible to analyze the usefulness of serology as an indicator of action against the vector.

Another indication for etiological treatment in secondary prevention is to prevent the reactivation of a chronic infection. Immunosuppression due to the application of immunosuppressive therapies [93] or HIV/acquired immune deficiency syndrome [94] increases the risk of reactivation in patients with chronic infection. Even though the effectiveness of etiological treatment for the clinical control of reactivation episodes has been demonstrated, it is necessary to gather evidence as to whether preventive treatment is effective in patients with no signs of clinical reactivation and with abnormal immunological parameters [95]. In this regard, some protocols recommend treating organ donors infected with *T. cruzi* to reduce the risk of transmission by transplant [96]. In this case, treatment should be considered as a primary prevention action.

- *Tertiary prevention level*—The use of etiological treatment against *T. cruzi* infection in order to reduce the negative effect of established disease has been evaluated through randomized clinical trials for assessing its efficacy in patients with stablished cardiac disease [97, 98]. These trials have assessed the efficacy of benznidazole for preventing the progression of cardiac disease, and no clinical benefit was demonstrated up to date, when treatment is prescribed in patients with stablished cardiac disease [99–101].

  Several observational studies have been published showing the effects of etiological treatment on patients infected with *T. cruzi* with respect to preventing the progression of chronic chagasic cardiomyopathy [102–104]. These studies achieved quality of evidence type II and strength of recommendation B and C. The prognosis of patients with heart failure or advanced stages of chagasic cardiomyopathy is poor [105], but similar to that of other patients who develop heart failure. Since the disease is chronic and heart damage develops over decades, it is very important to recognize the factors that determine the progression of the disease in the early stages [106]. The evolution of knowledges addressed to the current scenario with a new paradigm regarding benefit of trypanocide treatment [107]. The etiological treatment should be considered as a protective factor in the model of the physiopathology of chagasic cardiomyopathy.

As mentioned above, the effectiveness of etiological treatment in the control of reactivation episodes has been proved, showing recovery from severe manifestations of reactivation such as meningoencephalitis, myocarditis, and panniculitis [93, 94].

Based on our current understanding of the disease, there is a consensus that every patient infected with *T. cruzi* must (children and young) or should be treated (adults). Treatment can cure infection and reduce or prevent the progression of Chagas-related heart disease/cardiomyopathy.

Incorporating etiological treatment as a public health strategy useful at the primary, secondary, and tertiary prevention levels is essential to reduce the burden of the disease and to eliminate Chagas disease as a public health issue.

There are guidelines advising regarding management and care of people infected with *T. cruzi*. A precocious and adequate clinical management is required to all acute or chronic cases. A timely and adequate clinical, pharmacological, and chirurgical management if necessary is essential. The specific treatment has to be considered currently among other therapies [45]. For instance and according to several longitudinal studies, timely management of initial arrhythmia, heart failure or megaesophagus is fundamental to avoid or delay the appearance of severe or irreversibl situations [105, 106].

Chagas disease cannot be eradicated due to the multiple scenarios that contribute to the occurrence of new cases and mainly because of the demonstrated existence of infected wild triatomines in permanent contact with domestic cycles. However, it is possible to interrupt the transmission in humans.

Acknowledgements "The Drugs for Neglected Diseases *initiative* (DNDi)" is grateful to its donors, public and private, who have provided funding to DNDi since its inception in 2003. A full list of DNDi's donors can be found at http://www.dndi.org/donate/donors/.

References

1. World Health Organization. Control of Chagas Disease. WHO Technical Report Series 905; 2002.
2. BID/OPS/IDRC/CNZ. Programa Regional para el Control de la Enfermedad de Chagas en América Latina. Lineamientos y recomendaciones técnicas y de política pública para la enfermedad de Chagas. Montevideo, Uruguay: BID; 2010. 80 pp.
3. Noireau F, Diosqui P, Jansen AM. *Trypanosoma cruzi*: adaptation to its vectors and its hosts. Vet Res. 2009;40:26.
4. Cortez MR, Monteiro F, Noireau F. New insights on the spread of *Triatoma infestans* from Bolivia – implications for Chagas disease emergence in the Southern Cone. Infect Genet Evol. 2010;10:350–3.
5. Cortez MR, editor. Triatominos de Bolivia y la enfermedad de Chagas. Bolivia: Programa Nacional de Chagas, Ministerio de Salud y Deportes; 2007. 352 pp.
6. BID/OPS/IDRC/CNZ. Programa Regional para el Control de la Enfermedad de Chagas en América Latina. Iniciativa de Bienes Públicos Regionales. Montevideo, Uruguay: BID; 2010. 242 pp.
7. Cortez MR, Jansen AM, Gurtler R, Noireau F. Ecología de *Triatoma infestans* y *Trypanosoma cruzi* y la implicación de los reservorios silvestres en los valles andinos de Bolivia. In: Libro de resúmenes "Actualización de la Tripanosomiasis Americana". Proyecto SSA. Paraguay: ATU/Comunidad Europea, Dirección General de Investigación Científica y Tecnológica, Universidad Nacional de Asunción Paraguay; 2007. p. 135–40.
8. Oliveira Filho AM. New alternatives for Chagas disease control. Mem Inst Oswaldo Cruz. 1984;79(Suppl):117–23.

9. Silveira AC. Programa del Escudo Epidemiológico Boliviano. Contrato préstamo BID 1031/ SF-BO. Informe de consultoría en control, prevención, diagnóstico y tratamiento de la enfermedad de Chagas; 2005. 20 pp.
10. Noireau F, Cortez MR, Monteiro F, Jansen AM, Torrico F. Can wild *Triatoma infestans* foci in Bolivia jeopardize Chagas disease control efforts? Trends Parasitol. 2005;21:7–10.
11. Gürtler RE, Cecere MC, et al. Monitoring house reinfestation by vectors of Chagas disease: a comparative trial of detection methods during a four-year follow-up. Acta Trop. 1999;72(2):213–34.
12. Gurtler R. Eco-epidemiologia regional de la transmisión vectorial: enfermedad de Chagas en el Gran Chaco. In: La enfermedad de Chagas, a la Puerta de los 100 años del conocimiento de una endemia Americana ancestral. Buenos Aires: OPS/Fundación Mundo Sano; 2007. p. 137–55.
13. Silveira AC. O manejo da doença de Chagas como problema de Saude Publica. In: La enfermedad de Chagas, a la Puerta de los 100 años del conocimiento de una endemia Americana ancestral. Buenos Aires: OPS/Fundación Mundo Sano; 2007. p. 119–28.
14. Salvatella R. Una visión de la enfermedad de Chagas desde su propia historia. In: La enfermedad de Chagas, a la Puerta de los 100 años del conocimiento de una endemia Americana ancestral. Buenos Aires: OPS/Fundación Mundo Sano; 2007. p. 19–22.
15. Dias JCP. O controle da doença de Chagas no Brasil. In: Silveira AC, editor. El control de la enfermedad de Chagas en los países del Cono Sur de América Historia de una iniciativa internacional, 1991/2001. Uberaba, MG: OPAS/OMS, Faculdade de Medicina do Triângulo Mineiro; 2002. p. 146–250.
16. Dias JC. Southern Cone Initiative for the elimination of domestic populations of *Triatoma infestans* and the interruption of transfusional Chagas disease. Historical aspects, present situation, and perspectives. Mem Inst Oswaldo Cruz. 2007;102:11–8.
17. Yadón ZE, Gürtler RE, Tobar F, Medici AC. Descentralización y Gestión del Control de las Enfermedades Transmisibles en América Latina. Buenos Aires: OPS/TDR; 2006. 350 pp.
18. Silveira AC. El impacto de la descentralización de los sistemas de salud en la prevención y control de la enfermedad de Chagas: el caso de Brasil. In: Descentralización y Gestión del control de las Enfermedades Transmisibles en América Latina. Buenos Aires: OPS/TDR; 2006. p. 203–14.
19. Picollo MI, Vassena C, et al. High resistance to pyrethroid insecticides associated with ineffective field treatments in *Triatoma infestans* (Hemiptera: Reduviidae) from northern Argentina. J Med Entomol. 2005;42(4):637–42.
20. Cortez MR, Emperaire L, et al. Sylvatic *Triatoma infestans* (Reduviidae, Triatominae) in the Andean valleys of Bolivia. Acta Trop. 2007;102(1):47–54.
21. Gorla D, Hashimoto K. Control strategies against Triatominae. In: American Trypanosomiasis Chagas disease. Amsterdam: Elsevier; 2017. p. 223–42. https://doi.org/10.1016/B978-0-12-801029-7.00010-1.
22. World Health Organization. Global vector control response. Geneva: WHO; 2017. Available from: http://www.who.int/malaria/areas/vector_control/Draft-WHO-GVCR-2017-2030-esp.pdf?ua=1.
23. PNUD, editor. Cambio Climático y el Desafío de la Salud en Bolivia. Programa de las Naciones Unidas para el Desarrollo. Bolivia: PNUD; 2013. 161 pp. Available from http://www.bo.undp.org/content/dam/bolivia/docs/MedioAmbiente/undp-bo-salud-2014.pdf.
24. OPS/OMS. Plan de acción para la eliminación de las enfermedades infecciosas desatendidas y las medidas posteriores a la eliminación 2016–2022. In: 55th Directing Council, 68th Session of the Regional Committee of WHO for the Americas; Sept 26–30. Washington, DC: PAHO; 2016. Available from: https://www.paho.org/hq/dmdocuments/2016/CD55-15-s.pdf.
25. OPS/OMS. Enfermedad de Chagas en las Américas: una revisión de la situación actual de salud pública y su visión para el futuro. Informe. Washington, DC: Conclusiones y Recomendaciones; 2018. Available from: https://www.paho.org/hq/index.php?option=com_content&view=article&id=14399:enfermedad-de-chagas-en-las-americas-una-revision-de-la-situacion-actual-de-salud-publica-y-su-vision-para-el-futuro&Itemid=72315&lang=es.

26. PAHO/WHO. Plan of action on entomology and vector control 2018–2023. 56th Directing Council, 70th Session of the Regional Committee of WHO for the Americas; Sept 23–27. Washington, DC: PAHO/WHO; 2018. Available from: https://www.paho.org/hq/index.php?option=com_docman&task=doc_download&gid=45774&Itemid=270&lang=en.
27. Abad-Franch F, Vega MC, Rolón MS, Santos WS, Rojas de Arias A. Community participation in Chagas disease vector surveillance: systematic review. PLoS Negl Trop Dis. 2011;5(6):e1207. https://doi.org/10.1371/journal.pntd.0001207.
28. Lardeux F, Depickerea S, Aliaga C, Chavez T, Zambrana L. Experimental control of *Triatoma infestans* in poor rural villages of Bolivia through community participation. Trans R Soc Trop Med Hyg. 2015;109:150–8.
29. Noya González O, de Noya BA, Robertson L. Chapter 5. Epidemiological factors related to food borne transmission of Chagas Disease. In: de Noya BA, Noya González O, Robertson L, editors. *Trypanosoma cruzi* as a food borne pathogen. London: Springer; 2015. p. 41–51.
30. Noya O, Ruiz-Guevara R, Díaz-Bello Z, Alarcón de Noya B. Epidemiología y clínica de la transmisión oral de *Trypanosoma cruzi*. Rev Esp Epidem: XI. Workshop on Chagas disease, Barcelona, España; 2015. p. 23–34.
31. Alarcón de Noya B, Noya O. An ecological overview on the factors that drives to *Trypanosoma cruzi* oral transmission. Acta Trop. 2015;151:94–102.
32. Díaz-Bello Z, Zavala-Jaspe R, Reyes-Lugo M, Colmenares C, Noya-Alarcón O, Noya O, Herrera L, Alarcón de Noya B. Urban *Trypanosoma cruzi* oral transmission: from a zoonotic founder focus to the largest microepidemic of Chagas disease. SOJ Microbiol Infect Dis. 2016;4(1):1–9.
33. Reyes-Lugo M. *Panstrongylus geniculatus* Laterille 1811. (Hemiptera: Reduviidae: Triatominae), vector de la enfermedad de Chagas en el ambiente domiciliario del centro-norte de Venezuela. Rev Biomed. 2009;20:180–205.
34. Salazar R, Castillo-Neyra R, Tustin AW, Borrini-Mayori K, Naquira C, Levy MZ. Bed bugs (*Cimex lectularius*) as vectors of *Trypanosoma cruzi*. Am J Trop Med Hyg. 2015;92: 331–5.
35. Xavier SC, Roque ALR, Bilac D, de Araujo VAL, Neto SF, da Silva LFC, Jansen AM. Distantiae transmission of *Trypanosoma cruzi:* a new epidemiological feature of acute Chagas disease in Brazil. PLoS Negl Trop Dis. 2014;8(5):e2878. https://doi.org/10.1371/journal.pntd.0002878.
36. Añez N, Crisante G, Romero M. Supervivencia e infectividad de formas metacíclicas de *Trypanosoma cruzi* en alimentos experimentalmente contaminados. Boletín de Malariología y Salud Ambiental. XLIX(1), Ene–Jul; 2009.
37. Cardoso AVN, Lescano SAZ, Amato Neto V, Gakiya E, Santos SV. Survival of *Trypanosoma cruzi* in sugar cane used to prepare juice. Rev Inst Med Trop Sao Paulo. 2006;28:287–9.
38. Barbosa RL, Dias VL, Pereira KS, Schmidt FL, Franco RM, Guaraldo AM, Alves DP, Passos LA. Survival *in vitro* and virulence of *Trypanosoma cruzi* in açai pulp in experimental acute Chagas Disease. J Food Prot. 2012;75:601–6.
39. Robertson L, Noya O. Chapter 8. Prophylactic measures and implementation of control measures in food borne Chagas disease. In: de Noya BA, Noya González O, Robertson L, editors. *Trypanosoma cruzi* as a food borne pathogen. London: Springer; 2015. p. 81–8.
40. Thomas ME, Rasweiler JJ IV, D'Alessandro A. Experimental transmission of the parasitic flagellates *Trypanosoma cruzi* and *Trypanosoma rangeli* between triatomine bugs or mice and captive Neotropical bats. Mem Inst Oswaldo Cruz. 2007;102:559–65.
41. Malaquias Dias MGB, Gruendling AP, Araújo SM, Gomes ML, Toledo MJ. Evolution of infection in mice inoculated by the oral route with different developmental forms of *Trypanosoma cruzi* I and II. Exp Parasitol. 2013;135:511–7.
42. Sosa-Estani S, Segura EL. Integrated control of Chagas disease for its elimination as public health problem-a review. Mem Inst Oswaldo Cruz. 2015;110(3):289–98. https://doi.org/10.1590/0074-02760140408.
43. Schmunis GA, Cruz JR. Safety of the blood supply in Latin America. Clin Microbiol Rev. 2005;18:12–29. Erratum in Clin Microbiol Rev 18:582.

44. Howard EJ, Xiong X, Carlier Y, Sosa-Estani S, Buekens P. Frequency of the congenital transmission of *Trypanosoma cruzi*: a systematic review and meta-analysis. BJOG. 2014;121(1):22–33. https://doi.org/10.1111/1471-0528.12396. Epub 2013 Aug 7. Review.
45. Organización Panamericana de la Salud. Guía para el diagnóstico y el tratamiento de la enfermedad de Chagas. Washington, DC: OPS; 2018. http://iris.paho.org/xmlui/handle/123456789/49653.
46. Carlier Y, Sosa-Estani S, Luquetti AO, Buekens P. Congenital Chagas disease: an update. Mem Inst Oswaldo Cruz. 2015;110(3):363–8. https://doi.org/10.1590/0074-02760140405. Epub Mar 6. Review.
47. Blanco SB, Segura EL, Cura EN, Chuit R, Tulián L, Flores I, Garbarino G, Villalonga JF, Gürtler RE. Congenital transmission of *Trypanosoma cruzi*: an operational outline for detecting and treating infected infants in north-western Argentina. Tropical Med Int Health. 2000;5:293–301.
48. Fabbro DL, Danesi E, Olivera V, Codebó MO, Denner S, Heredia C, Streiger M, Sosa-Estani S. Trypanocide treatment of women infected with *Trypanosoma cruzi* and its effect on preventing congenital Chagas. PLoS Negl Trop Dis. 2014;8(11):e3312. https://doi.org/10.1371/journal.pntd.0003312. eCollection 2014 Nov.
49. Carlier Y, Torrico F, Sosa-Estani S, Russomando G, Luquetti A, Freilij H, Albajar Vinas P. Congenital Chagas disease: recommendations for diagnosis, treatment and control of newborns, siblings and pregnant women. PLoS Negl Trop Dis. 2011;5(10):e1250. https://doi.org/10.1371/journal.pntd.0001250. Epub 2011 Oct 25. Review. No abstract.
50. Picado A, Cruz I, Redard-Jacot M, et al. The burden of congenital Chagas disease and implementation of molecular diagnostic tools in Latin America. BMJ Glob Health. 2018;3:e001069. https://doi.org/10.1136/bmjgh-2018-001069.
51. Kransdorf EP, Czer LS, Luthringer DJ, Patel JK, Montgomery SP, Velleca A, et al. Heart transplantation for Chagas cardiomyopathy in the United States. Am J Transplant. 2013;13(12):3262–8.
52. Fores R, Sanjuan I, Portero F, Ruiz E, Regidor C, Lopez-Velez R, et al. Chagas disease in a recipient of cordblood transplantation. Bone Marrow Transplant. 2007;39(2):127–8.
53. Rodriguez-Guardado A, González ML, Rodriguez M, Flores-Chavez M, Boga JA, Gascon J. T*rypanosoma cruzi* infection in a Spanish liver transplant recipient. Clin Microbiol Infect. 2015;21(7):687.e1–3.
54. Flores-Chávez M, Fernández B, Puente S, Torres P, Rodríguez M, Monedero C, Cruz I, Gárate T, Cañavate C. Transfusional Chagas disease: parasitological and serological monitoring of an infected recipient and blood donor. Clin Infect Dis. 2008;46(5):e44–7.
55. Gray EB, La Hoz RM, Green JS, Vikram HR, Benedict T, Rivera H, Montgomery SP. Reactivation of Chagas disease among heart transplant recipients in the United States, 2012–2016. Transpl Infect Dis. 2018;20(6):e12996.
56. Huprikar S, Bosserman E, Patel G, Moore A, Pinney S, Anyanwu A, Neofytos D, Ketterer D, Striker R, Silveira F, Qvarnstrom Y, Steurer F, Herwaldt B, Montgomery S. Donor-derived *Trypanosoma cruzi* infection in solid organ recipients in the United States, 2001–2011. Am J Transplant. 2013;13(9):2418–25.
57. Steele LS, MacPherson DW, Kim J, Keystone JS, Gushulak BD. The sero-prevalence of antibodies to *Trypanosoma cruzi* in Latin American refugees and immigrants to Canada. J Immigr Minor Health. 2007;9(1):43–7.
58. Slot E, Hogema BM, Molier M, Bart A, Zaaijer HL. Risk factors and screening for *Trypanosoma cruzi* infection of Dutch blood donors. PLoS One. 2016;11(3):e0151038. https://doi.org/10.1371/journal.pone.0151038.
59. Angheben A, Anselmi M, Gobbi F, et al. Chagas disease in Italy: breaking an epidemiological silence. Euro Surveill. 2011;16:19969.
60. Gabrielli S, Girelli G, Vaia F, et al. Surveillance of Chagas disease among at-risk blood donors in Italy: preliminary results from Umberto I Polyclinic in Rome. Blood Transfus. 2013;11:558–62.

61. Piron M, Vergés M, Muñoz J, Casamitjana N, Sanz S, Maymó RM, Hernández JM, Puig L, Portús M, Gascón J, Sauleda S. Seroprevalence of *Trypanosoma cruzi* infection in at-risk blood donors in Catalonia (Spain). Transfusion. 2008;48(9):1862–8.
62. Kitchen AD, Hewitt PE, Chiodini PL. The early implementation of *Trypanosoma cruzi* antibody screening of donors and donations within England: preempting a problem. Transfusion. 2012;52(9):1931–9.
63. Requena-Méndez A, Albajar-Viñas P, Angheben A, Chiodini P, Gascón J, Muñoz J. Chagas Disease COHEMI Working Group. Health policies to control Chagas disease transmission in European countries. PLoS Negl Trop Dis. 2014;8(10):e3245.
64. Angheben A, Boix L, Buonfrate D, Gobbi F, Bisoffi Z, Pupella S, Gandini G, Aprili G. Chagas disease and transfusion medicine: a perspective from non-endemic countries. Blood Transfus. 2015;13:540–50.
65. Pinazo MJ, Miranda B, Rodríguez-Villar C, Altclas J, Brunet Serra M, García-Otero EC, de Almeida EA, et al. Recommendations for management of Chagas disease in organ and hematopoietic tissue transplantation programs in nonendemic areas. Transplant Rev (Orlando). 2011;25(3):91–101.
66. Muñoz J, Portus M, Corachan M, Fumado V, Gascon J. Congenital *Trypanosoma cruzi* infection in a non-endemic area. Trans R Soc Trop Med Hyg. 2007;101(11):1161–2.
67. Jackson Y, Myers C, Diana A, Marti HP, Wolff H, Chappuis F, et al. Congenital transmission of Chagas disease in Latin American immigrants in Switzerland. Emerg Infect Dis. 2009;15(4):601–3.
68. Gastañaga Holguera T, Llorente-Gomez B, Merino P, Illescas T, Villar G, Herraiz MA. Hydrops fetalis in a congenital Chagas case in a non-endemic area. J Obstet Gynaecol. 2016;36(5):672–3.
69. Muñoz J, Coll O, Juncosa T, Verges M, del Pino M, Fumado V, et al. Prevalence and vertical transmission of *Trypanosoma cruzi* infection among pregnant Latin American women attending 2 maternity clinics in Barcelona, Spain. Clin Infect Dis. 2009;48(12):1736–40.
70. Francisco-González L, Gastañaga-Holguera T, Jiménez Montero B, Daoud Pérez Z, Illán Ramos M, Merino Amador P, Herráiz Martínez MÁ, Ramos Amador JT. Seroprevalence and vertical transmission of Chagas disease in a cohort of Latin-American pregnant women in a tertiary hospital in Madrid. An Pediatr (Barc). 2018;88(3):122–12.
71. Barona-Vilar C, Giménez-Martí MJ, Fraile T, González-Steinbauer C, Parada C, Gil-Brusola A, Bravo D, Gómez MD, Navarro D, Perez-Tamarit A, Fernandez-Silveira L, Fullana-Montoro A, Borrás R. Prevalence of *Trypanosoma cruzi* infection in pregnant Latin American women and congenital transmission rate in a non-endemic area: the experience of the Valencian Health Programme (Spain). Epidemiol Infect. 2012;140(10):1896–903.
72. Rodari P, Angheben A, Gennati G, Trezzi L, Bargiggia G, Maino M, Ruggeri M, Rampello S, Soavi L, Rizzi M. Congenital Chagas disease in a non-endemic area: results from a control programme in Bergamo province, Northern Italy. Travel Med Infect Dis. 2018;25:31–4.
73. Basile L, Oliveira I, Ciruela P, Plasencia A. Working Group For Developing The Catalonian Screening Programme For Congenital Transmission of Chagas Disease. The current screening programme for congenital transmission of Chagas disease in Catalonia, Spain. Euro Surveill. 2011;16(38):19972.
74. Murcia L, Simón M, Carrilero B, Roig M, Segovia M. Treatment of infected women of childbearing age prevents congenital *Trypanosoma cruzi* infection by eliminating the parasitemia detected by PCR. J Infect Dis. 2017;215(9):1452–8.
75. Stillwaggon E, Perez-Zetune V, Bialek SR, Montgomery SP. Congenital Chagas disease in the United States: cost savings through maternal screening. Am J Trop Med Hyg. 2018;98(6):1733–42.
76. Sicuri E, Munoz J, Pinazo MJ, Posada E, Sanchez J, Alonso PL, et al. Economic evaluation of Chagas disease screening of pregnant Latin American women and of their infants in a non endemic area. Acta Trop. 2011;118(2):110–7.
77. Lee BY, Bacon KM, Bottazzi ME, Hotez PJ. Global economic burden of Chagas disease: a computational simulation model. Lancet Infect Dis. 2013;13:342–8.

78. Bartsch SM, Avelis CM, Asti L, Hertenstein DL, Ndeffo-Mbah M, Galvani A, Lee BY. The economic value of identifying and treating Chagas disease patients earlier and the impact on *Trypanosoma cruzi* transmission. PLoS Negl Trop Dis. 2018;12(11):e0006809. https://doi.org/10.1371/journalpntd.0006809. eCollection 2018 Nov.
79. Sosa-Estani S, et al. Therapy of Chagas disease: implications for levels of prevention. J Trop Med. 2012;2012:292138. 10 pages. https://doi.org/10.1155/2012/292138.
80. Sosa-Estani S, Cura E, Velazquez E, Yampotis C, Segura EL. Etiological treatment of young women infected with *Trypanosoma cruzi*, and prevention of congenital transmission. Rev Soc Bras Med Trop. 2009;42(5):484–7.
81. Moscatelli G, Moroni S, García-Bournissen F, et al. Prevention of congenital Chagas through treatment of girls and women of childbearing age. Mem Inst Oswaldo Cruz. 2015;110:507–9.
82. Alvarez MG, Vigliano C, Lococo B, Bertocchi G, Viotti R. Prevention of congenital Chagas disease by Benznidazole treatment in reproductive-age women. An observational study. Acta Trop. 2017;174:149–52.
83. Herwaldt BL. Laboratory-acquired parasitic infections from accidental exposures. Clin Microbiol Rev. 2001;14(4):659–88.
84. de Andrade AL, Zicker F, de Oliveira RM, Almeida Silva S, Luquetti A, Travassos LR, Almeida IC, de Andrade SS, de Andrade JG, Martelli CM. Randomised trial of efficacy of benznidazole in treatment of early *Trypanosoma cruzi* infection. Lancet. 1996;348(9039):1407–13.
85. Sosa Estani S, Segura EL, Ruiz AM, Velazquez E, Porcel BM, Yampotis C. Efficacy of chemotherapy with benznidazole in children in the indeterminate phase of Chagas' disease. Am J Trop Med Hyg. 1998;59(4):526–9.
86. Schijman AG, Altcheh J, Burgos JM, Biancardi M, Bisio M, Levin MJ, Freilij H. Aetiological treatment of congenital Chagas' disease diagnosed and monitored by the polymerase chain reaction. J Antimicrob Chemother. 2003;52(3):441–9. Epub 2003 Aug 13.
87. Streiger ML, Del Barco ML, Fabbro DL, Arias ED, Amicone NA. Longitudinal study and specific chemotherapy in children with chronic Chagas' disease, residing in a low endemicity area of Argentina. Rev Soc Bras Med Trop. 2004;37(5):365–75.
88. Yun O, Lima MA, Ellman T, et al. Feasibility, drug safety, and effectiveness of etiological treatment programs for Chagas disease in Honduras, Guatemala, and Bolivia: 10-year experience of Médecins Sans Frontiéres. PLoS Negl Trop Dis. 2009;3(7):e488.
89. Sguassero Y, Roberts KN, Harvey GB, Comandé D, Ciapponi A, Cuesta CB, Aguiar C, Castro AM, Danesi E, et al. Course of serological tests in treated subjects with chronic *Trypanosoma cruzi* infection: a systematic review and meta-analysis of individual participant data. Int J Infect Dis. 2018;73:93–101. https://doi.org/10.1016/j.ijid.2018.05.019. Epub 2018 Jun 4. Review.
90. de Oliveira W Jr. Depression and quality of life in Chagas patients. Rev Soc Bras Med Trop. 2006;39(Suppl 3):130–2.
91. Villa L, Morote S, Bernal O, Bulla D, Albajar-Vinas P. Access to diagnosis and treatment of Chagas disease/infection in endemic and non-endemic countries in the 21st century. Mem Inst Oswaldo Cruz. 2007;102(1):87–93.
92. Silveira AC. O controle da doença de Chagas nos países do Cone Sul da América. História de uma iniciativa internacional, 1991/2001. In: Silveira AC, de Arias AR, Segura E, Guillén G, Russomando G, Schenone H, Dias JCP, Valdes J, Lorca M, Salvatella R, editors. El control de la enfermedad de Chagas enlos países del Cono Sur de América: historia de una iniciativa internacional, 1991/2001. Uberaba: Facultad de Medicina do Triangulo Mineiro; 2002. p. 15–43.
93. Diez M, Favaloro L, Bertolotti A, Burgos JM, Vigliano C, Lastra MP, Levin MJ, Arnedo A, Nagel C, Schijman AG, Favaloro RR. Usefulness of PCR strategies for early diagnosis of Chagas disease reactivation and treatment follow-up in heart transplantation. Am J Transplant. 2007;7:1633–40.
94. Cordova E, Boschi A, Ambrosioni J, Cudos C, Corti M. Reactivation of Chagas disease with central nervous system involvement in HIV-infected patients in Argentina, 1992–2007. Int J Infect Dis. 2008;12:587–92.

95. Nishioka S. Benznidazol na quimioprofilaxia primária da reativação de doença de Chagas em chagásicos crônicos em uso de corticosteroides em doses imunodepressoras: há evidência suficiente para a recomendação do seu uso? Ver Soc Bras Med Trop. 2000;33:83–5.
96. Altclas J, Sinagra A, Dictar M, Luna C, Verón MT, de Rissio AM, García MM, Salgueira C, Riarte A. Chagas disease in bone marrow transplantation: an approach to preemptive therapy. Bone Marrow Transplant. 2005;36:123–9.
97. Reyes PA, Vallejo M. Trypanocidal drugs for late stage, symptomatic Chagas disease (*Trypanosoma cruzi* infection). Cochrane Database Syst Rev. 2005;(4):CD004102.
98. Marin-Neto JA, Rassi A Jr, Morillo CA, Avezum A, Connolly SJ, Sosa-Estani S, Rosas F, Yusuf S, BENEFIT Investigators. Rationale and design of a randomized placebo-controlled trial assessing the effects of etiologic treatment in Chagas' cardiomyopathy: the Benznidazole Evaluation For Interrupting Trypanosomiasis (BENEFIT). Am Heart J. 2008;156:37–43.
99. Morillo CA, Marin-Neto JA, Avezum A, Sosa-Estani S, Rassi A Jr, Rosas F, Villena E, Quiroz R, Bonilla R, Britto C, Guhl F, Velazquez E, Bonilla L, Meeks B, Rao-Melacini P, Pogue J, Mattos A, Lazdins J, Rassi A, Connolly SJ, Yusuf S, BENEFIT Investigators. Randomized trial of benznidazole for chronic chagas' cardiomyopathy. N Engl J Med. 2015;373(14):1295–306. https://doi.org/10.1056/NEJMoa1507574. Epub 2015 Sep 1.
100. Pecoul B, Batista C, Stobbaerts E, Ribeiro I, Vilasanjuan R, Gascon J, Pinazo MJ, Moriana S, Gold S, Pereiro A, Navarro M, Torrico F, Bottazzi ME, Hotez PJ. The BENEFIT trial: where do we go from here? PLoS Negl Trop. 2016;10(2):e0004343. https://doi.org/10.1371/journal.pntd.0004343. eCollection 2016 Feb.
101. Schmidt A, Dias Romano MM, Marin-Neto JA, Rao-Melacini P, Rassi A Jr, Mattos A, Avezum Á Jr, Villena E, Sosa-Estani S, Bonilla R, Yusuf S, Morillo CA, Maciel BC, BENEFIT Investigators. Effects of trypanocidal treatment on echocardiographic parameters in Chagas cardiomyopathy and prognostic value of wall motion score index: a BENEFIT Trial Echocardiographic Substudy. J Am Soc Echocardiogr. 2018;2(2):286–295.e3. https://doi.org/10.1016/j.echo.2018.09.006. [Epub ahead of print].
102. Viotti R, Vigliano C, Lococo B, et al. Long-term cardiacoutcomes of treating chronic Chagas disease with benznidazole versus no treatment: a nonrandomized trial. Ann Intern Med. 2006;144(10):724–34.
103. Fabbro DL, Streiger ML, Arias ED, Bizai ML, Del Barco M, Amicone NA. Trypanocide treatment among adults with chronic Chagas disease living in Santa Fe City (Argentina), over a mean follow-up of 21 years: parasitological, serological and clinical evolution. Rev Soc Bras Med Trop. 2007;40(1):1–10.
104. Cardoso CS, Ribeiro ALP, Oliveira CDL, Oliveira LC, Ferreira AM, Bierrenbach AL, Silva JLP, Colosimo EA, Ferreira JE, Lee TH, Busch MP, Reingold AL, Sabino EC. Beneficial effects of benznidazole in Chagas disease: NIH SaMi-Trop cohort study. PLoS Negl Trop Dis. 2018;12(11):e0006814. https://doi.org/10.1371/journalpntd.0006814. eCollection 2018 Nov.
105. Rassi A Jr, Rassi A, Little WC, Xavier SS, Rassi SG, Rassi AG, Rassi GG, Hasslocher-Moreno A, Sousa AS, Scanavacca MI. Development and validation of a risk score for predicting death in Chagas' heart disease. N Engl J Med. 2006;355:799–808.
106. Viotti R, Vigliano C, Lococo B, Petti M, Bertocchi G, Álvarez MG. Indicadores clínicos de progresión de la miocarditis chagásica crónica. Rev Esp Cardiol. 2005;58:1037–44.
107. Viotti R, Alarcón de Noya B, Araújo-Jorge T, Grijalva MJ, Guhl F, López MC, Ramsey JM, Ribeiro I, Schijman AG, Sosa-Estani S, Torrico F, Gascon J, Latin American Network for Chagas Disease, NHEPACHA. Towards a paradigm shift in the treatment of chronic Chagas disease. Antimicrob Agents Chemother. 2014;58:635–9.

A Social Approach to Chagas Disease: A First Step to Improve Access to Comprehensive Care

4

Mariana Sanmartino, Andrea Avaria Saavedra, and Leonardo de la Torre Ávila

4.1 Introduction

The increasing complexity of the problematic of Chagas,[1] product of the multiple realities where it is currently present, poses the need of recognizing that we are facing an intricate socio-environmental health problem related to both rural and urban environments in not only the Latin American context but also worldwide [1]. These current and diverse scenarios imply crucial challenges for health teams, researchers from multiple disciplines, and public policy makers because they evidence the existence and needs of populations usually disregarded. In this context, it is pertinent to ask ourselves "what are we talking about when we talk about Chagas?" For this question, relevant answers should be provided for each of the scenarios where Chagas exists. The committed and permanent search for these answers will lead inevitably to recognize the need of

[1] We will use the *Chagas* or *problematic of Chagas* denominations to avoid circumscribing the subject to the biomedical aspects (Chagas disease) and therefore we recognize the multiple causality and the multiple implications for people in different territories, and social and institutional realities that address it.

M. Sanmartino (✉)
Grupo de Didáctica de las Ciencias - IFLYSIB (CONICET-UNLP),
La Plata, Provincia de Buenos Aires, Argentina
e-mail: mariana.sanmartino@conicet.gov.ar

A. Avaria Saavedra
Facultad de Ciencias Sociales y Humanidades, Universidad Autónoma de Chile – Proyecto FONIS SA18I0056, Providencia, Región Metropolitana, Chile
e-mail: andrea.avaria@uautonoma.cl

L. de la Torre Ávila
InSPIRES Project, The Barcelona Institute for Global Health and International Health Service, Hospital Clínic i Provincial de Barcelona, Barcelona, Spain
e-mail: leonardo.delatorrea@isglobal.org

M.-J. Pinazo Delgado, J. Gascón (eds.), *Chagas Disease*,
https://doi.org/10.1007/978-3-030-44054-1_4

developing comprehensive approaches including the experiences and knowledge of all the actors involved in one way or another. When we talk about "all the actors involved," we refer to those who provide academic knowledge from the diverse disciplines necessary to understand Chagas beyond the biomedical area, and those who are affected directly or indirectly by the disease. Furthermore, the search for answers to this question should involve all the actors who might provide knowledge from the areas of communication and education, widely and openly understood.

According to estimates from the World Health Organization, between six and seven millions of people are currently infected by *Trypanosoma cruzi* [2]. This figure evidences complexity not only because of the magnitude it represents but also because it is estimated that barely 10% out of these people are aware of the fact that they carry the parasite, and only 1% out of that population access to the available therapies [1]. The evidence speaks for itself and, in front of it, the words spoken by Carlos Chagas and Emmanuel Dias more than a 100 years ago retain their full validity: "besides the technical innovations, the definite overcoming of the human Chagas disease involves, above all, political willingness and social responsibility" [3]. We humbly add to the masters' reflection that, besides biomedical and technological advances, the definite overcoming of the problematic of Chagas involves, above all, the incorporation of issues that have not been considered in the traditional approaches, such as respect to the peasant and indigenous' knowledge, the particularities of migrant and traveling contexts in the current globalization framework, and the weight of stigmatization and, thereby, of social exclusion (a frequent experience, stronger and more tangible than the physical consequences of *T. cruzi* infecting an organism), among others. We agree, in this sense, with the paradox formulated by Ventura-Garcia et al.: "Although the importance of social and cultural factors is broadly acknowledged, current approaches to NTDs[2] almost always neglect aspects of the sociocultural—biological—environmental triad. This results in a narrower understanding of Chagas disease that hampers sustainable prevention and control" [4:1].

In this scenario and considering the complexity of Chagas, we consider crucial to address the actions of intervention, care, and research related to the topic from a new paradigm. To accomplish this, it is essential to return to some pending discussions and incorporate new theoretical frameworks to understand the problematic.

This chapter will allow from a perspective that considers and values the questioning to the classical epistemologies of health and illness, enhancing the hegemonic approach perspective by the incorporation of new actors and essential knowledge to embrace the diversity of the current scenarios. We question the "place" from where we talk and address the problematic of Chagas because it evidences not only underlying epistemological frameworks but also the ways of addressing the relationship with the Other and the mechanisms and practices from which we face health interventions and acknowledge power relationships and the knowledge at stake. From these considerations, we would like to propose reflections and alternative epistemological scenarios from which it is possible, from our perspective, to address the problematic of Chagas in a contextualized and comprehensive way.

[2] NTDs: Neglected Tropical Diseases.

4.2 About the Social and Inclusive Approach

4.2.1 What Are We Talking About When We Talk About Chagas?

From a biomedical perspective, Chagas disease has been described as an illness caused by the parasite *Trypanosoma cruzi*, which is transmitted to human beings and other mammals by an insect (depending on the place, known as *vinchuca*, *barbeiro*, *kissing bug*, *chinche besucona* and *chinchorro*, among other names) or through other ways such as transplacental and oral transmissions, blood transfusions, organ transplantation, or, to a lesser extent, laboratory accidents.

For decades, Chagas disease has been associated to the populations with scarce economic resources from rural and semirural areas of the American continent. However, today more than ever, we know that it is not an "exclusive" disease of certain human groups and geographical contexts. Currently, it is essential to recognize that we are facing an extremely complex problem that overcame the geographical, cultural, social, and disciplinary traditional borders a long time ago. The present challenge implies acknowledging that "Chagas disease requires an explicitly multidimensional approach in which prevention, control, and care strategies and programs are designed and implemented jointly, and in which the social and biomedical sciences, together with the experience of those affected, are incorporated and articulated" [4]. In other words, it is crucial to support the fact that, beyond the biomedical and epidemiological aspects traditionally (and almost exclusively) considered, in the current and past configurations of Chagas disease converge an intricate web of elements related to the social, cultural, economic, political, and environmental aspects, among others. Today, more than 100 years after its "discovery," it is evident that the classical scenario in which it was described has changed and that, in a globalized context, we must overcome the biomedical dimension to improve people's health conditions.

Therefore, to adopt a comprehensive approach, we define four large dimensions, whose dynamic combination of elements conveys the complexity of the problem. These dimensions are combined, metaphorically, in a *kaleidoscopic puzzle* [5], where the parts become meaningful when considered in mutual dependence and interrelationship within the whole and, in turn, depend on the analysis perspectives from which they are observed. The relationships established within and among the dimensions provoke that today, at a certain place in the world, the problematic of Chagas emerges with particular features, in contrast with some other regions and historical moments. Basically, these four proposed dimensions could be defined as follows:

- The biomedical dimension includes features ranging from the biology of the causal agent and vectors to medical issues regarding the disease manifestation, diagnosis, treatment, and transmission.
- The epidemiologic dimension concerns the aspects that characterize the situation from a population scale, using parameters such as prevalence and incidence, distribution, and infestation rates. The phenomenon of the growing

migration that influences the configurations of the problem is also considered in this dimension.
- The sociocultural dimension is related to cultural patterns and cosmovisions, home conditions, environmental management, the distinctive features of both rural and urban contexts, and social representations and prejudices that reproduce, for example, discrimination and stigmatization.
- The political-economic dimension involves, apart from the economic and macroeconomic conditions affecting the problem, the features related to public management and health, educational, legislative, and economic decisions at local, regional, and global levels. Furthermore, this dimension includes the position and decisions that we, as citizens and professionals, assume when thinking of Chagas disease in different environments (research, teaching, communication, health care, etc.).

When recognizing this multidimensionality of intervening elements and factors, it is evident the need of counting on the contributions from all the fields of knowledge and incorporating to the approach new and contextualized theoretical frameworks that might enlighten the comprehension of this problematic. In this sense, we are particularly interested in the contributions of the so-called Epistemologies of the South [6], which presents a way of proceeding that allows appraising, legitimating, and validating the knowledge born in the fight of social groups that have systematically suffered the exclusions, injustices, and discriminations caused by capitalism, colonialism, and patriarchy. Regarding those who live *systematically the exclusions, injustices, and discriminations caused by Chagas disease*, we find with (high) frequency the reproduction of characterizations that stigmatize and configure *an Other* affected by Chagas disease as *poor, silent, victim, patient*; this Other is generally a passive recipient of solutions designed from a great geographical, ideological, and cultural distance.

Because of this, it is time to revise critically the hegemonic discourses and recognize the obstacles they imply for the understanding of the topic. Identifying the resistance faced by people "affected by Chagas disease" and knowing and acknowledging those Others as active individuals, holders of knowledge and creative answers, is the point from where it will emerge uncountable learnings that must be considered for the collective search of comprehension and solutions.

The criticisms and foundations on which the Epistemologies of the South are built, as Boaventura de Sousa Santos himself claims [6], demand thinking about the unthinkable, assume the surprise as the constitutive act of the theoretical labor recognizing that, in the current context of social and political transformation, we do not need *cutting-edge theories*—which, by definition, do not allow themselves to be surprised—but *rearguard theories* instead. That is, we must accompany closely the transforming labor of the social movements to question and understand the possibilities in which collective implies creating articulations, translations, alliances among the diverse actors, their positions, and experiences. In other words, the Epistemologies of the South poses more the work of an implicated witness and less the one of a clairvoyant leadership; they require the development of approximations to what is new for some people and old for some other people; they lead to open analytical spaces for *surprising* realities (because they are new or because, so far, they have been produced as nonexistent) where transforming emergencies and liberating alternatives might sprout. In particular, for the topic we are considering here, the Epistemologies of the South are one of the most fertile frameworks to incorporate new realities and analytical spaces to collect from there transforming and liberating scenarios that, within the context of health care and biomedical hegemonic approach of Chagas, have been forgotten or simply disregarded.

4.2.2 Current Context(s) of Chagas Disease: Diverse Scenarios, Diverse Challenges

The current scenario of the problematic of Chagas, considered as a global health problem, defies us at various levels. On the one hand, it requires the comprehension of the *territory*, the multiple geographical, social, political, institutional, personal, familiar, and communitarian contexts in which the disease and its consequences are (re)produced. It is essential to identify the mechanisms used to address the problematic, that is, intervention methodologies, epistemological frameworks, and paradigms from which Chagas is understood and approached. It is also crucial to consider the structures used and activated to deal with Chagas, identify the actors involved in these processes, and recognize the complexity and the challenges implied by the new scenarios, both for researchers and health teams and for the affected people, their families and communities, to act accordingly.

These challenges imply the awareness of this complex problem as regards not only the global framework where it is currently manifested due to human mobility, but also the way in which Chagas affects the individuals and their families, transnational families, and individuals who travel because of their mobility projects and that, independently of their mobility, should be assisted in their needs to guarantee their rights [7]. The epidemiological change implies recognizing a territory that has widened; the disease localization becomes mobile through the traveling individuals. From this perspective, we have to observe the way in which Chagas is approached in these multiple scenarios, identify the institutions, and characterize the structures recognizing that the individuals come from diverse health and health care experiences and that they are in a different scenario. We should recognize that a located approach, limited to a context and to one way of intervention, is no longer possible. Instead, we should consider the complexity of intervening and identify the opportunities to be opened: we should go from national to transnational, from biomedical to multidimensional, from individual to familiar and collective, abandoning the idea of the Other as a *patient*, dependent, and recognize him/her as an agent, that is, as a subject able to identify his/her position in the structure, the properties of that position, and the politicization of his/her practices [8].

Transforming the paradigms through which we comprehend (and describe) Chagas and the *affected people* implies to transit the overcoming of positivist, biomedical, and Eurocentric epistemologies that sustain the objectivity and the distance as related and comprehensive principles and withdraw from the individual's experiences and knowledge. As we mentioned, we should also approach to epistemologies that allow an emancipation process, that is, a transformation not only in the academic, institutional, and medical performances but also in the search for the transformation of the inequality structural conditions.

A change in the methodologies used to intervene in the problematic is crucial; the strictly biomedical approach fragments the understanding of the topic, reduces and inactivates the individual. Chagas should be faced from approaches that lay their foundations in the questioning to the colonialist practices through which knowledge and learnings concerning the topic are reproduced, as well as in the subordination, invisibility, and inequity practices that support the general health performance [9]. We should go forward, sharpen our vision, and enhance the listening and, therefore, our understanding.

In the current scenario of human mobility, diversity challenges health teams; it calls them to consider the cultural, social, and epidemiological diversity in the care ordinariness, which does not necessarily imply a questioning to the subordination or subalternity logic reproduced in the health care systems or to the ways of approaching the health issue. Within this framework, interculturality, as a relational strategy in these contexts, constitutes a proposal that emerges from the questioning to the epistemological frameworks through which individuals intervene in health to solve the implications of a disease that burst in new scenarios and health care teams and systems, which are not always trained or prepared to face the problem, and the tensions that imply the learnings and conceptions of health and illness. To open to the Other entails the recognition of our own and the Other's cosmovisions, an exercise that is only possible through empathy and dialogue [10]. Not recognizing the differences closes the systems and encapsulates them to reality constructions where the own system is reproduced. In general, the State opens to incorporate the differences, but neither from the recognition nor from the agency co-construction processes; on the contrary, diversity is approached from an opening that, in practice, incorporates the differences from assimilationist and domesticating practices [11]. On the other hand, the development of cultural competences is also an opportunity for the health teams because it allows them to open themselves and incorporate the Other through a dialogic process that resignifies and enriches the practice itself.

We should understand that biomedicine is a cultural construction, associated to cosmovisions, generally ethnocentric, and Eurocentric; it places the knowledge in the medical practitioner (generally, a man). Populations and cultures elaborate constructions of health and illness that emerge from their cosmovisions; this is the way in which the notions of care, health, and illness are built and the mechanisms intervening in this process are developed [12, 13]. In this context, assuming oneself as being ill from Chagas disease has characteristics related to the subjective construction of the disease and the context where it is experimented, elements that are intertwined with the gender, the institutionality, and the objective and subjective conditions the individuals face because of, for example, migration or belonging to a transnational family [14, 15].

Recognizing the social, cultural, and gender components associated to health problems has allowed widening the approach perspective and acknowledging the inside of the problematic and the multiplicity of elements to be considered in the description and understanding of Chagas. In this sense, rurality stresses significantly the presence of the problematic. In this context, the housing issue constitutes a relevant tension for the people affected, not only because the house is associated to the presence of the insect vectors but also because in this scenario a series of aspects (such as local practices in housing building and use) coalescences with mechanisms through which the problematic is solved (fumigation, house or belonging burning, house relocation, etc.). Hence, from the intervention in health, we have activated health care practices that burst in the most intimate individuals' space, their homes, depriving, discrediting the people and families' efforts, and underestimating the value of ancestral practices. The action of the States through public policies, from fumigation to relocation of families and demolition/building or disposal of houses, entails the loss of intimacy and resolution capacity and agency of individuals and families. Thus, it has been observed that these practices of eradicating adobe brick and thatched roof houses do not necessarily translate, in fact, in a decrease of the risk of contracting Chagas disease through the vectorial way because the presence of insect vectors in a building is more related with structural issues than with the sort of material implied [5].

On the other hand, in the context of people's growing spatial mobility the individuals themselves and their bodies become currently the mobile depositaries of *the disease*. People face the lack of knowledge as well as the lack of opportunities of solving health problems, especially in the scenarios in which the professional and institutional capacities do not approach or consider Chagas disease. Through people's mobility, the fragility of the guaranteed health rights is evidenced in both the countries of origin and destination as well as in the social and institutional issues. In this scenario, the formation structures of the health care personnel, the resolution capacities of unknown problematics and, mainly, the capacity of dialogue and recognition of the Other and the legitimate knowledge he/she has as regards his/her body are confronted. Understanding the problematic of Chagas as a public health problem at a global level challenges countries and health care systems to interact and develop strategies that altogether allow guaranteeing rights, accessibility, and elimination of stigmatization associated to it. Chagas disease is an opportunity to open ourselves to emancipating learnings and practices.

4.2.3 Present and Absent Actors: Toward the Inclusion of Experiences

As we have been pointing out, considering the person affected by Chagas as an *actor* (and not as a *patient* or *passive recipient* of interventions designed from institutional and ideological distances) allows recognizing and promoting opportunities for these people's participation in initiatives to solve the complex social-environmental problematic that Chagas constitutes.

How an actor becomes an actor? Thanks to the agency, the ability to act, the ability to transform. According to Pierre Bourdieu, among other constructivist thinkers, the agency is the power of influencing in the structures, despite the historical traction through which these structures condition, the actions and their scope possibilities [16].

Probably, it is impossible to draw an exhaustive map of the actors involved in the complex web of Chagas; moreover, when this map and the lines through which we could analyze the influence of ones over the others vary according to the context. Anyway, thinking about the parts involved in any complex problem from the theory of social action [16] might promote that the problem could be understood and critically observed. After this analysis, it could be faced with inclusive initiatives.

We could ask ourselves, for example: which actors have had the true agency in the intervention planning proposed so far against Chagas disease? Those who have not had and/or do not have this agency? Furthermore, it should be necessary to stimulate asking these questions for the concrete case in each initiative (prevention, research, political incidence, etc.). To question ourselves is an exercise that, from the Epistemologies of the South viewpoint [6], allows us to widen and deepen the initiative scope and demand engagements, as it will be seen later.

The experience of rural communities related to their coexistence with the vector insect might be regarded or disregarded at the time of designing contemporary monitoring and control programs in the so-called *endemic areas*. The same can be decided after confirming the influence ability of certain actors from Latin American populations, for example in Europe, at the time of activating the demand of care and treatment strategies of Chagas disease considering their contacts in social networks and their relatives in origin and destination places. To ignore or add those actors to the understanding or transforming table is, as we have indicated, the first one of the decisions that might support (or not) a change of paradigm in the approach of the problematic of Chagas.

Bordieu's reflections have never been purely theoretical; on the contrary, they have been built after diverse research in the field. If one could imagine the author trying to understand the problem of Chagas disease, probably he would have proposed to start with a conversation with the first affected ones, as actors. Maybe, he would have relied upon the "spontaneous sociology" of the social actors, which is something more complex than discarding the scholar knowledge and retaining the ordinary one [16]. He would agree with de Sousa Santos in affirming that, from the "sociology of absences," before working with experts one should work with all the experiences, especially with the ones that have been excluded or actively produced as nonexistent, that is, as unbelievable alternatives to what exists [6].

Understanding the affected people's multiple experiences and learnings is one of the ways that urges to be explored to comprehend the problematic and evidence the anxieties and the diverse learnings. Hence, knowing the experiences of the teams of chemical control against the vector, the primary health care physicians, the specialists, the local authorities, the decision makers, etc., would allow a more global and comprehensive understanding of the problematic. To interpret all the actors, Bordieu would say, it is not necessary to comprehend them from our own conceptualizations but to figure out the "logic of practice," that is, the logic explained from the performance of acts displayed in their respective times [17].

Furthermore, the stories heard during the specialized care in Chagas disease in contexts characterized as *nonendemic* in Europe approximate us to the actors from experiences that are even more invisible than the ones experienced in the traditional rural and Latin American scenarios. Their narrations talk about the past time (the childhood in the rural community of origin, the migrant trajectories, etc.) and reveal, in the same way, the coexistence with the vector and the lack of information about the real risks they faced or, even worse, the lack of ability to change the material conditions associated to the disease. Far from their birthplace, these people identify a change that might be linked with the elimination of the disease. However, the presence of Chagas disease at the *endemic areas*, and at *nonendemic areas*, on the one hand, indicates that certain conditions are perpetuated and, on the other hand, evidences that a lot has been achieved from one generation to the next one as regards the vector control and the access to diagnosis and medication. Several of these transformations are related to a change in the population's welfare conditions, housing modifications, access to treatment, etc. These and other actions allow cutting—socially and within families—the *T. cruzi* transmission.

Thus, we can claim that there is agency conquest in those who, through multiple processes of social mobility, have accessed to better conditions and structural opportunities. The associativity and the social mobility also contribute in the conquest of rights. We can see that phenomenon in those who decide to generate a change in their communities through the participation in peasant organizations that fights for violated rights in rural areas or in those who take part in associations of people affected by Chagas disease that proposes to activate, in *endemic* and *nonendemic* contexts, the access to diagnosis and treatment or propose to look for the customs release of second-hand cardiac pacemakers coming from abroad.

The scenarios are diverse, the people's actions and reactions are multiple, as diverse as their learnings and experiences. Within the framework of these diversities, we can comprehend that the working conditions limit the access to health care. We need to identify the stigmas associated to the disease, and understand the complexity of each social scenario, for example, for people in the European context and people in the current Latin American rural context. At the same time, this positioning implies to understand that—close or far from the *endemic* zones—people try to move away from those constructions that impoverish them or represent them as *ill people*, which produce them as *nonexistent* in terms of the *sociology of absences* proposed by de Sousa Santos [6].

It should be stressed that the study of perceptions and social representations of Chagas disease should not be approached only toward the knowledge and representations of the affected people, because the way in which the other actors see and understand the problem also reproduces the features of the topic complexity. We should recognize in all the actors the presence of stereotypes, assessments, attitudes toward the problematic and the people affected, the modes of transmission and the contexts in which the disease is spread (or not), and its causes and consequences [18]. In this sense, we agree with some other authors that approach the need of incorporating and unveiling the complexities in the perceptions and social representations about Chagas [5, 18, 19]. We understand that it is key to deepen the learnings and experiences of all the actors involved in the complexity of Chagas (health teams, researchers, communicators, educators, decision makers, etc.), considering besides that all of them are stained with reality constructions produced in each context. As we shall explain below, these studies constitute a concrete proposal so that the interventions in Chagas disease lay their foundations in a contextualized and inclusive beginning, tending to equity among the parties involved.

4.3 About the Comprehensive Care

4.3.1 What Are We Talking About When We Talk About "Comprehensive Care"?

As we have stated along these pages, we agree with Ventura-Garcia et al. when they propose that Chagas disease is embedded in a web of relationships marked by biological, sociocultural, political, economical, historical, and environmental circumstances that shape its incidence and prevalence, as well as the population's response [4]. In the complexity and multidimensionality of this *web of relationships* the need of approaches that meet expectations is evidenced, whether from research and care, or from other instances. Because of this fact, the development of the so-called (and sometimes overused) *comprehensive care* leads us to challenges that cannot be underestimated.

Beyond some worthy exceptions, we generally observe that one of the main failures of the implemented strategies lies in the predominance of a fragmented and biased viewpoint. This viewpoint is oriented toward Chagas as a disease, ignoring that it constitutes a complex problematic characterized by factors of different nature, which make sense when they are considered in mutual dependence and relationship [5]. In this sense, when exploring, for example, the social representations about the problematic of Chagas in members of the health care team, a restrictive viewpoint on the biomedical aspects is evidenced; moreover, that viewpoint is biased by stereotypes (poverty disease, silent disease, rural disease, a disease associated to housing precariousness, etc.) [18, 19] that hinder the development of the pursued *comprehensive approach*. The reproduction of prejudices leads to stick to attitudes of blaming and stigmatization toward the people affected by Chagas disease themselves, preventing in turn to enquire about the underlying causes that bring together in the problem persistence, whereas the association between rurality and poverty, established by most of the biomedical professionals, reflects the prejudices associated to the nonurban way of life. This viewpoint ignores, and usually disregards, the value of traditions, habits, and practices of certain groups affected by the problematic so that the *experts* become the only ones trained to provide solutions to improve the affected people's lives [18, 20].

Therefore, at the moment of thinking about the spaces where *comprehensive care* should be displayed, it is inevitable to consider the existence of diverse scenarios provided that, as it has been already mentioned, the migrant and urbanization phenomena allow locating Chagas also in urban contexts and worldwide. Due to these reasons, we understand the urgent need of putting into practice strategies promoting health that consider the topic as a public health issue not only at regional and national scales but also at international level.

Chagas has become a paradigmatic example of the challenges that the current territorial changes and globalization pose for health research and public health care. In this sense, it is crucial to aim at the development of an integral, updated, and contextualized approach, oriented not only to preventing the disease but also to promoting health as a means of improving people's lives.

Understood this way, comprehensive care includes both the necessary diversity of health professionals to *treat* properly the requirements of each case and the implementation of multidisciplinary and multisectorial devices to guarantee the approach of the violated human rights, the cultural, social, labor, and legal aspects, and the materialization of true knowledge exchange nourished by the involved actors' experiences and knowledge. In particular, we consider that a worthy contribution in this sense is constituted by the proposal of the *Ecology of knowledge* provided by the Epistemologies of the South [6], which begins by assuming that all the relationship practices among human beings, as well as between human beings and nature, imply more than one way of knowledge and, thus, of ignorance. It consists, on the one hand, of exploring alternative scientific practices that have become visible through the plural epistemologies of scientific practices and, on the other hand, of promoting the interdependence of scientific and nonscientific knowledge. In other words, it involves an ecology based in the "recognition of the plurality of heterogeneous knowledge (being one of them modern science) and the continuous and dynamic interconnections among them without compromising its autonomy" [6:10]. Hence, against a rooted monocultural conception of knowledge, the *Ecology of knowledge* does not consider knowledge abstractly, but as learning practices that allow or prevent certain interventions in the real world.

4.3.2 Difficulties and Tensions as Regard Access: Objective and Subjective Barriers

The processes involved in the population's health care entail the interaction among health care teams, professionals, and people and their families. Usually, this relationship takes place in an institutional context of hegemonic knowledge considered, in general, legitimate. In addition to this, the individuals in the interaction put into dialogue or confrontation their own cosmovisions, that is, the way through which people and systems understand health and illness and, consequently, construct care modes and mechanisms used to elaborate resolution strategies for health problems. In these spaces, there is an interaction of care, health, and illness models that must be evidenced, especially when we approach Chagas disease from hegemonic models [21].

At the same time, these interactions are constructed within sociopolitical frameworks where health is a right, as well as it is people's mobility in most of the countries around the world. In order to understand health and illness in the context of Chagas, it becomes essential to territorialize the problematic firstly, which involves the comprehension of the environment, its changes and adaptations as well as the scenarios (institutional, sociopolitical, local, etc.) associated to the transformations in the individuals' life conditions. These changes are related to the neoliberal exploitation methods and consumption; within this framework, the individuals interact among them through the means because they construct transnational families, and they organized themselves according to international organization structures.

In local contexts, human mobility has shown the fragility of the universal health care systems and, paradoxically, the acceptance of diversity. The current social scenarios reveal the social tensions based on xenophobia and racism as well as the gender inequalities. In which way Chagas disease, as a health problem, is perceived in these contexts? Within the framework of the health system, Chagas disease allows, for example, being aware of the women's position in care practices against health problems, and the responsibility attributed to them as regards health care of the family in general and the children in particular. The way in which women assume the responsibility in the transmission of Chagas places them as high vulnerability individuals because of stigmatization, social rejection, gender violence, etc. In this background, women must be guaranteed health conditions for themselves, such as preconception controls to diminish the chances of infection of Chagas through pregnancy and, at the same time, improve their health conditions and quality of life [15]. However, we should take care about the mechanisms used to approach them, preventing the reproduction of prejudices and discriminatory behavior and being aware of the social exclusion processes due to the relationship between gender and the disease transmission mechanisms.

Physical conditions, ethnicity, social class, and religion account for the need of constructing an approach that implies the revision of relationship strategies [19]. Thus, for instance, in migrant population contexts, both in research and in the development of care and detection policies of Chagas disease, it is essential to revise the strategies carefully to sensitize, educate, and implement interventions with native and foreign populations. On the other hand, as we stated in the previous section, the learnings and experiences of the affected people should be considered to promote the dialogue among the health care team members and the other involved social actors. We also stated that this dialogue will lead to the development of strategies oriented toward guaranteeing the right to health disregarding people's origin and context.

Health should be acknowledged as a fundamental right for all people in the diverse global scenarios. This means that States should guarantee the care, treatment, and monitoring of Chagas disease in *endemic* and *nonendemic* contexts as well as in the current context where this problematic constitutes a global health problem and the population has not been considered yet the leading and main actor to whom all the actions related to the disease approach should be addressed.

Currently, the access barriers experimented by the population affected by Chagas reflect relationships with objective elements that condition the access to both health care and work due to presence of positive serology. On the other hand, these barriers imply that the individuals, although being holders of universal rights, face situations of discrimination and social exclusion, product of their national, ethnical, and gender. Both in the societies where the people affected by Chagas disease are *traditionally* located and in the *new contexts* where the disease is present, people face social exclusion constructions associated to discriminatory practices based on a diversity of prejudices and stigmas that constitute barriers of structural and subjective or symbolic inequality.

One of the risks faced by health teams is to attribute themselves the Others' *voice* when recognizing situations of exploitation, undermining, and social exclusion, among others, suffered by the people affected by Chagas. Although this assertion states the need of contributing in the restitution of the individuals' rights (because health teams, as State representatives, have the duty of guaranteeing the population's right to health), it is crucial to remember that this contribution should take place acknowledging the individuals' agency. Thus, professionals may support organization processes in which the individuals themselves might be represented and acknowledge individual agencies that favor intercultural interchange processes, but it is not acceptable for them to arrogate *the voice* of those who live Chagas disease as an everyday life reality. Men, women, and children, who within the framework of the problematic have been displaced or excluded, must be listened; but they themselves should articulate and develop mechanisms of social representation.

Furthermore, the presence of active individuals is key, those who, on the basis of their own experiences, might contribute to the development of culturally pertinent interventions and actions of intercultural character that imply an input to the improvement of the conditions of the violation of rights. We refer, for example, to interventions such as the programs "Expert patient" [22] or "Spread the voice" [23] that, somehow, acknowledge the individuals' agency within the framework of the problematic of Chagas. Agency that articulates with the medical performance provides not only a better understanding of the problem from the affected individuals' viewpoint and experiences but also a privileged learning space for the involved health teams.

4.3.3 Initiatives Designed in an Inclusive Relationship

Research findings and concrete intervention experiences performed in different places of the world present eloquent results about the importance of considering diverse actors as authentic partners to generate transformations inside the complex problem of Chagas. In the last few years, several innovative approaches have pioneered, both individually and collectively, the treatment improvement of individuals directly or indirectly affected by Chagas by combining biomedical, psychological, social, and anthropological elements and seeking to overcome stigma and barriers [7, 22–25]. At the same time, information, education, and communication initiatives have been developed in a similar spirit, including strategies and resources designed to address the issue holistically and incorporate constructive and innovative perspectives, both for the benefit of the individuals affected and for society [5, 24]. Although the initiatives are still not numerous, they set a precedent to be considered and allow enlightening adaptations in broader areas (at local, rural, urban, cosmopolitan, and transnational contexts).

Analyzing all the implications of the previously cited experiences exceeds the aim of these pages. However, it is possible to recognize common ground to point out that, beyond the particularities of each case, in all the situations some transversal elements are highlighted, such as the acknowledgement of the agency or the potentiality of agency of all the actors, the integration (to a greater or lesser extent) of the diverse actors at the moment of designing the initiative, the perspective of rights, the consideration of the multiplicity of voices and aspects of the individuals' everyday life (beyond the *disease*), and the consciousness about costs and engagements of this kind of approach.

However, it should be clarified that an inclusive and dialogic beginning does not guarantee in itself the intervention success but, maybe, it allows delineating scenarios in which the eventual conflict due to the disagreement of conceptions and expectations of the diverse actors might be known in time, and the tensions might be channeled toward understanding and mutual cooperation. We might draw away from, for example, the "mismatch between the experience of those affected and the medical classifications of Chagas disease" indicated by Ventura-Garcia et al. [4]. The dialogue between the health systems and the individuals, and the opportunities displayed with the aim of understanding the diversity of conceptions of health and illness would improve the scenarios and the practices of the medical systems, not only because they allow a better agreement and approach to the affected population, but also because in the relativization of their own conceptions of health and the world, new possibilities are open to perform in health and health care [19]. Then, almost as a first measure, we suggest to reinforce research about the social representations of Chagas disease influencing the behavior, the agency, of both the affected individuals and the other involved actors [5, 18]. Hence, it will be possible to obtain practical and contextualized recommendations to co-design the best interventions as more information becomes available about the different experiences and the diversity of knowledge related to the problematic of Chagas and all the elements emerging at the moment of asking oneself *What are we talking about when we talk about Chagas?* in each particular context.

As we have already stated, from the possibilities opened when we recognize the affected individuals' experience and acknowledge their agency, cultural and intercultural pertinent interventions such as the "Expert Patient" [22] or "Spread the voice" [23] have been generated. Another type of agency whose potentialities have been evidenced is the one of the migrant people in the so-called *transnational social spaces*, that is, those spaces emancipated from the national borders. The transnational practices—those ones regularly held, with systematization and strong adherence between the origin and destination migrant communities—explain that in the transnational spaces flow not only economic remittances but also social remittances, understood as "the normative structures (ideas, values, and beliefs), the practice systems, and the social capital flowing from the resident families into the host society toward its origin society" [26]. In this transnational context, data are already collected about changes in the searching paths of solutions of the affected individuals by Chagas disease in *endemic* areas that might be explained from the appropriation of ideas and practices influenced by their relatives living abroad [27]. These opportunities broaden and become complex if we consider that the mobility of several families can be ascribed in circular or cyclic mobility tradition from which the actors demand the health team's capacity of adaptation, maybe toward the offering of a transnational care in Chagas disease, although that type of intervention could have been held only in structures financed by specific cooperation projects [27].

Mentions should also be made of forms of social mobilization spearheaded by associations of people affected by Chagas disease, medical workers, and researchers all over the world. They have been crucial in terms of raising awareness, promoting access to diagnosis and care, and boosting the development of applied research [28]. One extremely important initiative in this area was the creation in 2009 of the International Federation of Associations of People Affected by Chagas Disease (FINDECHAGAS),[3] which currently brings together more than 20 associations in the Americas, Europe, and the Western Pacific. This instance collects the experience and needs of women and men affected by Chagas disease worldwide.

Considering another type of actors, it is worthy to highlight the existence of initiatives such as the "Global Coalition of Chagas Disease"[4] that gathers researchers, public health professionals, and private donors who articulate with affected people and some other members of the civil society. Among their principal challenges, the Coalition leads efforts to approach aspects different from the biomedical dimension of Chagas such as the incidence and the access to diagnosis and treatment.

Finally, from the World Health Organization (WHO) as a possible scenario of confluence and spreading of validated initiatives, we underline the recent creation of the Technical Group No. 6 on Information, Education, and Communication (TG6 IEC)[5] of the WHO Control Program of Chagas Disease. This instance is presented as a key opportunity for a group of professionals—specialists from different areas without conflicts of interest—to counsel the WHO from multidimensional viewpoints to promote, on a larger scale, the change of approach necessary to account for the challenges posed in the new epidemiologic scenarios and the essential epistemological changes we have been stating.

[3] http://findechagas.org/

[4] http://www.coalicionchagas.org/

[5] www.beatchagas.org/

4.4 An Overview: Some Keys to Facilitate the Desired Access

As a corollary of the words shared in these pages, we come back to the reflection leading us to think about the numerous elements that should be incorporated systematically and rigorously to the search for answers to understand the current problematic of Chagas and reach contextualized and long-lasting solutions. In this sense, we agree on that the new paradigm we referred to at the beginning of these chapter must consider the fundamental and unavoidable importance of subjectivity as it is stated by de Sousa Santos [6], who considers that objectivity depends on the quality of its subjective dimension and the development and consolidation of social transformation collective actions.

Therefore, it is necessary to systematize and assess the interventions from integrative perspectives as well as to develop research tools to expand the focus of these activities incorporating the sociocultural aspects of health. "Furthermore, both research focused on incorporating people's experience and needs into policies and interventions in endemic and non-endemic countries and the development of preventive and/or control actions conducted with attention to affected individuals beyond medical spaces are crucial" [4:6]. However, we have to be cautious because the call is not only to add these perspectives, as we have already stated; on the contrary, it is about articulating dialogic relationships that favor supporting acknowledgement processes of the distances of knowledge and power from which we face Chagas. These distances translate into barriers for the access to an adequate and prompt health care and the stigmatization and violation of the affected people's rights. Thereby, the revision of processes that in practice exclude, assimilate, and undermine is imperative to promote, from that revision, a dialogue introducing new social and institutional learnings and practices.

The epistemological transformation will allow enhancing the delimitation of the problematic of Chagas and, at the same time, will favor processes to acknowledge emergent actors and practices in order to incorporate diversities as regard gender, age, ethnicity, and class and to include other actors such as social, national, and international organizations, the media, etc. From research, its practices and the diverse ways of constructing knowledge, it must be evidenced the need of incorporating from the considerations of the Epistemologies of the South [4] the dialogic articulation required today in the understanding of Chagas disease. In other words, from these new perspectives, we might be alert to the changes and transformations raised from the individuals who consult (or not) and to the modifications from the health care systems manifested in those new relationships, to go beyond the current unequal framework of relationships and transform society. There we will find the signals to overcome the dichotomies accounting for practices of disqualification and *nonexistence*, in the end, of social exclusion: healthy/ill, ignorant/expert, poor/not poor, endemic/nonendemic, patient/agent, with a voice/without a voice, etc. By being aware of these dichotomies, we evidence the ideological frameworks through which we approach health and illness in general and Chagas disease in particular. With these evidences on the table, we will begin to walk with steady steps the collective way toward the development of true comprehensive approaches.

References

1. Coura JR, Viñas PA. Chagas disease: a new worldwide challenge. Nature. 2010;465(7301 Suppl):S6–7.
2. WHO. Chagas disease (American trypanosomiasis). 2018. http://www.who.int/es/news-room/fact-sheets/detail/chagas-disease-(american-trypanosomiasis). Accessed 9 Nov 2018.
3. Dias JCP. Present situation and future of human Chagas disease in Brazil. Mem Inst Oswaldo Cruz. 1997;92:13–5.
4. Ventura-Garcia L, Roura M, Pell C, Posada E, Gascón J, Aldasoro E, et al. Socio-cultural aspects of Chagas disease: a systematic review of qualitative research. PLoS Negl Trop Dis. 2013;7(9):e2410. https://doi.org/10.1371/journal.pntd.0002410. Accessed 26 Aug 2019.
5. Sanmartino M (Coordinación). Hablamos de Chagas. Aportes para (re)pensar la problemática con una mirada integral. Contenidos: Amieva C, Balsalobre A, Carrillo C, Marti G, Medone P, Mordeglia C, Reche VA, Sanmartino M, Scazzola MS. Buenos Aires: CONICET; 2015.
6. De Sousa Santos B. Descolonizar el saber, reinventar el poder. Montevideo: Trilce, Extensión Universitaria UDELAR; 2010.
7. Velarde-Rodríguez M, Avaria Saavedra A, Gómez-i-Prat J, Jackson Y, de Oliveira Junior WA, Camps-Carmona B, Albajar-Vinas P. Need of comprehensive health care for *T. cruzi* infected immigrants in Europe. Rev Soc Bras Med Trop. 2010;42(Suppl 2):111–5.
8. Fernández P. Elementos para una teoría de la acción social. In: Liedo MC, Fernández P, editors. Saberes y emancipaciones desde el sur. Córdoba, Argentina: Villa María, Eduvim; 2016.
9. Lander E (Comp). La colonialidad del saber: eurocentrismo y ciencias sociales. Buenos Aires: CLACSO. 2000. http://bibliotecavirtualclacsoorgar/clacso/sur-sur/20100708034410/lander-pdf. Accessed 25 Jan 2020.
10. Walsh C. Interculturalidad crítica y (de) colonialidad, Ensayos desde Abya Yala. Serie pensamiento descolonial. Quito: Abya Yala; 2012.
11. Avaria A. El Estado y la incorporación de las diferencias. ¿Problema Resuelto? In: Fundación CIDOB. Pensar las dinámicas interculturales. Aproximaciones y perspectivas. Foro de doctorandos, Documentos CIDOB, N° 10. Dinámicas interculturales. Barcelona: CIDOB; 2007. p 122–140.
12. Kleinman A, Eisenberg L, Good B. Culture, illness, and care: clinical, lessons from anthropologic and cross-cultural research. Focus. 2006;4:140–9.
13. Good B. Medicina, racionalidad y experiencia. Una perspectiva antropológica. Barcelona: Edicions Bellaterra; 2003.
14. Avaria A. Migrar, enfermar-sanar lejos de casa. Bolivianos en Barcelona, experiencia hecha carne. In: Agar L (coord.) Migraciones, salud y globalización: entrelazando miradas. Santiago: Edición OIM, OPS, MINSAL; 2010. p. 199–213
15. Avaria A, Gómez i Prat J. Si tengo Chagas es mejor que me muera. El desafío de incorporar una aproximación sociocultural a atención de personas afectadas por la enfermedad de Chagas. Revista multidisciplinar de SIDA, Tuberculosis, drogodependencia, y otras enfermedades emergentes. IV Taller sobre la enfermedad de Chagas importada: tratamiento y transmisión vertical. 2008;10:40–5.
16. Corcuff P. Las nuevas sociologías. Madrid: Alianza Editorial; 1995.
17. Giorgis M. La virgen prestamista. Buenos Aires: Antropofagia; 2004.
18. Sanmartino M, Amieva C, Medone P. Representaciones sociales sobre la problemática de Chagas en un servicio de salud comunitaria del área de Gran La Plata, en Buenos Aires, Argentina. Glob Health Promotion. 2018;25(3):102–10. https://doi.org/10.1177/1757975916677189. Accessed 25 Jan 2020.
19. Avaria A. Un cuerpo vale más que mil palabras. Mujeres y hombres bolivianos en Barcelona. Corporización de la migración: cuerpo migrante, cuerpo trabajador, cuerpo enfermo. Tesis Doctoral, Departament - Antropologia Cultural i Història d'Amèrica i d'Àfrica, Barcelona; 2014. http://hdl.handle.net/2445/62705. Accessed 25 Jan 2020.

20. Padilla Velázquez R. Conocimiento epidemiológico de la enfermedad de Chagas por los médicos familiares de la UMF 66 del Instituto Mexicano del Seguro Social. Tesis de Epidemiología, Universidad Veracruzana, Veracruz; 2014.
21. Menéndez E. Modelos de atención de los padecimientos, de exclusiones teóricas y articulaciones prácticas. Ciência y Súde colectiva. 2003;8(1):185–297.
22. Claveria Guiu I, Caro Mendivelso J, Ouaarab Essadek H, et al. The Catalonian Expert Patient Programme for Chagas Disease: an approach to comprehensive care involving affected individuals. J Immigr Minority Health. 2017;19:80–90. https://doi.org/10.1007/s10903-016-0345-y. Accessed 26 Aug 2019.
23. Muñoz E. Facilitar el acceso a la atención en Chagas a inmigrantes bolivianos en el Servicio de Salud Internacional del Hospital Clínico de Barcelona. Barcelona: Universidad Autónoma de Barcelona, Mimeo; 2018.
24. Sanmartino M, Avaria Saavedra A, Gomez i Prat J, Parada C, Albajar-Viñas P. Que no tengan miedo de nosotros: el Chagas según los propios protagonistas. Interface (Botucatu) 2015. 2015;19(55):1063–75.
25. Oliveira Júnior WA. Atenção integral ao paciente chagasico: uma proposta para o cuidar. Arq Bras Cardiol. 2005;84(1):1–2.
26. Levitt P. The transnational villagers. Los Ángeles: California University Press; 2001.
27. Pinazo M-J, Pinto J, Ortiz L, Sánchez J, García W, Saravia R, et al. A strategy for scaling up access to comprehensive care in adults with Chagas disease in endemic countries: The Bolivian Chagas Platform. PLoS Negl Trop Dis. 2017;11(8):e0005770. https://doi.org/10.1371/journal.pntd.0005770.
28. Médicos Sin Fronteras (MSF). Movilización popular y enfermedad de Chagas. Río de Janeiro: MSF; 2012.

Diagnosis of *Trypanosoma cruzi* Infection: Challenges on Laboratory Tests Development and Applications

5

Montserrat Gállego, Alejandro G. Schijman, and Julio Alonso-Padilla

5.1 Laboratory Techniques for Diagnosis of the Acute or Chronic Stage of Chagas Disease

The range of techniques to diagnose a *Trypanosoma cruzi* (*T. cruzi*) infection includes parasitological, molecular, and serological methods. The use of one type or another is mostly related to the stage of the disease: acute or chronic [1]. This is so because parasitemia varies between the two stages being elevated during the acute phase and low and intermittent during the chronic phase. Thus, parasitological and molecular methods are generally used for the diagnosis of acute Chagas disease, whereas serological methods are the preferred choice for the diagnosis of chronic Chagas disease [1, 2].

Despite the great advances achieved in the last few decades, the absence of consensus on the techniques to be used, the different criteria regarding which diagnostic

M. Gállego
Section of Parasitology, Department of Biology, Healthcare and the Environment, Faculty of Pharmacy, Universitat de Barcelona, Barcelona, Spain

Chagas, Parasitic and Imported Disease, Barcelona Institute for Global Health (ISGlobal), Hospital Clínic - University of Barcelona, Barcelona, Spain
e-mail: mgallego@ub.edu

A. G. Schijman
Laboratory of Molecular Biology of Chagas Disease (LaBMECh), Instituto de Investigaciones en Ingeniería Genética y Biología Molecular "Dr Hector Torres" (INGEBI-CONICET), Buenos Aires, Argentina
e-mail: schijman@dna.uba.ar

J. Alonso-Padilla (✉)
Chagas, Parasitic and Imported Disease, Barcelona Institute for Global Health (ISGlobal), Hospital Clínic - University of Barcelona, Barcelona, Spain
e-mail: julio.a.padilla@isglobal.org

M.-J. Pinazo Delgado, J. Gascon (eds.), *Chagas Disease*,
https://doi.org/10.1007/978-3-030-44054-1_5

algorithms to follow, and the inexistence of governmental guidelines in some countries, both endemic and non-endemic, still complicate the disease diagnosis scenario. Besides, the genetic and antigenic variability among *T. cruzi* strains poses another challenge to deal with, as it plays a role in the distinct geographical performances of the tests [1]. On the other hand, biomarkers for the disease prognosis and the early assessment of treatment efficacy are also being investigated, but any test based on them is still far from commercialization [3].

5.1.1 Parasitological Techniques

Parasitological methods were the earliest developed and still include the standard method recommended by the WHO for acute Chagas disease diagnosis: microscopy-based visualization of *T. cruzi* trypomastigotes in freshly collected peripheral blood [4]. In its simplest format, a drop of blood, which can be taken by digital puncture, is placed between a slide and a coverslip and searched for the presence of motile trypomastigote forms using a 40× objective lens upon which these can be recognized by their movement among the red blood cells [4]. The sensitivity and limit of detection vary accordingly to the volume of blood examined [5], and thin and thick blood smears stained with Giemsa could also serve to the diagnosis [4].

With the aim to increase the sensitivity of microscopy-based direct observation of trypomastigotes, concentration methods such as the Strout or the micromethod were developed [6, 7]. The Strout requires a greater volume of blood (5 mL), and it is thus indicated for the diagnosis of acute Chagas in adults when there is no chance to perform serology, i.e., when there has not been sufficient time to mount the antiparasitic humoral response. For the Strout, blood is obtained without anticoagulant and left at room temperature (RT) to coagulate. Serum can then be obtained following a double centrifugation procedure, and trypomastigotes are searched in its sediment under the microscope. In contrast, the micromethod is the technique of choice for the diagnosis of congenital Chagas disease in newborns due to the smaller volume of blood required (~0.3 mL may suffice). In this case, 4–6 heparinized capillary tubes containing 50 μL of blood are centrifuged and the parasites presence is searched at the interface between the blood cells lower phase and the plasma upper phase [6]. Its detection limit is very variable (40–500 parasites/mL) [6, 7], and it is recommended to examine the samples immediately after their collection or within 24 h from it [8].

Both Strout and micromethod have in common that they require highly trained personnel [1]. In terms of their throughput limitation, in order to be able to perform the examination of a large number of blood samples, a computational method has been described for the automatic detection of *T. cruzi* parasites based on machine learning and image processing algorithms [9].

Although nowadays not as common as the former, there are other methods for the parasitological diagnosis of *T. cruzi* infections like the xenodiagnosis and the hemoculture. These are more laborious than direct microscopy-based observation, and despite they may yield a greater sensitivity, it can take 15–60 days to know their outcome [1]. In addition, they require of specific laboratory conditions and entail safety concerns as both involve growing the parasites. For instance, xenodiagnosis requires of uninfected laboratory-reared triatomine vectors (order Hemiptera; family Reduviidae) which are allowed to feed on the subject to be diagnosed and then examined for trypanosomes that may be found in the intestinal contents of the insects 1 or 2 months later. Hemocultures are as well examined weeks after inoculating culture media (LIT, Novy-MacNeal-Nicolle agar (NNN), or Evans' modified Tobie's medium) with the suspected subject's blood and its detection limit is 5 parasites/mL [10].

5.1.2 Molecular-Based Diagnostics

The Polymerase Chain Reaction (PCR) has been used for decades for the amplification and detection of pathogen-specific nucleic acid sequences so as to diagnose infectious diseases [11]. *T. cruzi* parasites have nuclear and kinetoplast DNA, which contain repetitive sequences that are suitable for PCR amplification [12]. PCR was used for the first time to detect *T. cruzi* DNA on both the triatomine vector and the mammalian host in 1989 [13]. Amplification of the parasite DNA has been described from many types of samples like blood buffy coat, umbilical cord blood, serum, cerebrospinal fluid, as well as tissue samples from the placenta, heart, esophagus, colon, brain, or skin [14–19]. Nonetheless, for practical reasons, the preferred sample is non-coagulated periphery blood, generally collected in tubes carrying EDTA as anticoagulant.

Normally, a DNA extraction procedure must always precede the PCR. In order to increase the blood's half-life at RT, 1:1 (vol.:vol.) addition of 6 M guanidine hydrochloride to 0.2 M EDTA blood collecting tubes can be done before the DNA extraction step [20]. Although this treatment was shown to increase the subsequent PCR sensitivity in comparison with the use of untreated whole blood as sample [21], there are yet no commercial tubes including it. Moreover, its cost is high and its use is usually limited to reference diagnostic centers [16]. It must be noted that guanidine EDTA-blood (GEB) samples have been described to be unsuitable for DNA extractions based on magnetic particles, at least when samples are not processed immediately [22].

Several PCR protocols have been now described for *T. cruzi* DNA detection. In them, two repetitive sequences in the parasite's genome are the most used amplification targets: satellite DNA (satDNA), which is a conserved nuclear mini-satellite region designated TCZT with the highest number of repetitions (~10 thousand copies per genome organized in series); and kinetoplastid DNA (kDNA), composed by networks of concatenated DNA maxi- and mini-circles that may reach between 5000 and 10,000 copies [23]. Conventional PCRs allow the detection of a single diagnostic band of a known size in an agarose gel and being the simplest they were the first to be used [12, 13], sometimes followed by DNA hybridization [24, 25]. Nested PCR (nPCR) protocols, which provide a higher sensitivity than conventional PCRs, have also been used for the clinical diagnosis of the disease [26, 27]. However, they entail a high risk of false positive results due to contaminating amplicons, so real-time PCR (rtPCR) technology is the preferred PCR assay at present. It provides a rapid throughput of results (amplification and detection in one step) with a reduced risk of carry-over contamination, and it can be optimized as a quantitative assay given appropriate standards are included (quantitative PCR or qPCR) [17, 28–31].

The described sensitivity and specificity values of qPCR methods are variable and depend on the epidemiological characteristics of the population, as well as on many technical factors such as the volume of the clinical sample, the conditions of its storage, the method used to isolate the DNA, the parasite target sequences, the set of primers and probes chosen for the amplification, the reagents used for it, and the thermo-cycling conditions followed [22, 23]. Nonetheless, molecular-based methods provide a more sensitive alternative to classical parasitological techniques, and they are thus very useful for the diagnosis of acute Chagas disease, such as congenital and oral transmission cases or those due to parasite reactivation in immunosuppressed individuals [2, 23]. In the case of congenital infection, detection of the parasite DNA in the newborn's blood by PCR could greatly help to improve the early diagnosis of the infection and preclude the high rate of loss to follow-up observed with current algorithms [32]. Molecular methods have also proved very useful to monitor *T. cruzi* infection status of solid organ transplantation recipients as PCR can be used independently of the host immune status allowing the parasite detection even before the appearance of symptoms [33–35].

In contrast, molecular tests should not be ordered for diagnosis of the chronic stage because low and intermittent parasitemia will preclude any sensitive detection [19, 36]. Nonetheless, in the absence of other approved biomarkers, and considering that patients' positive serological status can take many years to revert, qPCRs have been shown to be very helpful in the follow-up of treated chronically infected adults, especially during the performance of clinical trials [37, 38]. In this context, one of the limitations of DNA-based amplification methods is whether a positive result reflects the presence of intact parasites or circulating DNA derived from lysed ones [39]. Cancino-Faure and co-workers [40] have developed a qPCR with a propidium monoazide (PMA) dye that allows the differentiation between viable and nonviable *T. cruzi* epimastigotes, which could be very useful for the assessment of lytic drugs. Moreover, Juiz and co-workers [41] have developed a rtPCR procedure based on *T. cruzi* 18s ribosomal RNA (rRNA) to detect transcriptionally active parasites in tissue specimens.

Another drawback posed on molecular detection of *T. cruzi* is the lack of homogeneity between protocols. That is why multi-centric studies to analytically validate and standardize the best methodologies were performed [19, 31, 37]. A recent article has reviewed the clinical sensitivity and specificity from PCR studies of at least 100 subjects with suspicion or previous diagnosis of *T. cruzi* infection [23]. In summary, kDNA targeted qPCRs appeared to be slightly more sensitive than satDNA-based qPCR protocols [2]. The analytical sensitivity was more uniform among different Discrete Typing Units (DTUs) for kDNA qPCRs than for satDNA qPCRs, being the latter less sensitive for some TcI and TcIV strains due to a lower number or satellite DNA units in their genomes [23]. Nonetheless, a duplex satDNA qPCR performed in peripheral blood samples from patients residing in Colombia, where genotype I (TcI) is the prevailing DTU, showed high specificity in both acute and chronic Chagas disease patients and high sensitivity in acute patients, whereas in chronic patients clinical sensitivity reached 64.2% [16]. This degree of PCR positivity to detect circulating *T. cruzi* in chronically infected patients has been observed as well in many other studies [23]. Besides, kDNA-PCR specificity is questioned in areas of *Trypanosoma rangeli* (*T. rangeli*) and *T. cruzi* co-circulation due to the similarities of their mini-circle sequences, and satDNA qPCRs should be used there [2, 23]. Notably, the recent characterization of satellite sequences from a higher number of strains has allowed the improvement of primer/probe design, and consequently sensitivity [42].

5.1.3 Serology-Based Diagnostics

Several serology-based techniques have been developed for the diagnosis of chronic Chagas disease including indirect immunofluorescence antibody test (IFAT), agglutination tests (AT), indirect hemagglutination assay (HAI), enzyme-linked immunosorbent assay (ELISA), immunoprecipitation (IP) or particle gel immunoassay (PaGIA), radioimmunoprecipitation assay (RIPA), immunochromatographic test (ICT), in vitro enzyme strip assay (ESA), chemiluminescent microparticle immunoassay (CMIA), indirect chemiluminescent immunoassays (CLIA), and western blotting (WB). There are commercialized tests for detecting anti-*T. cruzi* antibodies of all these techniques except RIPA test.

WHO criteria for the serological diagnosis of chronic Chagas disease recommends the use of two tests based on distinct antigen sets [4], and this recommendation is followed in the last inform of the Pan American Health Organization (PAHO) [43]. In case of discordance, a third test should be available (Fig. 5.1). Correct use of two of the conventional techniques most frequently applied (ELISA, IFAT, or HAI) allows the diagnosis of >98% of those chronic patients screened, and the sensitivity and specificity of the commercial techniques may approach 99% [44]. On the other hand, due to the high sensitivity (~100%) shown by some of the latest developed techniques (i.e., CMIA), it has been proposed the use of a single technique for the diagnosis of chronic Chagas disease (at least in non-endemic areas), though increasing the cut-off established by the manufacturer [45, 46]. However, despite the cost savings that the proposed change would provide, the most recently published report from the PAHO [43] does not consider it and remains similar to the WHO recommendations from 2002 [4].

In spite of the great advances achieved, serological methods yet present some disadvantages like potential cross reactivity with other related protozoan diseases, especially leishmaniasis [47]. Besides, since serological markers may take several years to disappear in drug treated patients, a positive serology does not necessarily mean an active infection [48]. On the other side, although reversion to a negative serology result will indicate cure, due to the long time span to arrive to it, serology cannot currently be used in a time practical manner to inform on drug treatment

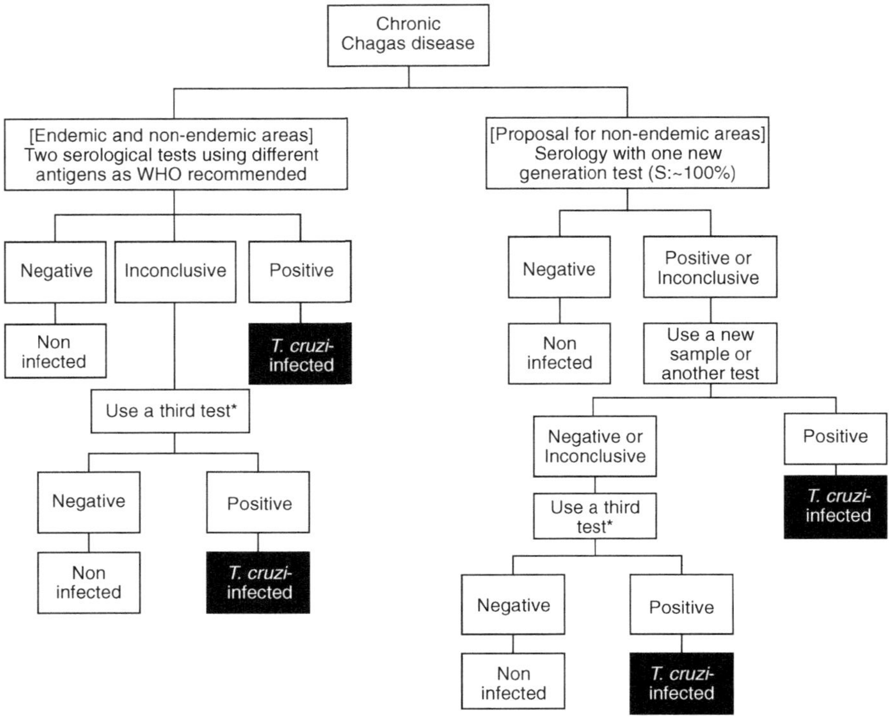

Fig. 5.1 Box scheme representation of chronic Chagas disease diagnosis algorithms (currently used and proposed). * Third test can be either based on a distinct set of antigens or be a different diagnostic technique; *S* sensitivity

response. In addition, the majority of the serological techniques aforementioned need experienced staff and expensive large sized pieces of equipment (immunofluorescence microscope, spectrophotometer reader,…), which makes them not suitable for the diagnosis of the disease in ill-equipped laboratories in primary healthcare centers of many regions of endemic countries. Rapid diagnostic tests (RDTs) were developed as point-of-care (POC) diagnostics for these settings as they require little or no equipment, have few manipulation steps, can be read visually and provide a quick turnaround of results (see Sect. 5.2.2.3 below). A Target Product Profile (TPP) addressing the characteristics required for such POC tests has been published [49].

Serology is also a fundamental tool in the algorithms for the diagnosis of congenital Chagas disease (Fig. 5.2). At birth, most of children are seropositive due to the transmission of maternal IgG antibodies, which will lead to false positive results. On the other hand, the detection of anti-*T. cruzi* IgM antibodies in newborns is controversial, because it may also lead to false positive cases due to abnormal placental passage of maternal IgM or antibodies type rheumatoid factor directed against IgM allotypes [50, 51]. Besides, the kinetics of IgM in the infected newborn is different

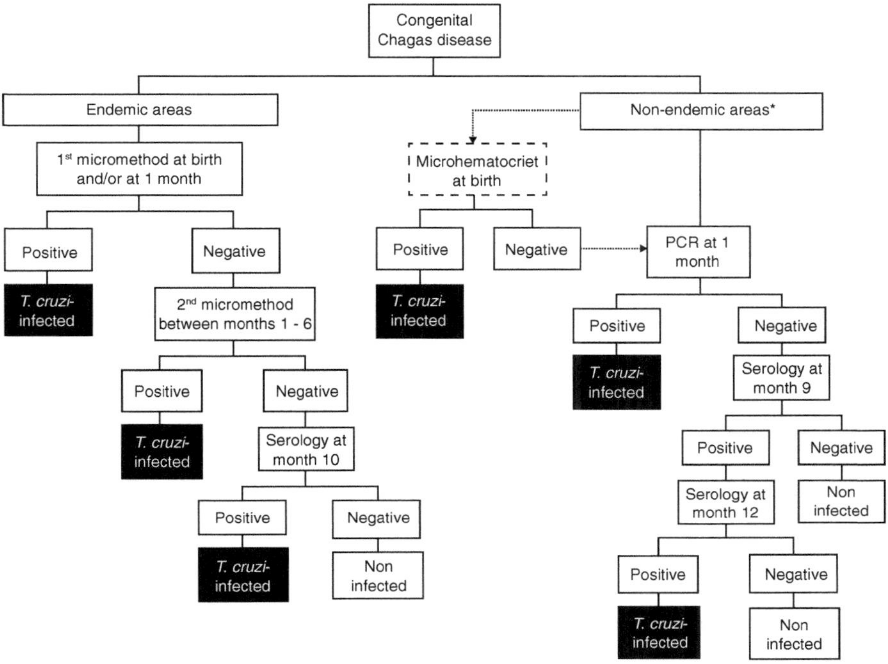

Fig. 5.2 Box scheme representation of congenital Chagas disease diagnosis showing the most used algorithms in endemic and non-endemic settings. * This algorithm is based on what is performed in the Catalunya (Spain) at present; box with dashed line indicates that the procedure is optional as microhematocrit is solely performed with fresh blood and in those centers with trained personnel available

from that of acute non-congenital Chagas, and there are few commercial kits to detect IgM. The detection of anti-*T. cruzi* IgGs in newborns must be performed by 9 months of age onwards as it is when mother-derived ones wane [50, 51]. If CMIA is used, in case of positivity at 9 months of age, it has been suggested to perform another test at 12 months of age due to the high sensitivity of the technique, which could pick up some remaining mother-derived IgGs potentially leading to confounding results [52] (Fig. 5.2).

5.2 Advances in Chagas Disease Diagnosis During the Last 10 Years

Although research on Chagas disease has been traditionally neglected, in the last few years there has been a shift in that drift and many advances are being made. For instance, there are several clinical trials ongoing with new drugs or evaluating new regimens of currently available ones (benznidazole (BNZ) and nifurtimox (NFX)) [53]. Bearing in mind that increasing access to diagnosis (and treatment), and finding reliable tests of cure are yet key pending subjects in the disease control path, research on new diagnostic methodologies has been a very active field in the last decade. Here we provide an overview of this matter, dealing first with molecular-based diagnostics and later on with serological methods.

5.2.1 Advancements in Molecular Diagnostics

5.2.1.1 Polymerase Chain Reaction (PCR)

Novel primer and probe sequences as well as sampling strategies are currently being designed and evaluated in the field to improve accuracy of available PCR assays [19, 39, 43, 44, 54]. Also, an external quality assurance program for the evaluation of *T. cruzi* qPCRs performance in molecular laboratories has been designed and implemented mostly in the framework of clinical trials with trypanomicidal drugs [38, 55]. The diagnosis potential of a digital droplet (dd)PCR platform has also been recently evaluated for the detection and quantification of *T. cruzi* satDNA in a panel of 192 previously characterized DNA specimens, extracted from blood samples of individuals with and without Chagas disease [56]. The ddPCR assay performed well, showing a limit of detection of 5 copies/μL or 1 parasite/mL with excellent agreement in clinical samples in comparison to standardized qPCR [56].

Of most interest in relation to the previously discussed lack of standardization is that two *T. cruzi* qPCR tests have become commercially available [22, 54]. If they were not so expensive, their introduction could mean a major improvement in the Chagas disease diagnosis scenario. However, it must be noted that despite commercial qPCR kits might be available, close attention should still be paid to the steps preceding them (blood volume, sample storage and treatment, and DNA extraction method followed) in order to obtain reliable and comparable results [22]. It must not be forgotten either that any qPCR protocol would still need of trained personnel and equipped laboratories, which place them in no good position for their widespread use in many regions highly endemic to the disease.

5.2.1.2 Loop Mediated Isothermal Amplification (LAMP)

The loop mediated isothermal amplification assay (LAMP) is an easy-to-do molecular technique less exigent in terms of equipment requirements than PCRs [57]. It is able to amplify large amounts of DNA within 30–60 min of incubation at 60–65 °C, employing a complex design of primer sequences and strand displacement *Bst* DNA polymerase [57]. Furthermore, its reagents are stable at room temperatures up to 37 °C avoiding the need of a cold chain, no thermocycler is needed for the reaction, and the reaction product visualization can be by naked eye or followed in real time by turbidity or fluorescence using intercalating dyes [57].

A first *T. cruzi* LAMP procedure targeting 18 s rRNA gene that was evaluated in triatomine feces showed a sensitivity of 100 fentograms (fg) of DNA per test but was cross reactive with *Leishmania* spp. DNA [58], whereas detection level sensitivity in human blood was 50 parasites/mL [59]. A recent prototype kit for detection of *T. cruzi* satDNA in human blood samples has been developed by Eiken Company

[60]. It contains dried reagents on the inside of the microtube caps, and it detected 1×10^{-2} parasite equivalents/mL in blood samples anticoagulated with EDTA and spiked with known concentrations of culture parasites when DNA extraction was done using commercial columns or rapid "boil and spin" method [60]. Remarkably, it did not amplify *Leishmania* spp. or *T. rangeli* DNAs, and the method appears highly sensitive for congenital Chagas disease, for immunosuppressed patients with Chagas reactivation, and for detection of orally infected cases [61]. The attributes desired for a congenital POC test have been described [49], and larger field studies are required to determine whether the *T. cruzi*-LAMP prototype by Eiken fulfills with them.

Other isothermal amplification technologies are also being investigated as it is the case of the recombinase polymerase amplification (RPA). This even exceeds LAMP in terms of lower amplification temperature (i.e., lower consumption of electricity for the reaction or simpler thermal systems to achieve it) and shorter amplification times. A RPA assay has been described for *T. cruzi* DNA detection, and it showed a very good performance comparable to that of qPCR in dogs' samples from Mexico [62].

5.2.1.3 Nucleic Acid Aptamers

Nucleic acid aptamers are nucleic acid species that have been engineered through repeated rounds of in vitro selection or SELEX (systematic evolution of ligands by exponential enrichment) to bind to various molecular targets such as small molecules, proteins, nucleic acids, and even cells, tissues and organisms [63]. Aptamers are useful in biotechnological and therapeutic applications as they offer molecular recognition properties that rival that of the commonly used antibodies [63]. In addition to their discriminate recognition, aptamers offer advantages over antibodies as they can be engineered completely in a test tube, are readily produced by chemical synthesis, have desirable storage properties, and elicit little or no immunogenicity in therapeutic applications.

Nagarkatti et al. [64] have applied RNA aptamers to detect biomarkers of *T. cruzi* infection in the plasma of infected mice. Aptamers were generated against *T. cruzi* trypomastigote excreted/secreted antigens (TESA) purified from in vitro culture supernatants of infected host cells and used as specific ligands in enzyme-linked aptamer assays. TESA molecules could be detected in the blood of infected mice during both the acute and the chronic phases.

5.2.2 Advancements in Serology-Based Diagnostics

Numerous studies with recombinant antigens have observed that there are antigens with ability to distinguish between acute and chronic Chagas disease [65, 66], as well as to discriminate heart or digestive clinical forms [67, 68], or the variation of sensitivity in different patient populations [69]. Incorporation of these recombinant antigens and/or synthetic peptides in serodiagnostic assays has certainly improved their specificity, but their use as single assays has yet reduced sensitivity compared with conventional assays based on whole protein lysates [47]. In this regard, the design of cocktails with a battery of recombinant antigens, mixtures of synthetic peptides, or multi-epitope antigens aims to overcome this limitation.

5.2.2.1 Multi-antigenic Tests

The development of tests based on several parasite antigens, or chimeric antigens built with epitopes coming from different proteins is a subject of thorough research [70–72]. For example, Granjon et al. [70] have described the application of a cutting edge technology that allows printing several different recombinant antigens in a single 96-well plate micro-well following a pattern that includes inner well performance controls. The selection of the antigens ultimately included in the test was computationally assisted [70]. Similarly, an in silico process has been used by Mucci et al. [71] to identify the short peptides to be included in their proposed next generation ELISA. On the other hand, Santos et al. [72, 73] designed and evaluated four chimeric *T. cruzi* proteins using both ELISA and a liquid bead microarray (LMA) format. In sight of the test performance, these authors have proposed that its use would revoke the need for two different serological techniques for Chagas disease diagnosis [74].

Multi-antigen assays have already been evaluated in several Latin American countries with satisfactory performance in chronic Chagas disease diagnosis [65, 67, 75, 76]. Importantly, different research laboratories have identified, cloned, employed, and labeled *T. cruzi* antigens that actually corresponded to the same protein, thus generating redundant information. The equivalence among parasite antigens has been reported elsewhere [77], whereas those that worked better were the cytoplasmic antigens CRA, the trypomastigote surface molecule B13, trans-sialidase family SAPA, cytoskeleton FRA, microtubule-associated protein (MAP), acidic ribosomal protein JL5, and flagellar proteins [78].

5.2.2.2 Novel *T. cruzi* Antigens for Diagnosis Purposes

High concentrations of antibodies against an antigen can be an immunological response stimulated by the existence of tandem repeats (TR), which are defined by possessing one or more copies of an amino acid pattern and are characteristic of multigene families [79]. These TR are located in highly conserved regions and their high exposure to the immune system induces the production of specific antibodies with high affinity for the epitopes of the TR [77]. Their immunodominance is due mainly by the molar ratio and/or their particular spatial structure [80], which is by the way a characteristic of B lymphocytes' response to *T. cruzi* [78–81].

TRs of trypanosomatids are located within very long genes (10,000 bp approximately) that encode proteins with identified functions such as the "zinc finger proteins" motifs [79]. About 45% of TR protein sequences harbor a signal peptide and 53% have transmembrane domains and thus are expressed as surface proteins. They belong to the three main large gene families: trans-sialidase (TS), mucins, and mucin-associate surface proteins (MASPs) [79]. For example, one of the antigens containing TRs is the TS "shed acute-phase antigen" (SAPA), made up of a TR of 12 residues (DSSAHSTPSTPA) located in its C-terminal extension [82]. It is associated to the induction of neutralizing antibodies [83, 84]. This SAPA epitope has been characterized as an acute-phase antigen that induces the production of high titers of IgM and IgG antibodies in the initial phases of the disease, whereas during chronic infection its sensitivity is only 60% [82]. Accordingly, its use has been proposed for the diagnosis of congenital infection [77, 85]. Indeed, anti-SAPA response in sera from infected women and their infants showed that most infected neonates presented higher titers than their mothers, whereas in non-infected neonates titers were lower than those detected in maternal sera [86]. A positive correlation was observed between parasitic loads in mothers and infants evaluated by qPCR and anti-SAPA ELISA, so it has been proposed as biomarker for early diagnosis of newborns in healthcare facilities and hospitals where DNA amplification techniques are not available [86].

Other antigens under investigation are members of MASP family, which is composed of more than 1300 genes and 400 pseudo-genes arranged in arrays of up to 600 kb [87]. Many of the proteins in this family are expressed in the mammalian infective free-swimming trypomastigote stage, and many have been located to the trypomastigote membrane and some may in the secretome [88, 89].

In relation to non-peptidic antigens, the tri-saccharide Galα(1,3)Galβ(1,4) GlcNAcα is found on glycosylphosphatidylinositol-anchored mucins of the trypomastigote stage and it has been shown to elicit high levels of protective antibodies [90]. Neoglycoproteins containing it have been synthesized and evaluated as serodiagnostic antigens, which proved useful for the diagnosis of chronic Chagas disease by means of a chemiluminescent enzyme-linked immunosorbent assay [90, 91].

5.2.2.3 Non-conventional Recombinant Serological Commercial Methods

The most employed one is chemiluminescence, which is commercially available and used in many blood banks (CMIA). Its sensitivity is ~100% but its specificity may be lower, so it is advisable to use it together with a conventional one for ascertain diagnosis of a case. As it has been mentioned before, given its very high sensitivity, its use as single test for exclusion purposes in the diagnosis of chronic stage has been suggested [45] (Fig. 5.1).

RIPA (radioimmunoassay), which is not commercially available and used only in United States of America (USA), is another non-conventional test. It consists in the co-precipitation of two glycoproteins (of 72 and 90 kDa) specifically bound by IgGs in the patients sera that can provide very good sensitivity and specificity as it was shown upon comparison with commercial tests [92]. However, it is a complex procedure that uses radioactivity, which linked to its high cost likely results in lower applicability potential than other methodologies. Other non-conventional tests that may have similar disadvantages towards a widespread application, especially in endemic settings, are the Western blot TESA-blot test and lytic assays that involve the use of flow cytometry (non-commercially available) [93].

Last but not least in this section, we must not forget of RDTs, which were developed with remarkable POC characteristics and may revolutionize chronic Chagas disease diagnosis in the future [94]. In general, any RDT consists of a membrane sensitized with several recombinant antigens. When a drop of serum or blood is placed in contact in a few minutes, a reaction may be seen as a band if the test is positive. These rapid tests have a number of advantages: they do not require cold storage, nor any specific equipment for incubation or reading of results; some of them work with a very little volume of whole blood that can be obtained by finger prick greatly facilitating logistics and patients' inclination to the test; and very importantly, they have a quick results turnaround of <1 h, which involves that the person can walk away from the consultation with a result [94, 95]. Several works have been published showing reasonable specificity and sensitivity of some of them [95–98]. Nevertheless they still have precise indications and should not label an individual as infected unless a second, conventional test is used in parallel. In fact, RDTs are mainly in use as screening tools because current policies still enforce that the diagnosis of the disease is dictated by conventional serological methods [95]. In order to overcome this limitation, the combinatory use of RDTs has been proposed and tested successfully, in the same way that two serological tests based on distinct antigen sets are needed for a concluding Chagas diagnosis [95, 99]. But RDTs performance has been deemed to be geographically variable and therefore they should be geographically validated before suggesting a wider use [69].

5.2.3 Nanotechnology-Based Diagnostics

Nanotechnology-based tests can be adapted to POC and cost-effective detection of microbial agents. Castro-Sesquen [100] and co-workers demonstrated that a Chagas urine nanoparticle assay (Chunap) detects congenital Chagas disease in a single urine specimen at 1 month of life with more than 90% sensitivity and more than 95% specificity. The study also showed that poly-(NIPAm) particles coupled with trypan blue dye efficiently capture and concentrate *T. cruzi* antigens in urine, and under experimental conditions these particles were shown to protect *T. cruzi* antigens in the urine from enzymatic degradation. In addition, the non-invasive nature of the test will greatly enhance parental acceptability.

Although it is less sensitive than PCR, this disadvantage is perhaps balanced by the fact that the nanoparticles assay is simpler, it costs two times less in terms of consumables, and it can be completed within only 30 min. Furthermore, the fact that it incorporates four nanoprobes minimizes the need for confirmation of the specificity of positive results. However, upon concentration of the parasite antigens they must be detected by Western blot which may preclude Chunap widespread use in ill-equipped facilities of endemic regions.

5.2.4 Methodologies to Monitor Treatment Responses

Until more amenable and less expensive tests become available, blood-based qPCR techniques are being consistently used to detect therapeutic response or failure in clinical trials with traditional and novel drugs [37, 55, 101–103]. However, clearance of parasitic loads exerted by drugs can be transient and lead to misleading conclusions when follow-up is performed at the short term [55]. Ideally, molecular methods used for monitoring chronic patients should be performed for several years after treatment to confirm or discard available data. Even more now that recent findings of dormant amastigote subpopulations, refractory to benznidazole action, may represent a key factor leading to treatment failure [104].

5.3 Challenges in the Use of the Available Tools in Endemic and Non-endemic Regions

The gap towards getting diagnosed and treated for Chagas disease is due to a series of factors that encompass: lack of awareness and misinformation on the disease; failures in the drug supply chain; and unavailability of a timely diagnosis. The latter can be related to the fact that Chagas disease pathogenesis progresses in a silent manner until tissue damage is overt [1], but also to the fact that current diagnostics and/or diagnosis algorithms are impractical in many highly endemic regions that lay distant from reference laboratories [1, 2, 95].

The specific weight of the aforementioned features in precluding the access to treatment varies between endemic and non-endemic settings. In the latter (i.e., areas of Europe, Japan, or Australia), there is generally a better healthcare structure and an easier access to diagnosis and treatment. Therefore, in these settings it is mostly the lack of information about the disease the main reason why no diagnosis and treatment are eventually provided [105]. In contrast, in those countries with highly endemic regions for Chagas disease, obstacles to get access to treatment are more related to the absence of adequate healthcare structures that hinder the achievement of a timely diagnosis (and treatment) [105]. Furthermore, currently available methodologies are not practical to use under the field conditions present in those regions that frequently lay distant from reference laboratories. In these areas, it may happen that there are no properly equipped labs, nor trained personnel to run them. Moreover, application of current algorithms entails more than one consultation to get a confirmed diagnosis due to a delayed results turnaround of several weeks. As a consequence, this often means the loss of many patients to treatment and follow-up.

At present, the algorithm to diagnose chronic Chagas disease in endemic and non-endemic regions is very similar (Fig. 5.1). The use of two (or three when necessary) serological tests has proved very sensitive and specific. Nonetheless, as explained above, such tests cannot be used in many endemic regions that are distantly located from referential hospitals. It is there where the availability of POC RDTs can really make a difference, especially if algorithms fully relying on them are ever set on place. In this regard, the combinatory use of two RDTs based on distinct antigen sets has been proposed as an alternative to conventional serological methods [95, 99].

On the other hand, the high performance of new generation diagnostics such as CMIA has prompted the proposal of using just one test, at least to triage negative cases and further confirm only those cases inconclusive or positive (Fig. 5.1). In fact, work from two studies already sustain the use of CMIA as single test as far as its signal-to-cut off (S/CO) is over a certain threshold (>6 or ≥3.8, respectively, in [45, 46]). It would be ideal that such conclusions were ever made with a RDT, but their variable geographical performance and the likely relation of this with the regions' seroprevalence rates still casts some doubts on their widespread implementation [69].

Differences in the diagnostic algorithm between endemic and non-endemic settings are more marked in the case of congenital Chagas diagnosis, where the use of molecular methods is starting to become widely accepted in non-endemic countries (Fig. 5.2). For instance, recent consensus within the Generalitat de Catalunya Working Group on Chagas disease has agreed to perform PCR from blood collected at 1 month of age when the parasitic load is at its peak [106, 107], and potential false positive results derived from the transmission of *T. cruzi* DNA from the mother to the fetus are minimized [50]. In laboratories from endemic regions, parasitological techniques like the micromethod are yet at the frontline to detect the infection in newborns, despite their lower sensitivity and specificity in comparison to molecular methods [32]. Implementation of the latter would definitely mean a breakthrough considering the poor sensitivity of the micromethod and the great impact that the loss to follow-up has on achieving the serological diagnosis of infant Chagas by the time it can be made. There is little doubt that novel methodologies that enable molecular detection in ill-equipped labs from endemic areas like LAMP could make a huge difference in the management of congenital Chagas disease.

5.4 Final Remarks

- Laboratorial diagnosis of Chagas disease is well established. For acute phase, direct parasitological tests should be employed, which are rather simple but need skilled technicians to perform them. For chronic phase, the use of two serological techniques in parallel is necessary to ascertain a positive result or exclude the infection in a patient.
- In order to obtain a good performance, kits of recognized quality and good laboratory practices are necessary. To fulfill these needs, internal and external quality controls are imperative.
- Implementation of molecular diagnostics would improve acute phase diagnosis. Towards it, the standardization of currently available procedures is paramount for reliable inter-lab and inter-studies comparisons, especially in the following scenarios: early detection of congenital infection, assessment of infection reactivation in immune-suppressed patients, and for close follow-up of drug treatments.
- Field deployment and implementation of point-of-care serological tests like RDTs and molecular diagnostics like LAMP would mean a major breakthrough, generalizing access to diagnosis in currently neglected areas.
- Target product profiles (TPPs) of strategies for diagnosis of *T. cruzi* infection have been addressed, pointing to the need of developing point-of-care assays.

Their evaluation in field studies is needed to predict their usefulness in the clinical practice and for public health applications.

- The participation of the industry in the development of commercial and standardized tests is suitable. These tests have to be easy to implement, cheap and should be evaluated independently by reference laboratories.

References

1. Balouz V, Agüero F, Buscaglia CA. Chagas disease diagnostic applications: present knowledge and future steps. Adv Parasitol. 2017;97:1–45.
2. Alonso-Padilla J, Gallego M, Schijman AG, et al. Molecular diagnostics for Chagas disease: up to date and novel methodologies. Expert Rev Mol Diagn. 2017;17:669–710.
3. Pinazo MJ, Thomas MC, Bua J, et al. Biological markers for evaluating therapeutic efficacy in Chagas disease, a systematic review. Expert Rev Anti Infect Ther. 2014;12:479–96.
4. World Health Organization Expert Committee. Control of Chagas disease. WHO technical report series number 905. Brazilia: WHO; 2002.
5. Flores-Chávez M, de Fuentes I, Gárate T, et al. Diagnóstico de laboratorio de la enfermedad de Chagas importada. Enferm Infecc Microbiol Clin. 2007;25:29–37.
6. Feilij H, Muller L, Gonzalez Cappa SM. Direct micromethod for diagnosis of acute and congenital Chagas' disease. J Clin Microbiol. 1983;18:327–30.
7. Torrico MC, Solano M, Guzman JM, et al. Estimation of the parasitemia in *Trypanosoma cruzi* human infection: high parasitemias are associated with severe and fatal congenital Chagas disease. Rev Soc Bras Med Trop. 2005;38(Suppl 2):58–61.
8. Carlier Y, Torrico F. Congenital infection with *Trypanosyma cruzi*: from mecahnisms of transmission to strategies for diagnosis and control. Rev Soc Bras Med Trop. 2003;36:767–71.
9. Uc-Cetina V, Brito-Loeza C, Ruiz-Piña H. Chagas parasite detection in blood images using AdaBoost. Comput Math Methods Med. 2015;2015:139681.
10. Torrico MC, Solano MA, Córdova M, et al. Diagnóstico parasitológico de la enfermedad de Chagas: de la teoría a la práctica. Enf Emerg. 2011;13(Suppl 1):33–8.
11. Persing DH. Polymerase chain reaction: trenches to benches. J Clin Microbiol. 1991;29:1281–5.
12. Virreira M, Torrico F, Truyens C, et al. Comparison of polymerase chain reaction methods for reliable and easy detection of congenital *Trypanosoma cruzi* infection. Am J Trop Med Hyg. 2003;68:574–82.
13. Moser DR, Kirchhoff LV, Donelson JE. Detection of *Trypanosoma cruzi* by DNA amplification using the polymerase chain reaction. J Clin Microbiol. 1989;27:1477–82.
14. Burgos JM, Diez M, Vigliano C, et al. Molecular identification of *Trypanosoma cruzi* discrete typing units in end-stage chronic Chagas heart disease and reactivation after heart transplantation. Clin Infect Dis. 2010;51:485–95.
15. Diez M, Favaloro L, Bertolotti A, et al. Usefulness of PCR strategies for early diagnosis of Chagas' disease reactivation and treatment follow-up in heart transplantation. Am J Transplant. 2007;7:1633–40.
16. Hernández C, Teherán A, Flórez C, et al. Comparison of parasite loads in serum and blood samples from patients in acute and chronic phases of Chagas disease. Parasitology. 2018;145:1837–43.
17. Qvarnstrom Y, Schijman AG, Veron V, et al. Sensitive and specific detection of *Trypanosoma cruzi* DNA in clinical specimens using a multi-target real-time PCR approach. PLoS Negl Trop Dis. 2012;6:e1689.
18. Russomando G, Figueredo A, Almiron M, et al. Polymerase chain reaction-based detection of *Trypanosoma cruzi* DNA in serum. J Clin Microbiol. 1992;30:2864–8.

19. Schijman AG, Bisio M, Orellana L, et al. International study to evaluate PCR methods for detection of *Trypanosoma cruzi* DNA in blood samples from Chagas disease patients. PLoS Negl Trop Dis. 2011;5:e931.
20. Wincker P, Britto C, Pereira JB, et al. Use of a simplified polymerase chain reaction procedure to detect *Trypanosoma cruzi* in blood samples from chronic chagasic patients in a rural endemic area. Am J Trop Med Hyg. 1994;51:771–7.
21. Brasil PE, De Castro L, Hasslocher-Moreno AM, et al. ELISA versus PCR for diagnosis of chronic Chagas disease: systematic review and meta-analysis. BMC Infect Dis. 2010;10:337.
22. Abras A, Ballart C, Llovet T, et al. Introducing automation to the molecular diagnosis of *Trypanosoma cruzi* infection: a comparative study of sample treatments, DNA extraction methods and real-time PCR assays. PLoS One. 2018;13:e0195738.
23. Schijman AG. Molecular diagnosis of *Trypanosoma cruzi*. Acta Trop. 2018;184:59–66.
24. Avila HA, Pereira JB, Thiemann O, et al. Detection of *Trypanosoma cruzi* in blood specimens of chronic chagasic patients by polymerase chain reaction amplification of kinetoplast minicircle DNA: comparison with serology and xenodiagnosis. J Clin Microbiol. 1993;31:2421–6.
25. Schijman AG, Altcheh J, Burgos JM, et al. Aetiological treatment of congenital Chagas' disease diagnosed and monitored by the polymerase chain reaction. J Antimicrob Chemother. 2003;52:441–9.
26. Ochs DE, Hnilica VS, Moser DR, et al. Postmortem diagnosis of autochthonous acute chagasic myocarditis by polymerase chain reaction amplification of a species-specific DNA sequence of *Trypanosoma cruzi*. Am J Trop Med Hyg. 1996;54:526–9.
27. Marcon GEB, Andrade PD, De Albuquerque DM, et al. Use of a nested polymerase chain reaction (N-PCR) to detect *Trypanosoma cruzi* in blood samples from chronic chagasic patients and patients with doubtful serologies. Diagn Microbiol Infect Dis. 2002;42: 39–43.
28. Duffy T, Cura CI, Ramirez JC, et al. Analytical performance of a multiplex real-time PCR assay using TaqMan probes for quantification of *Trypanosoma cruzi* satellite DNA in blood samples. PLoS Negl Trop Dis. 2013;7:e2000.
29. Duffy T, Bisio M, Altcheh J, et al. Accurate real-time PCR strategy for monitoring bloodstream parasitic loads in Chagas disease patients. PLoS Negl Trop Dis. 2009;3:e419.
30. Piron M, Fisa R, Casamitjana N, et al. Development of a real-time PCR assay for *Trypanosoma cruzi* detection in blood samples. Acta Trop. 2007;103:195–200.
31. Ramírez JC, Cura CI, da Cruz Moreira O, et al. Analytical validation of quantitative real-time PCR methods for quantification of *Trypanosoma cruzi* DNA in blood samples from chagas disease patients. J Mol Diagn. 2015;17:605–15.
32. Picado A, Cruz I, Redard-Jacot M, et al. The burden of congenital Chagas disease and implementation of molecular diagnostic tools in Latin America. BMJ Glob Health. 2018;3: e001069.
33. Cura CI, Lattes R, Nagel C, et al. Early molecular diagnosis of acute chagas disease after transplantation with organs from *Trypanosoma cruzi*-infected donors. Am J Transplant. 2013;13:3253–61.
34. Benvenuti LA, Roggério A, Cavalcanti MM, et al. An autopsy-based study of *Trypanosoma cruzi* persistence in organs of chronic chagasic patients and its relevance for transplantation. Transpl Infect Dis. 2017;19(6).
35. Chin-Hong PV, Schwartz BS, Bern C, et al. Screening and treatment of Chagas disease in organ transplant recipients in the United States: recommendations from the chagas in transplant working group. Am J Transplant. 2011;11:672–80.
36. Brasil PE, Castro R, Castro LD. Commercial enzyme-linked immunosorbent assay versus polymerase chain reaction for the diagnosis of chronic Chagas disease: a systematic review and meta-analysis. Mem Inst Oswaldo Cruz. 2016;111:1–19.
37. Moreira OC, Ramírez JD, Velázquez E, et al. Towards the establishment of a consensus real-time qPCR to monitor *Trypanosoma cruzi* parasitemia in patients with chronic Chagas disease cardiomyopathy: a substudy from the BENEFIT trial. Acta Trop. 2013;125:23–31.

38. Parrado R, Ramirez JC, de la Barra A, et al. Real-time PCR for the evaluation of treatment response in clinical trials of adult chronic Chagas disease: usefulness of serial blood sampling and qPCR replicates. Antimicrob Agents Chemother. 2018. pii: AAC.01191-18.
39. Britto CC. Usefulness of PCR-based assays to assess drug efficacy in Chagas disease chemotherapy: value and limitations. Mem Inst Oswaldo Cruz. 2009;104:122–35.
40. Cancino-Faure B, Fisa R, Alcover MM, et al. Detection and quantification of viable and non-viable *Trypanosoma cruzi* parasites by a propidium monoazide real-time polymerase chain reaction assay. Am J Trop Med Hyg. 2016;94:1282–9.
41. Juiz NA, Solana ME, Acevedo GR, et al. Different genotypes of *Trypanosoma cruzi* produce distinctive placental environment genetic response in chronic experimental infection. PLoS Negl Trop Dis. 2017;11:e0005436.
42. Ramírez JC, Torres C, Curto MLA, et al. New insights into *Trypanosoma cruzi* evolution, genotyping and molecular diagnostics from satellite DNA sequence analysis. PLoS Negl Trop Dis. 2017;11:e0006139.
43. Organización Panamericana de la Salud (OPS). Guía para el diagnóstico y el tratamiento de la enfermedad de Chagas. Washington, DC: Estados Unidos de América; 2018.
44. World Health Organization. Anti-*Trypanosoma cruzi* ASSAYS: operational characteristics. Report 1. Geneva: WHO; 2010.
45. Abras A, Gállego M, Llovet T, et al. Serological diagnosis of chronic Chagas disease: is it time for a change? J Clin Microbiol. 2016;54:1566–72.
46. Pérez-Ayala A, Fradejas I, Rebollo L, et al. Usefulness of the ARCHITECT Chagas® assay as a single test for the diagnosis of chronic Chagas disease. Trop Med Int Health. 2018;23:634–40.
47. Caballero ZC, Sousa OE, Marques WP, et al. Evaluation of serological tests to identify *Trypanosoma cruzi* infection in humans and determine cross-reactivity with *Trypanosoma rangeli* and *Leishmania* spp. Clin Vaccine Immunol. 2007;14:1045–9.
48. Viotti R, Alarcón De Noya B, Araujo-Jorge T, et al. Towards a paradigm shift in the treatment of chronic Chagas disease. Antimicrob Agents Chemother. 2014;58:635–9.
49. Porrás AI, Yadon ZE, Altcheh J, et al. Target product profile (TPP) for Chagas disease point-of-care diagnosis and assessment of response to treatment. PLoS Negl Trop Dis. 2015;9:e0003697.
50. Carlier Y, Sosa-Estani S, Luquetti AO, et al. Congenital Chagas disease: an update. Mem Inst Oswaldo Cruz. 2015;110:363–8.
51. Carlier Y, Torrico F, Sosa-Estani S, et al. Congenital Chagas disease: recommendations for diagnosis, treatment and control of newborns, siblings and pregnant women. PLoS Negl Trop Dis. 2011;5:e1250.
52. Abras A, Muñoz C, Ballart C, et al. Towards a new strategy for diagnosis of congenital *Trypanosoma cruzi* infection. J Clin Microbiol. 2017;55:1396–407.
53. United States National Library of Science. "Search of Condition or disease": "Chagas disease". https://clinicaltrials.gov/.
54. Seiringer P, Pritsch M, Flores-Chavez M, et al. Comparison of four PCR methods for efficient detection of *Trypanosoma cruzi* in routine diagnostics. Diagn Microbiol Infect Dis. 2017;88:225–32.
55. Torrico F, Gascon J, Ortiz L, et al. Treatment of adult chronic indeterminate Chagas disease with benznidazole and three E1224 dosing regimens: a proof-of-concept, randomised, placebo-controlled trial. Lancet Infect Dis. 2018;18:419–30.
56. Ramírez JD, Herrera G, Hernández C, et al. Evaluation of the analytical and diagnostic performance of a digital droplet polymerase chain reaction (ddPCR) assay to detect *Trypanosoma cruzi* DNA in blood samples. PLoS Negl Trop Dis. 2018;12:e0007063.
57. Notomi T, Okayama H, Masubuchi H, et al. Loop-mediated isothermal amplification of DNA. Nucleic Acids Res. 2000;28:E63.
58. Thekisoe OM, Rodriguez CV, Rivas F, et al. Detection of *Trypanosoma cruzi* and *T. rangeli* infections from *Rhodnius pallescens* bugs by loop-mediated isothermal amplification (LAMP). Am J Trop Med Hyg. 2010;82:855–60.

59. Rivero R, Bisio M, Velázquez EB, et al. Rapid detection of *Trypanosoma cruzi* by colorimetric loop-mediated isothermal amplification (LAMP): a potential novel tool for the detection of congenital Chagas infection. Diagn Microbiol Infect Dis. 2017;89:26–8.
60. Besuschio SA, Llano Murcia M, et al. Analytical sensitivity and specificity of a loop-mediated isothermal amplification (LAMP) kit prototype for detection of *Trypanosoma cruzi* DNA in human blood samples. PLoS Negl Trop Dis. 2017;11:e0005779.
61. Besuschio SA, Muñoz-Calderón A, Fernández M, et al. LAMP y Enfermedad de Chagas: detección de ADN de *T. cruzi* y monitoreo de tratamiento en brote por transmisión oral y reactivación por inmunocompromiso. In: Libro de Resúmenes, XXX Reunión Annual de la Sociedad Argentina de Protozoología; 2018. p. 91.
62. Jimenez-Coello M, Shelite T, Castellanos-Gonzalez A, et al. Efficacy of recombinase polymerase amplification to diagnose *Trypanosoma cruzi* infection in dogs with cardiac alterations from an endemic area of Mexico. Vector-Borne Zoonotic Dis. 2018;18:417–23.
63. Kanwar JR, Roy K, Maremanda NG, et al. Nucleic acid-based aptamers: applications, development and clinical trials. Curr Med Chem. 2010;22:2539–57.
64. Nagarkatti R, de Araujo FF, Gupta C, et al. Aptamer based, non-PCR, non-serological detection of Chagas disease biomarkers in *Trypanosoma cruzi* infected mice. PLoS Negl Trop Dis. 2014;8:e2650.
65. Umezawa ES, Luquetti AO, Levitus G, et al. Serodiagnosis of chronic and acute Chagas' disease with *Trypanosoma cruzi* recombinant proteins: results of a collaborative study in six Latin American countries. J Clin Microbiol. 2004;42:449–52.
66. Flechas ID, Cuellar A, Cucunubá ZM, et al. Characterising the KMP-11 and HSP-70 recombinant antigens' humoral immune response profile in chagasic patients. BMC Infect Dis. 2009;9:186.
67. Longhi SA, Brandariz SB, Lafon SO, et al. Evaluation of in-house ELISA using *Trypanosoma cruzi* lysate and recombinant antigens for diagnosis of Chagas disease and discrimination of its clinical forms. Am J Trop Med Hyg. 2012;87:267–71.
68. Vasconcelos RH, Amaral FN, Cavalcanti MG, et al. Increased levels of IgA antibodies against CRA and FRA recombinant antigens of *Trypanosoma cruzi* differentiate digestive forms of Chagas disease. Hum Immunol. 2010;71:964–7.
69. Verani JR, Seitz A, Gilman RH, et al. Geographie variation in the sensitivity of recombinant antigen-based rapid tests for chronic *Trypanosoma cruzi* infection. Am J Trop Med Hyg. 2009;80:410–5.
70. Granjon E, Dichtel-Danjoy ML, Saba E, et al. Development of a novel multiplex immunoassay multi-cruzi for the serological confirmation of Chagas disease. PLoS Negl Trop Dis. 2016;10:e0004596.
71. Mucci J, Carmona SJ, Volcovich R, et al. Next-generation ELISA diagnostic assay for Chagas disease based on the combination of short peptidic epitopes. PLoS Negl Trop Dis. 2017;11:e0005972.
72. Santos FL, Celedon PA, Zanchin NI, et al. Performance assessment of four chimeric *Trypanosoma cruzi* antigens based on antigen-antibody detection for diagnosis of chronic chagas disease. PLoS One. 2016;11:e0161100.
73. Santos FL, Celedon PA, Zanchin NI, et al. Accuracy of chimeric proteins in the serological diagnosis of chronic Chagas disease—a phase II study. PLoS Negl Trop Dis. 2017;11:e0005433.
74. Santos FL, Celedon PA, Zanchin NI, et al. Performance assessment of a *Trypanosoma cruzi* chimeric antigen in multiplex liquid microarray assays. J Clin Microbiol. 2017;55:2934–45.
75. Villagrán-Herrera ME, Sánchez-Moreno M, Rodríguez-Méndez AJ, et al. Comparative serology techniques for the diagnosis of *Trypanosoma cruzi* infection in a rural population from the state of Querétaro, Mexico. Mem Inst Oswaldo Cruz. 2014;109:967–72.
76. Vega Benedetti AF, Cimino RO, Cajal PS, et al. Performance of different *Trypanosoma cruzi* antigens in the diagnosis of Chagas disease in patients with American cutaneous leishmaniasis from a co-endemic region in Argentina. Trop Med Int Health. 2013;18:1103–9.

77. da Silveira JF, Umezawa ES, Luquetti AO. Chagas disease: recombinant *Trypanosoma cruzi* antigens for serological diagnosis. Trends Parasitol. 2001;17:286–91.
78. Frasch AC. Trans-sialidase, SAPA amino acid repeats and the relationship between Trypanosoma cruzi and the mammalian host. Parasitology. 1994;108(Suppl):S37–44.
79. Goto Y, Carter D, Reed SG. Immunological dominance of *Trypanosoma cruzi* tandem repeat proteins. Infect Immun. 2008;76:3967–74.
80. Schofield L. On the function of repetitive domains in protein antigens of *Plasmodium* and other eukaryotic parasites. Parasitol Today. 1991;7:99–105.
81. Alvarez P, Leguizamón MS, Buscaglia CA, et al. Multiple overlapping epitopes in the repetitive unit of the shed acute-phase antigen from *Trypanosoma cruzi* enhance its immunogenic properties. Infect Immun. 2001;69:7946–9.
82. Cazzulo JJ, Frasch AC. SAPA/trans-sialidase and cruzipain: two antigens from *Trypanosoma cruzi* contain immunodominant but enzymatically inactive domains. FASEB J. 1992;6:3259–64.
83. Buscaglia CA, Alfonso J, Campetella O, et al. Tandem amino acid repeats from *Trypanosoma cruzi* shed antigens increase the half-life of proteins in blood. Blood. 1999;93:2025–32.
84. Pitcovsky TA, Buscaglia CA, Mucci J, et al. A functional network of intramolecular cross-reacting epitopes delays the elicitation of neutralizing antibodies to *Trypanosoma cruzi* trans-sialidase. J Infect Dis. 2002;186:397–404.
85. Reyes MB, Lorca M, Muñoz P, et al. Fetal IgG specificities against *Trypanosoma cruzi* antigens in infected newborns. Proc Natl Acad Sci U S A. 1990;87:2846–50.
86. Volta BJ, Russomando G, Bustos PL, et al. Diagnosis of congenital *Trypanosoma cruzi* infection: a serologic test using shed acute phase antigen (SAPA) in mother-child binomial samples. Acta Trop. 2015;147:31–7.
87. El-Sayed NM, Myler PJ, Bartholomeu DC, et al. The genome sequence of *Trypanosoma cruzi*, etiologic agent of chagas disease. Science. 2005;309:409–15.
88. Bartholomeu DC, Cerqueira GC, Leão ACA, et al. Genomic organization and expression profile of the mucin-associated surface protein (masp) family of the human pathogen *Trypanosoma cruzi*. Nucleic Acids Res. 2009;37:3407–17.
89. De Pablos LM, González GG, Solano Parada J, et al. Differential expression and characterization of a member of the mucin-associated surface protein family secreted by *Trypanosoma cruzi*. Infect Immun. 2011;79:3993–4001.
90. Ashmus RA, Schocker NS, Cordero-Mendoza Y, et al. Potential use of synthetic α-galactosyl-containing glycotopes of the parasite *Trypanosoma cruzi* as diagnostic antigens for Chagas disease. Org Biomol Chem. 2013;11:5579–83.
91. Izquierdo L, Marques AF, Gállego M, et al. Evaluation of a chemiluminescent enzyme-linked immunosorbent assay for the diagnosis of *Trypanosoma cruzi* infection in a nonendemic setting. Mem Inst Oswaldo Cruz. 2013;108:928–31.
92. Otani MM, Vinelli E, Kirchhoff LV, et al. WHO comparative evaluation of serologic assays for Chagas disease. Transfusion. 2009;49:1076–82.
93. WHO Technical Report Series—Research priorities for Chagas disease, human African trypanosomiasis and leishmaniasis. WHO Special Programme for Research and Training in Tropical Diseases [TDR], 2012 (975): p. v–xii, 1–100.
94. Médecins Sans Frontières. Campaign for Access to Essential Medicines. International meeting: new diagnostic tests are urgently needed to treat patients with Chagas disease. Rev Soc Bras Med Trop. 2008;41:315–9.
95. Egüez KE, Alonso-Padilla J, Terán C, et al. Rapid diagnostic tests duo as alternative to conventional serological assays for conclusive Chagas disease diagnosis. PLoS Negl Trop Dis. 2017;11:e0005501.
96. Sánchez-Camargo CL, Albajar-Viñas P, Wilkins PP, et al. Comparative evaluation of 11 commercialized rapid diagnostic tests for detecting *Trypanosoma cruzi* antibodies in serum banks in areas of endemicity and nonendemicity. J Clin Microbiol. 2014;52:2506–12.
97. Shah V, Ferrufino L, Gilman RH, et al. Field evaluation of the InBios Chagas detect plus rapid test in serum and whole-blood specimens in Bolivia. Clin Vaccine Immunol. 2014;21:1645–9.

98. Roddy P, Goiri J, Flevaud L, et al. Field evaluation of a rapid immunochromatographic assay for detection of *Trypanosoma cruzi* infection by use of whole blood. J Clin Microbiol. 2008;46:2022–7.
99. Mendicino D, Colussi C, Moretti E. Simultaneous use of two rapid diagnostic tests for the diagnosis of Chagas disease. Trop Doct. 2018;27:49475518813792.
100. Castro-Sesquen YE, Gilman RH, Galdos-Cardenas G, et al. Use of a novel Chagas urine nanoparticle test (Chunap) for diagnosis of congenital Chagas disease. PLoS Negl Trop Dis. 2014;8:e3211.
101. Morillo CA, Marin-Neto JA, Avezum A, et al. Randomized trial of benznidazole for chronic Chagas' cardiomyopathy. N Engl J Med. 2015;373:1295–306.
102. Álvarez MG, Hernández Y, Bertocchi G, et al. New scheme of intermittent benznidazole administration in patients chronically infected with *Trypanosoma cruzi*: a pilot short-term follow-up study with adult patients. Antimicrob Agents Chemother. 2016;60:833–7.
103. Morillo CA, Waskin H, Sosa-Estani S, et al. Benznidazole and posaconazole in eliminating parasites in asymptomatic *T. cruzi* carriers: the STOP-CHAGAS trial. J Am Coll Cardiol. 2017;69:939–47.
104. Sánchez-Valdéz FJ, Padilla A, Wang W, et al. Spontaneous dormancy protects *Trypanosoma cruzi* during extended drug exposure. Elife. 2018;7:e34039.
105. Pinazo MJ, Gascon J. The importance of the multidisciplinary approach to deal with the new epidemiological scenario of Chagas disease (global health). Acta Trop. 2015;151:16–20.
106. Bua J, Volta BJ, Velazquez EB, et al. Vertical transmission of *Trypanosoma cruzi* infection: quantification of parasite burden in mothers and their children by parasite DNA amplification. Trans R Soc Trop Med Hyg. 2012;106:623–8.
107. Bua J, Volta BJ, Perrone AE, et al. How to improve the early diagnosis of *Trypanosoma cruzi* infection: relationship between validated conventional diagnosis and quantitative DNA amplification in congenitally infected children. PLoS Negl Trop Dis. 2013;7:e2476.

Diagnosis of Chagas Disease: Are Clinical Definitions of Heart Involvement Accurate Enough?

6

Andréa Silvestre de Sousa,
Fernanda de Souza Nogueira Sardinha Mendes,
Paloma Jordà, and Ana García-Álvarez

6.1 Chronic Chagas Cardiomyopathy: Definition, Stages, and Evolution of Chronic Chagas Disease (CCD)

Chronic Chagas cardiomyopathy (CCC) occurred when Chagas infectious disease has progressed to a chronic phase, characterized by at least one typical electrocardiographic (ECG) abnormality. A simple ECG is considered a definitive exam for diagnosing CCC in patients with two positive anti-*Trypanosoma cruzi* IgG serologies, even without myocardial dysfunction or symptoms. Ventricular or atrioventricular conduction disturbances, premature ventricular contractions, electrically inactive zones, primary repolarization alterations, and low QRS voltage may be present in isolation or associated in the same patient. Some patients have a progressive impairment of biventricular systolic function, with a segmental predominant pattern that can be detected by echocardiography, which is associated to arrhythmia genesis and cardioembolic phenomena [1].

6.1.1 Current Classifications

There are different classification systems to stratify Chagas disease or CCC. As they share some codes for strata classification, it is important to know the differences to avoid misunderstandings. Because of the importance of CCC in clinical follow-up and prognosis, current classifications are based on the abnormalities identified by ECG, chest X-ray, echocardiography, and clinical symptoms of heart failure.

A. Silvestre de Sousa · F. de Souza Nogueira Sardinha Mendes
Evandro Chagas National Institute of Infectious Diseases,
Oswaldo Cruz Foundation, Manguinhos, Rio de Janeiro, Brazil

P. Jordà · A. García-Álvarez (✉)
Institut Clínic Cardiovascular, Hospital Clínic—IDIBAPS, Barcelona, Spain
e-mail: PJORDAB@clinic.cat; ANAGARCI@clinic.cat

M.-J. Pinazo Delgado, J. Gascon (eds.), *Chagas Disease*,
https://doi.org/10.1007/978-3-030-44054-1_6

Digestive semiology is not evaluated to classify these patients. By definition, abnormal ECG findings are considered the diagnostic criteria for CCC.

Five main classifications are used: Kuschnir [2], Brazilian Consensus on Chagas Disease [3], modified Los Andes classification [4], Latin American Guidelines [5], and American Heart Association (AHA) Statement [6]. The Brazilian Consensus classification has the focus on CCC; all the others include the chronic indeterminate form. Only AHA classification cites chronic digestive form, excluding patients who present these symptoms in isolation. The description of these five classifications is presented below:

Kuschnir Classification (1985)
- **0:** Normal ECG findings and normal heart size (on chest X-ray)
- **I:** Abnormal ECG findings and normal heart size (on chest X-ray)
- **II:** Left ventricular (LV) enlargement
- **III:** Congestive heart failure (CHF)

Brazilian Consensus Classification (2005, Reviewed in 2015)
- **A:** Abnormal ECG findings, normal echocardiographic findings, no signs of CHF
- **B1:** Abnormal ECG findings, abnormal echocardiographic findings with left ventricular ejection fraction (LVEF) $\geq$ 45%, no signs of CHF
- **B2:** Abnormal ECG findings, abnormal echocardiographic findings with LVEF <45%, no signs of CHF
- **C:** Abnormal ECG findings, abnormal echocardiographic findings, compensated CHF
- **D:** Abnormal ECG findings, abnormal echocardiographic findings, refractory CHF

Modified Los Andes Classification (2007)
- **IA:** Normal ECG findings, normal echocardiographic findings, no signs of CHF
- **IB:** Normal ECG findings, abnormal echocardiographic findings, no signs of CHF
- **II:** Abnormal ECG findings, abnormal echocardiographic findings, no signs of CHF
- **III:** Abnormal ECG findings, abnormal echocardiographic findings, CHF

Latin American Guidelines (2011)
- **A:** No symptoms of heart failure and no structural heart disease (normal ECG and chest X-ray)
- **B1:** Asymptomatic patient with ECG changes (arrhythmias or conduction disorders); mild echocardiographic contractile abnormalities with normal global ventricular function
- **B2:** Decreased LVEF on patients who have never had any signs or symptoms of heart failure
- **C:** Patients with LV dysfunction and prior or current symptoms of heart failure
- **D:** Patient with symptoms of heart failure at rest, refractory to maximized medical therapy (NYHA functional class IV)

American Heart Association Statement (2018)

- **A:** Patients at risk for developing heart failure; positive serology for *T. cruzi*; neither structural cardiomyopathy or heart failure symptoms; normal ECG; no digestive changes (indeterminate form)
- **B1:** Patients with structural cardiomyopathy evidenced by electrocardiographic or echocardiographic changes, but with normal ventricular function and neither current or previous signs and symptoms of heart failure
- **B2:** Patients with structural cardiomyopathy characterized by global ventricular dysfunction and neither current or previous signs and symptoms of heart failure
- **C:** Patients with ventricular dysfunction and current or previous symptoms of heart failure New York Heart Association (NYHA) (functional class II, III or IV)
- **D:** Patients with refractory symptoms of heart failure at rest despite optimized clinical treatment requiring specialized interventions

The synthesis of the alterations in ECG, chest X-ray, echocardiography, and clinical heart failure symptoms according to each of the existing classifications can be visualized in Table 6.1. It is interesting to evaluate the discriminatory capacity of

Table 6.1 Classifications of Chagas disease and alterations in exams and symptoms

	Classifications				
Exam alterations or clinical symptoms1.	Kuschnir	Brazilian Consensus	Los Andes modified	Latin American Guidelines	AHA statement
Normal ECG and no apparent SHD	0	NP	IA	A	A[a]
Normal ECG with contractile abnormalities	NP	NP	IB	NP	B1
Abnormal ECG and no apparent SHD	I	A	NP	B1	B1
Abnormal ECG and contractile abnormalities with normal LVEF	NP	B1	II	B1	B1
Abnormal ECG and probable SHD (chest X-ray) or abnormal LVEF	II	B1 (LVEF ≥45%), B2 (LVEF <45%)	II	B2	B2
Compensated heart failure	III	C (compensated),	III	C (compensated)	C (compensated)
Refractory heart failure	NP	D (refractory)	NP	D (refractory)	D (refractory)

ECG electrocardiogram, *SHD* structural heart disease by chest X-ray (Kuschnir) or echocardiogram, *LVEF* left ventricular ejection fraction, *NP* not provided, *AHA* American Heart Association
[a]With no digestive alterations

each classification in relation to prognosis, theoretically decreasing gradually along the different evolutionary stages of the disease.

In Kuschnir classification, although some intermediate conditions like segmental contractility abnormalities cannot be evaluated due to the low-sensibility method used to define structural heart disease (chest X-ray), there is an increasing severity pattern between the stages (0, I–III) and cardiac disease progression.

The Brazilian Consensus Classification is based in ACC-AHA heart failure classification, but does not evaluate patients in indeterminate chronic form. As the objective was to identify distinct prognostic stages and there was some evidence that patients with normal ECG might have the same risk of death as the population without Chagas disease, this classification limited to assess patients with CCC. It was derived from a cohort study that observed an increase in 5-year mortality associated with progressive LV global dysfunction and the presence of heart failure symptoms: 13%, 45%, 91%, and 98%, associated with stages A, B, C, and D, respectively. To increase the prognostic stratification accuracy, group B was subdivided into B1 and B2, with a cut-off point at the LVEF of 45% on the echocardiography, with a progressive reduction of 5-year survival [7].

Los Andes classification initially used ventriculography to evaluate structural heart disease, substituted in 2007 by echocardiography. In this case, there is a detailed assessment of LV contractility abnormalities, even with normal ECG. Patients were subdivided into IA and IB, even with no additional prognostic information. On the other hand, patients with abnormal ECG and/or with progression of ventricular dysfunction were included in the same stage (II), with a very low capacity to distinguish the prognosis.

Considering the potential risk of progression to CCC, Latin American Guidelines Classification included patients with indeterminate chronic form in stage A (normal ECG), even anticipating survival equal to general population with negative serology to Chagas disease. They did not define a different subgroup with normal ECG and segmental contractility abnormalities, probably because they considered that prognosis is not modified in this situation. Systolic ventricular dysfunction and heart failure symptoms define different prognosis stages (B2 and C/D, respectively), whereas patients with ECG abnormalities with or without contractile abnormalities are included in the same subgroup (B1), with an apparent limitation for prognostic stratification.

AHA Statement is very similar to Latin American Guidelines classification. In particular, it explicitly describes stage A as the indeterminate chronic form, by excluding patients with isolated digestive symptoms. In addition, it further widens the subgroup B1 by including patients with segmental contractility abnormalities and normal ECG, which greatly limits its ability to stratify the prognosis: while subgroup A might have a prognosis similar to the general population, the subgroup B1 is very heterogeneous, ranging from normal to abnormal ECG, with or without structural heart disease. Beside this, B2 group comprises all patients with systolic dysfunction, ranging from mild to severe, which constitutes another very heterogeneous group, with a great leap of worsened prognosis for the C/D subgroups.

6.2 New Tests to Assess Heart Involvement. Benefits and Limitations of Use in Endemic and Non-endemic Areas

6.2.1 Myocardial Deformation Analysis and Diastolic Function Assessment by Echocardiography

6.2.1.1 Diagnostic Approach

Echocardiography has become the most important complementary test for the diagnosis and follow-up of patients with CCC as it provides valuable information on cardiac structure and function that complements the data obtained from the ECG and X-ray.

Echocardiographic findings range from a completely normal exam in the early stages to a severe deterioration in systolic and diastolic function. Cardiac involvement caused by Chagas disease has a typical segmental distribution with preference towards the apex and inferolateral walls [4, 8]. Apical aneurysm of the LV is the most characteristic lesion of CCC, which can range from a microaneurysm difficult to visualize without the use of contrast, to one of a larger size indistinguishable from a post-myocardial infarction aneurysm. In a study that evaluated 849 patients with Chagas disease using echocardiography, segmental motility abnormalities were observed in 13% of patients with normal ECG and chest X-ray, 33% of patients with ECG disturbances, and 70% of patients with cardiomegaly on chest X-ray [8]. These results, in accordance with other series, suggested that the presence of segmental wall motion abnormalities is often not associated with other signs detectable by ECG or increased cardiothoracic index on the X-ray.

Detection of regional contractility abnormalities in patients with conserved LVEF may have prognostic implications given its association with the progression of heart disease [9] and the potential risk of cardioembolic events and ventricular arrhythmias. However, identification of subtle visual motility abnormalities using conventional 2D echocardiography is not simple or precise as it is subjective and highly dependent on the skills of the interpreter. In the last years, evaluation of myocardial strain using tissue Doppler and speckle-tracking techniques have emerged as promising tools to allow a more precise and quantitative measurement of the regional myocardial function.

In a study by Silva et al. [10], tissue Doppler longitudinal and radial myocardial deformation in 39 patients showed differences between CCC patients (echocardiographic or ECG abnormalities), patients in the indeterminate form and controls. They found significantly reduced global radial and longitudinal strain in CCC patients as compared with control subjects and a significantly reduced radial strain at the mid inferolateral segment in the indeterminate group as compared with the control group. In comparison with tissue Doppler, the speckle-tracking method is less dependent on the angle of analysis, has less inter-observer variability and allows measuring 3D myocardial deformation: longitudinal, circumferential, and radial. In a study performed in a non-endemic area [11], 98 subjects were studied using speckle-tracking echocardiography: 22 with CCC defined as ECG or 2D echocardiographic abnormalities, 32 in the indeterminate form, and 44 control subjects. Global radial, circumferential, and longitudinal strain showed a significant decreasing trend across groups. Patients in the indeterminate form had significantly lower global radial strain and segmental radial strain in the mid-inferior segment compared with control subjects [11]. Subsequently, Barbosa et al. [12] applied speckle-tracking to 78 asymptomatic Chagas disease patients and 38 healthy subjects. They found significantly lower global longitudinal and radial strain in Chagas disease patients with no significant differences in circumferential strain. When assessing regional deformation, significant differences were found for the longitudinal strain in the basal inferior, basal inferoseptal, and apical inferolateral walls; radial strain in the basal segments of the anterior, inferior, and inferoseptal walls; and circumferential strain at the basal segment of the anterior wall. Similarly, Lima et al. [13] observed lower values of radial strain and strain rate in the inferior and inferolateral walls in CCC patients. Significantly lower global longitudinal strain was noted in the indeterminate form group. Interestingly, in the group with severe systolic dysfunction, a paradoxical increase in radial septal and anterior strain was demonstrated. In contrast, Gomes et al. [14] studied patients with Chagas disease with no evidence of cardiac involvement ($n = 83$) or at stage A of the cardiac form (changes limited to the ECG; $n = 42$) and 43 control subjects. They did not find differences in global longitudinal, circumferential, and radial LV strain across groups. However, global strains and radial strain in the inferoseptal wall were significantly lower in those patients who had fibrosis in cardiac magnetic resonance (CMR).

Diastolic dysfunction usually precedes systolic ventricular dysfunction in several cardiomyopathies. In Chagas disease, the studies published regarding the diastolic function pattern in patients in the indeterminate form have shown contradictory results, in some way influenced by the definition of the indeterminate form used in each study. Barros and colleagues observed prolongation of both the E-wave deceleration time of the mitral valve flow and the isovolumetric relaxation time in patients in the indeterminate form [15]. Similarly, Cianciulli et al. [16] reported that the evaluation of mitral valve flow by Doppler allowed detection of diastolic dysfunction in patients with normal ECG and 2D echocardiography. Conversely, in the study by Pazin-Filho et al. [17], that was limited to 43 patients, there was no evidence of diastolic dysfunction in the indeterminate form when it strictly included patients with neither ECG alterations nor global or segmental systolic dysfunction. However, in a meticulous review of this study, a trend towards higher atrial volumes, prolongation of the E-wave deceleration time, and reduction of the maximum E-wave velocity by tissue Doppler on the mitral annulus (Em) are observed, suggesting that insufficient statistical power may have contributed to the negative result. In a cohort of Chagas disease patients in a non-endemic area, patients in the indeterminate form showed significantly reduced Em and lengthened E-wave deceleration time of the LV inflow flow. Interestingly, half of patients in the indeterminate form had impaired relaxation patterns, whereas half of patients with ECG abnormalities suggestive of CCC had normal diastolic function. In the indeterminate form group, BNP levels were statistically higher in patients with diastolic dysfunction when compared with those with normal diastolic function and every patient with abnormally high levels of BNP had diastolic dysfunction [18]. These data suggest that diastolic function might have more sensitivity to detect early myocardial involvement than ECG changes. Globally, Em appeared to be the best parameter to identify the progressive worsening of LV diastolic dysfunction in CCC [14, 15].

6.2.1.2 Risk Stratification

Several echocardiographic variables have been associated with increased mortality: LV end-systolic and end-diastolic diameters; LVEF; segmental contractile abnormalities; RV contractility index; parameters of diastolic function and evaluation of filling pressures by Doppler; tissue Doppler; left atrial (LA) volumes; and LA strain [13–15, 19–25]. Of them, LVEF constitutes the most powerful and consistent independent echocardiographic predictor [26], although Em and the E/Em ratio have demonstrated to provide additive prognostic value [15, 20]. There is no robust data regarding the prognostic additional value of myocardial deformation techniques.

6.2.2 Cardiac Magnetic Resonance

6.2.2.1 Diagnostic Approach

CMR provides a comprehensive assessment of cardiac anatomy, global and regional biventricular function, and myocardial perfusion and viability. It is currently considered the gold standard technique for ventricular volumes quantification and for global and segmental motility analysis, overcoming certain limitations of echocardiography such as the quality of the acoustic window dependence or the requirement of geometric assumptions. Compared to echocardiography, CMR has the additional advantage of being able to assess tissue characteristics using the analysis of late gadolinium enhancement (LGE) sequences to identify fibrotic and/or necrotic segments and, more recently, T1 mapping sequences for the evaluation of diffuse fibrosis and extracellular volume.

In Chagas disease, the first series of patients in which myocardial fibrosis evaluated by LGE was analyzed in 2005 by Rochitte et al. [27]. They observed LGE in 69% of the global cohort and in 20% of the patients in the indeterminate disease, being the first to suggest that CMR is more sensitive to detect early myocardial involvement than the usual diagnostic techniques. In another study [28] evaluating by CMR a series of 67 patients living in a non-endemic area, LGE was observed in 7.4% of the indeterminate form group. Similarly, to that described by Rochitte et al., in the global population of Chagas disease patients, LGE was more prevalent in the inferolateral and apical segments, which correlated with the regional wall motion abnormalities, and showed a highly heterogeneous distribution, being subendocardial in 26.8%, midwall in 14.0%, subepicardial in 22.6%, and transmural in 36.0% of total segments with LGE.

As availability for CMR can be limited in certain endemic countries, the study of the correlation between LGE and new echocardiographic techniques such as LV strain and diastolic assessment is interesting. In a previously commented study of our group [28], CMR was performed in subset of 21 Chagas disease patients, 7 patients in each group (group 1 = indeterminate, group 2 = ECG abnormalities, group 3 = echocardiographic disturbances). Two (28%) patients in group 1, 1 (14%) patient in group 2, and 3 (43%) patients in group 3 had LGE. When these patients were classified according to the diastolic function pattern, none with normal diastolic function had LGE, whereas 40% of patients with impaired relaxation pattern and 50% with pseudonormal pattern showed delayed enhancement. In the previously commented study by Gomes et al. [14], patients with LGE (n = 14) had significantly lower global longitudinal, circumferential and radial LV strain, and lower radial strain in the basal inferoseptal wall despite similar LV ejection fraction.

6.2.2.2 Risk Assessment and Prognosis

In CCC patients, several previous studies had shown a significant association between the presence of LGE and LV dysfunction [27, 28] and sustained ventricular tachycardia (VT) [27, 29, 30]. Very recently, two studies simultaneously published in the Journal of the American College of Cardiology by Volpe et al. and Senra et al. [31, 32] have shown new data regarding the prognostic value of LGE in CCC. In the study by Volpe et al. [31], 100 out of 140 patients (71.4%) showed LGE, despite most patients were in NYHA functional class I (75%). Median follow-up was 2.75 years, during which 11 patients (7.8%) reached the primary endpoint (cardiovascular death and sustained VT). In the multivariable analysis, LGE emerged as an independent predictor for mortality and sustained VT and, which is more remarkable, no cardiovascular event was recorded among patients without LGE, despite low LVEF in half of them, pointing again the high negative predictive value of LGE. Senra et al. [32] assessed 130 patients with CCC. Similarly, despite most patients were in NYHA functional class I, myocardial LGE was detected in 76.1% of the population. After a median follow-up of 5.4 years, 58 patients (44.6%) reached the primary endpoint (all-cause mortality, heart transplantation, anti-tachycardia pacing or appropriate shock from an implanted cardiac defibrillator and aborted cardiac death). Again, on multivariate analysis LGE emerged as an independent predictor of the combined endpoint both as continuous or categorical variable. However, in this case, LVEF by CMR was not included as a covariate because, according to the authors, there was collinearity between LGE and LVEF. When the models were built including either LGE or CMR-LVEF, there were no differences in models performance, which underlines the high concordance between LV systolic function and LGE for global prognosis.

T1 mapping sequences emerge as promising markers of early myocardial involvement, but until now there are no published studies in Chagas disease patients.

6.2.3 Biomarkers

The activation of several pathophysiological mechanisms described in the CCC such as endothelial dysfunction, inflammation, and persistent myocardial fibrosis could be detected by measuring different biomarkers in blood.

Natriuretic peptides, particularly type B natriuretic peptide (BNP), are the biomarkers that have received more attention in CCC research, probably because the availability of economy and portable kits to be used in endemic areas. Ribeiro et al. demonstrated a high specificity with moderate sensitivity for BNP to detect LVEF<40% in infected patients with an abnormal ECG or chest X-ray [33]. In a second study, the same group reported that BNP levels correlated with LV dimensions and LVEF in patients with Chagas disease [34]. In a third study, they compared the diagnostic accuracy of the combination of BNP plasmatic levels and ECG vs. the standard strategy (ECG and chest X-ray) to detect LVEF <40% and demonstrated a significant improvement in specificity although the new strategy had less sensitivity [35]. Few studies have correlated BNP levels and diastolic function in Chagas disease [9, 36]; and most of them have been done in patients in CCC. Barbosa et al. [12] evaluated 59 patients with dilated cardiomyopathy due to Chagas disease and reported a marked elevated concentration of the amino-terminal portion proBNP specifically in patients with a restrictive diastolic pattern. Oliveira et al. [37] evaluated 36 patients, all of them with diffuse or segmental ventricular motion abnormalities, and described a significant correlation between BNP and E/E' ratio in the inferior wall. In a population of patients in earlier phases of the disease [18], the accuracy of BNP to detect any degree of diastolic dysfunction was good (area under curve of 0.73); and additionally, the ability to detect mild diastolic dysfunction in the group of patients in the undetermined form was also good (area under the curve of 0.69). The specificity of BNP levels above 37 pg/mL (upper limit of normal values) to detect mild diastolic dysfunction in patients in the indeterminate form of Chagas disease was 100%.

Several previous studies have also shown that C-reactive protein levels are elevated in children during the acute phase of the infection [38] as well as in patients in the chronic phase of CCC, both in advanced stages of the CCC [39, 40] and in the indeterminate form [41]. C-reactive protein is particularly attractive for clinical application because its biological characteristics make its determination easy and reproducible; however, the lack of specificity and evidence regarding its association with prognosis make that it is not currently recommended in the clinical basis.

Results regarding endothelin 1 levels, a potent vasoconstrictor, and TNFa have been contradictory [18, 35, 42].

6.3 Conclusion

Current classifications have limitations for identifying the risk of events other than heart failure, such as cardioembolic phenomena or sudden cardiac death. In fact, heart failure is the focus in question in all these scales [3]. This becomes clear when we note that most of them are based on the ABCD classification created by ACC-AHA for heart failure. However, in Chagas disease the sudden cardiac death is a relevant cause of death and manifests frequently without previous advanced symptoms or even without severe systolic dysfunction. It therefore should be considered on further classification tests to assess heart involvement that could predict fatal events, such as LGE-CMR or even the promising T1 mapping analysis, to improve the accuracy of Chagas definitions. Biomarkers and novel echocardiographic techniques such as speckle-tracking and detailed evaluation of diastolic function can detect early myocardial involvement. Although minor abnormalities in cardiac function or structure probably have not a global impact on mortality, they may help to individualize treatment and follow-up. Longitudinal studies with longer follow-up are needed to confirm that patients with worse strain values or diastolic dysfunction are more prone to develop overt CCC as well as their utility in risk stratification and response to therapy. From a diagnostic point of view, currently, in endemic areas ECG remains the main exam to screen for CCC and LVEF by echocardiogram continues to be the most important marker of global prognosis.

References

1. Consenso Brasileiro em Doença de Chagas. Epidemiol. Serv. Saude v. 15 nesp. Brasilia, jun. 2016.
2. Kuschnir E, Sgammini H, Castro R, Evequoz C, Ledesma R, Brunetto J. Evaluation of cardiac function by radioisotopic angiography, in patients with chronic Chagas cardiopathy. Arq Bras Cardiol. 1985;45(4):249–56.
3. Dias JC, Ramos AN, Gontijo ED, Luquetti A, Shikanai-Yasuda MA, Coura JR, et al. 2nd Brazilian consensus on Chagas disease, 2015. Rev Soc Bras Med Trop. 2016;49(Suppl 1):3–60.
4. Acquatella H. Echocardiography in Chagas heart disease. Circulation. 2007;115(9):1124–31.
5. Andrade JP, Marin-Neto JA, Paola AA, Vilas-Boas F, Oliveira GM, Bacal F, et al. I Latin American guidelines for the diagnosis and treatment of Chagas cardiomyopathy. Arq Bras Cardiol. 2011;97(2 Suppl 3):1–48.
6. Nunes MCP, Beaton A, Acquatella H, Bern C, Bolger AF, Echeverría LE, et al. Chagas cardiomyopathy: an update of current clinical knowledge and management: a scientific statement from the American Heart Association. Circulation. 2018;138(12):e169–209.
7. Xavier SS, Sousa AS, Hasslocher-Moreno A. Application of the new classification of cardiac insufficiency (ACC/AHA) in chronic Chagas cardiopathy: a critical analysis of the survival curves. Rev SOCER J. 2005;18:227–32.
8. Viotti RJ, Vigliano C, Laucella S, Lococo B, Petti M, Bertocchi G, et al. Value of echocardiography for diagnosis and prognosis of chronic Chagas disease cardiomyopathy without heart failure. Heart. 2004;90(6):655–60.
9. Pazin-Filho A, Romano MM, Almeida-Filho OC, Furuta MS, Viviani LF, Schmidt A, et al. Minor segmental wall motion abnormalities detected in patients with Chagas' disease have adverse prognostic implications. Braz J Med Biol Res. 2006;39(4):483–7.
10. Silva CE, Ferreira LD, Peixoto LB, Monaco CG, Gil MA, Ortiz J, et al. Evaluation of segmentary contractility in Chagas' disease by using the integral of the myocardial velocity gradient (myocardial strain) obtained through tissue Doppler echocardiography. Arq Bras Cardiol. 2005;84(4):285–91.
11. Garcia-Alvarez A, Sitges M, Regueiro A, Poyatos S, Jesus Pinazo M, Posada E, et al. Myocardial deformation analysis in Chagas heart disease with the use of speckle tracking echocardiography. J Card Fail. 2011;17(12):1028–34.
12. Barbosa MM, Nunes Mdo C, Ribeiro AL, Barral MM, Rocha MO. N-terminal proBNP levels in patients with Chagas disease: a marker of systolic and diastolic dysfunction of the left ventricle. Eur J Echocardiogr. 2007;8(3):204–12.

13. Lima MS, Villarraga HR, Abduch MC, Lima MF, Cruz CB, Bittencourt MS, et al. Comprehensive left ventricular mechanics analysis by speckle tracking echocardiography in Chagas disease. Cardiovasc Ultrasound. 2016;14(1):20.
14. Gomes VA, Alves GF, Hadlich M, Azevedo CF, Pereira IM, Santos CR, et al. Analysis of regional left ventricular strain in patients with Chagas disease and Normal left ventricular systolic function. J Am Soc Echocardiogr. 2016;29(7):679–88.
15. Barros MV, Machado FS, Ribeiro AL, Rocha MO. Diastolic function in Chagas' disease: an echo and tissue Doppler imaging study. Eur J Echocardiogr. 2004;5(3):182–8.
16. Cianciulli TF, Lax JA, Saccheri MC, Papantoniou A, Morita LA, Prado NG, et al. Early detection of left ventricular diastolic dysfunction in Chagas' disease. Cardiovasc Ultrasound. 2006;4:18.
17. Pazin-Filho A, Romano MM, Gomes Furtado R, de Almeida Filho OC, Schmidt A, Marin-Neto JA, et al. Left ventricular global performance and diastolic function in indeterminate and cardiac forms of Chagas' disease. J Am Soc Echocardiogr. 2007;20(12):1338–43.
18. Garcia-Alvarez A, Sitges M, Pinazo MJ, Regueiro-Cueva A, Posada E, Poyatos S, et al. Chagas cardiomiopathy: the potential of diastolic dysfunction and brain natriuretic peptide in the early identification of cardiac damage. PLoS Negl Trop Dis. 2010;4(9):e826.
19. Nascimento CA, Gomes VA, Silva SK, Santos CR, Chambela MC, Madeira FS, et al. Left atrial and left ventricular diastolic function in chronic Chagas disease. J Am Soc Echocardiogr. 2013;26(12):1424–33.
20. Nunes MP, Colosimo EA, Reis RC, Barbosa MM, da Silva JL, Barbosa F, et al. Different prognostic impact of the tissue Doppler-derived E/e' ratio on mortality in Chagas cardiomyopathy patients with heart failure. J Heart Lung Transplant. 2012;31(6):634–41.
21. Rassi Ddo C, Vieira ML, Arruda AL, Hotta VT, Furtado RG, Rassi DT, et al. Echocardiographic parameters and survival in Chagas heart disease with severe systolic dysfunction. Arq Bras Cardiol. 2014;102(3):245–52.
22. Sarabanda AV, Marin-Neto JA. Predictors of mortality in patients with Chagas' cardiomyopathy and ventricular tachycardia not treated with implantable cardioverter-defibrillators. PACE. 2011;34(1):54–62.
23. Theodoropoulos TA, Bestetti RB, Otaviano AP, Cordeiro JA, Rodrigues VC, Silva AC. Predictors of all-cause mortality in chronic Chagas' heart disease in the current era of heart failure therapy. Int J Cardiol. 2008;128(1):22–9.
24. Viotti R, Vigliano C, Lococo B, Petti M, Bertocchi G, Alvarez MG, et al. Clinical predictors of chronic chagasic myocarditis progression. Rev Esp Cardiol. 2005;58(9):1037–44.
25. Rassi A Jr, Rassi A, Little WC, Xavier SS, Rassi SG, Rassi AG, et al. Development and validation of a risk score for predicting death in Chagas' heart disease. N Engl J Med. 2006;355(8):799–808.
26. Bocchi EA, Bestetti RB, Scanavacca MI, Cunha Neto E, Issa VS. Chronic Chagas heart disease management: from etiology to cardiomyopathy treatment. J Am Coll Cardiol. 2017;70(12):1510–24.
27. Rochitte CE, Oliveira PF, Andrade JM, Ianni BM, Parga JR, Avila LF, et al. Myocardial delayed enhancement by magnetic resonance imaging in patients with Chagas' disease: a marker of disease severity. J Am Coll Cardiol. 2005;46(8):1553–8.
28. Regueiro A, Garcia-Alvarez A, Sitges M, Ortiz-Perez JT, De Caralt MT, Pinazo MJ, et al. Myocardial involvement in Chagas disease: insights from cardiac magnetic resonance. Int J Cardiol. 2013;165:107–12.
29. Mello RP, Szarf G, Schvartzman PR, Nakano EM, Espinosa MM, Szejnfeld D, et al. Delayed enhancement cardiac magnetic resonance imaging can identify the risk for ventricular tachycardia in chronic Chagas' heart disease. Arq Bras Cardiol. 2012;98(5):421–30.
30. Tassi EM, Continentino MA, Nascimento EM, Pereira Bde B, Pedrosa RC. Relationship between fibrosis and ventricular arrhythmias in Chagas heart disease without ventricular dysfunction. Arq Bras Cardiol. 2014;102(5):456–64.

31. Volpe GJ, Moreira HT, Trad HS, Wu KC, Braggion-Santos MF, Santos MK, et al. Left ventricular scar and prognosis in chronic Chagas cardiomyopathy. J Am Coll Cardiol. 2018;72(21):2567–76.
32. Senra T, Ianni BM, Costa ACP, Mady C, Martinelli-Filho M, Kalil-Filho R, et al. Long-term prognostic value of myocardial fibrosis in patients with Chagas cardiomyopathy. J Am Coll Cardiol. 2018;72(21):2577–87.
33. Ribeiro AL, dos Reis AM, Barros MV, de Sousa MR, Rocha AL, Perez AA, et al. Brain natriuretic peptide and left ventricular dysfunction in Chagas' disease. Lancet. 2002;360(9331):461–2.
34. Ribeiro AL, Teixeira MM, Reis AM, Talvani A, Perez AA, Barros MV, et al. Brain natriuretic peptide based strategy to detect left ventricular dysfunction in Chagas disease: a comparison with the conventional approach. Int J Cardiol. 2006;109(1):34–40.
35. Talvani A, Rocha MO, Cogan J, Maewal P, de Lemos J, Ribeiro AL, et al. Brain natriuretic peptide and left ventricular dysfunction in chagasic cardiomyopathy. Mem Inst Oswaldo Cruz. 2004;99(6):645–9.
36. Anez N, Carrasco H, Parada H, Crisante G, Rojas A, Fuenmayor C, et al. Myocardial parasite persistence in chronic chagasic patients. Am J Trop Med Hyg. 1999;60(5):726–32.
37. Oliveira BM, Botoni FA, Ribeiro AL, Pinto AS, Reis AM, Nunes Mdo C, et al. Correlation between BNP levels and Doppler echocardiographic parameters of left ventricle filling pressure in patients with Chagasic cardiomyopathy. Echocardiography. 2009;26(5):521–7.
38. Medrano NM, Luz MR, Cabello PH, Tapia GT, Van Leuven F, Araujo-Jorge TC. Acute Chagas' disease: plasma levels of alpha-2-macroglobulin and C-reactive protein in children under 13 years in a high endemic area of Bolivia. J Trop Pediatr. 1996;42(2):68–74.
39. Aparecida da Silva C, Fattori A, Sousa AL, Mazon SB, Monte Alegre S, Almeida EA, et al. Determining the C-reactive protein level in patients with different clinical forms of chagas disease. Rev Esp Cardiol. 2010;63(9):1096–9.
40. Lopez L, Arai K, Gimenez E, Jimenez M, Pascuzo C, Rodriguez-Bonfante C, et al. C-reactive protein and interleukin-6 serum levels increase as Chagas disease progresses towards cardiac failure. Rev Esp Cardiol. 2006;59(1):50–6.
41. Garcia-Alvarez A, Sitges M, Heras M, Poyatos S, Posada E, Pinazo MJ, et al. Endothelial function and high-sensitivity C-reactive protein levels in patients with Chagas disease living in a nonendemic area. Rev Esp Cardiol. 2011;64(10):891–6.
42. Salomone OA, Elliott PM, Calvino R, Holt D, Kaski JC. Plasma immunoreactive endothelin concentration correlates with severity of coronary artery disease in patients with stable angina pectoris and normal ventricular function. J Am Coll Cardiol. 1996;28(1):14–9.

Chronic Digestive Chagas Disease

7

Joffre Rezende Filho and Enio Chaves de Oliveira

7.1 Chronic Digestive Chagas Disease

The occurrence of innumerous cases with chronic dysphagia in Brazilian hinterlands, commonly named "chocking affection" ("mal de engasgo" in Portuguese), was known even before than the recognition of American Trypanosomiasis [1]. Chagas (1916) suggested for the first time that this endemic affection could be one of the manifestations of the disease caused by the infection he has described [2]. However, only 50 years later, the digestive involvement was accepted as a clinical form of Chagas disease (CD). Along this time, various observations contributed to the recognition of the chagasic etiology of the endemic megaviscera.

Vampré (1919), employing radiographies with bismuth as contrast media found esophageal dilation, demonstrated for the first time that the endemic affection causing dysphagia was associated with megaesophagus [3].

Amorim and Correia Neto (1932) found myenteric plexus lesions in cases of endemic megaesophagus, but no etiology could be demonstrated [4]. Etzel (1934) demonstrated myenteric plexus lesions also in the colon, showing the systemic involvement of the disease, suggesting that both megaesophagus and megacolon shared the same etiopathogenetic–myenteric neuropathy [5].

J. Rezende Filho (✉)
Division of Gastroenterology, Hospital das Clinicas, Universidade Federal de Goias, Goiânia, Goiás, Brazil

Departamento de Clinica Medica, Faculdade de Medicina, Universidade Federal de Goiás, Goiânia, Goiás, Brazil

E. Chaves de Oliveira
Chagas Disease Study Center, Hospital das Clinicas, Universidade Federal de Goias, Goiânia, Goiás, Brazil

Departamento de Cirurgia, Faculdade de Medicina, Universidade Federal de Goiás, Goiânia, Goiás, Brazil

M.-J. Pinazo Delgado, J. Gascon (eds.), *Chagas Disease*,
https://doi.org/10.1007/978-3-030-44054-1_7

Freitas (1947) found that more than 90% of patients with endemic megaesophagus presented positive serological reactions for CD, strongly suggesting more than a coincidental finding [6].

Koberle and contributors, in successive anatomopathological studies starting in 1955, demonstrated unquestionably that CD may present lesions of enteric nervous system in variable degrees along the digestive tract. In those cases, in whom more severe denervation occurred, megavisceras were present [7–9].

Rezende (1956), in a clinical-epidemiological study, based upon a large cohort of patients accepted definitively the etiology of the megaesophagus, named it "chagasic megaesophagus." In addition, he proposed that an indepedent clinical form of CD shoud be individualized—the digestive form, considering that chagasic patients may present only digestive manifestations, independently of cardiac or other manifestations [10, 11].

7.1.1 Characterization of the Digestive Form

The digestive form of CD is characterized by alterations in motor, secretory, and absorptive functions of the gastrointestinal tract seen in chronic phase of the disease [11, 12]. It is related to denervation of the enteric nervous system that may occur along the entire digestive tract [9]. The denervation occurs at a variable degree, being irregular and not continuous [9].

The exact mechanism of this denervation associated with the infection is still not entirely known, but immune mechanisms related to inflammation induced by the presence of the parasite may be involved [13, 14]. The persistence of *T. cruzi* DNA has been demonstrated in esophageal tissue even in patients with chronic phase of the disease [13, 14]. The action of nitric oxide (NO) induced by γ interferon during the inflammation process in the muscular layers and myenteric plexus might be relevant to this neuronal degeneration [15].

As a result of this denervation, digestive motility disturbances occur, leading to loss of coordination of peristalsis and relaxation of the sphincters, resulting in progressive dilation—megaformations, notably in the esophagus and colon, known as chagasic megaesophagus and megacolon [16]. The pathogenesis of the digestive form of CD has been described in Chap. 1, and it is summarized in Fig. 7.1.

7.1.2 Prevalence of Digestive Form of Chagas Disease

The prevalence of digestive manifestation varies considerably. In Central Brazil, around 10% of chagasic patients present the digestive form of the disease, either alone or associated with cardiac involvement [17]. Presently, in some places, like in Central Brazil, as a result of cessation of vectorial transmission, a decline in prevalence has been observed [18, 19]. The population affected with chagasic megaesophagus and megacolon is turning progressively older [19, 20]. In a recent

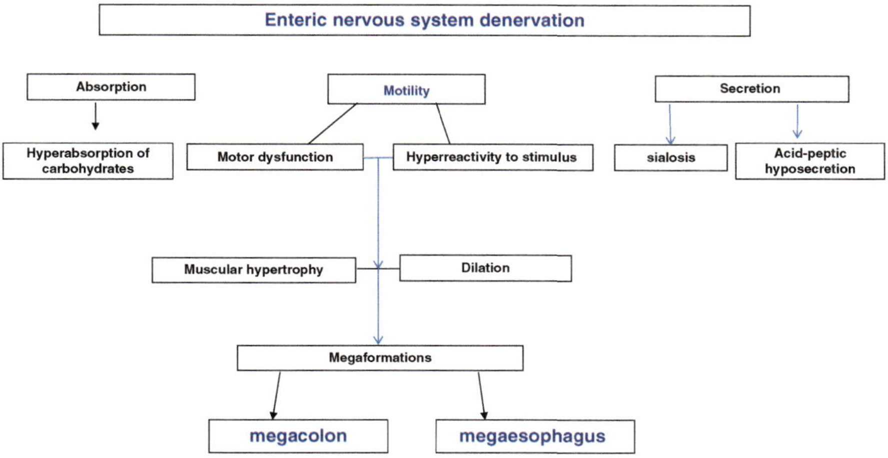

Fig. 7.1 Pathogenesis of digestive form of Chagas disease

publication, 68% of patients with *T. cruzi* infection from Cochabamba (Bolivia) show digestive disorders related to Chagas Disease [21].

In Northern part of South America, Central America, and Mexico, the digestive form of CD is rare [17, 22]. This regional variation in prevalence of chagasic megaesophagus and megacolon, probably, seems to be related to geographic distribution of distinct *Trypanosoma cruzi* subpopulations with different histotrophism [22–24]. The prevalence of digestive manifestations in patients living in nonendemic areas may vary considering the original location of the patients [25–27].

7.2 Chagasic Esophagopathy

The chagasic esophagopathy is characterized by a spectrum of motor disturbances in which megaesophagus represents the most severe involvement [12, 28, 29]. Alterations of peristalsis of the esophageal body and loss of relaxation of the lower esophageal sphincter occur, similar to idiopathic achalasia [28–31]. These result in esophageal transit disturbances and progressive esophageal dilation.

7.2.1 Clinical Manifestations

Chagasic megaesophagus is clinically indistinguishable from idiopathic achalasia of the esophagus. Positive serological reaction for Chagas disease and association with megacolon or other clinical manifestation such as cardiopathy may allow a differential diagnosis [28, 30]. Aprroximately 30% of patients with chagasic megaesophagus also present cardiopathy associated, named cardio-digestive form of CD. Usually, the esophageal symptoms appear early than the cardiac manifestations [28].

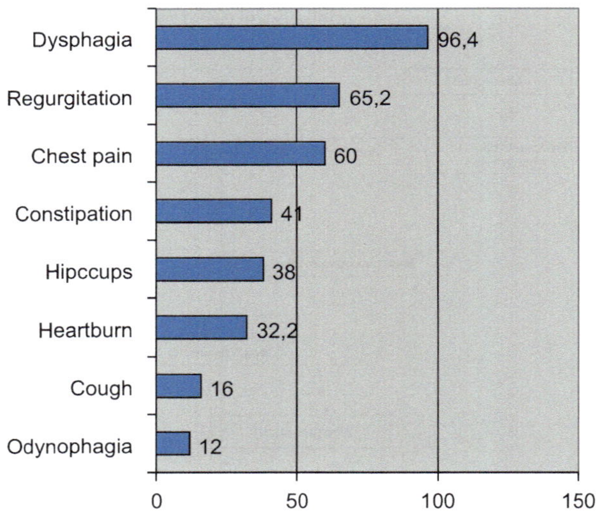

Fig. 7.2 Frequency of symptoms in chagasic megaesophagus in a cohort at the Hospital das Clinicas da UFG

The main esophageal symptom is slowly progressive dysphagia, even for years. Initially, the dysphagia is occasionally reported, more with cold foods and during emotional stress. Some patients live for years with an intermittent, mild dysphagia without further progression. In other cases, it progresses to become present more frequently, occurring with solids and liquids. Patients commonly report that they need to drink water during the meals to obtain relief. Sudden episodes of esophageal pain may appear. Usually, those episodes of chest pain are alleviated by drinking water [28]. In more advanced cases, passive supine regurgitation is common [28, 32]. Malnutrition may occur with progression of the esophageal dilation [32]. The frequency distribution of symptoms in patients with chagasic megaesophagus is shown in Fig. 7.2.

7.2.2 Diagnosis

The most common method used to evaluate chagasic esophagopathy has been radiological examination. The different morphofunctional aspects of the esophagus identify various stages of the disease and allow a radiological classification of the chagasic esophagopathy [33, 34]. In this classification, stage development of chagasic esophagopathy is divided into four groups, as follows (Fig. 7.3):

- Group I—Anectasic form. Normal diameter, minimal contrast retention, presence of a residual air column above the contrast
- Group II—Moderate dilatation, some contrast retention, increase in uncoordinated motor activity, relative hypertonia of the inferior third of the esophagus
- Group III—Large increase in diameter, great contrast retention, hypotonic esophagus with weak or absent motor activity

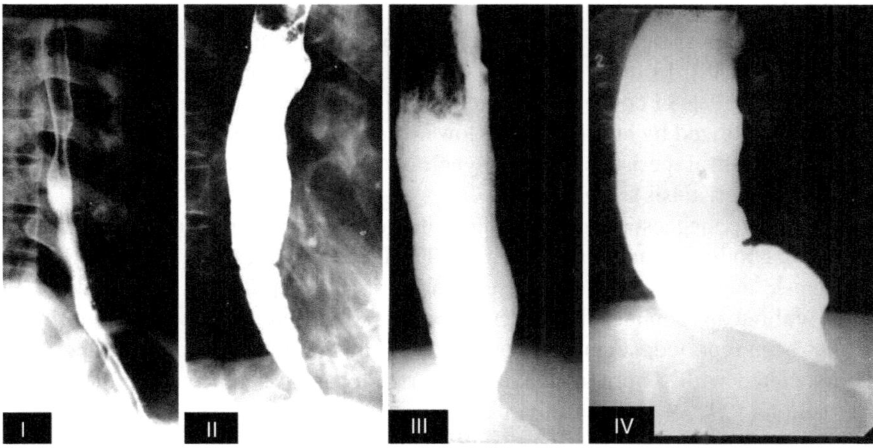

Fig. 7.3 Radiological classification of chagasic megaesophagus in four groups (stages), according to Rezende [33]

- Group IV—Large increase in volume, atonic, elongate esophagus, lying on the right diaphragmatic dome (dolicomegaesophagus)

In symptomatic cases without esophageal dilation, two radiographic images should be taken: the first immediately after swallowing 150 mL of barium meal, and the second 1 min later. In this second image, the following features make the correct diagnosis very likely: (a) normal diameter of the esophagus; (b) incomplete emptying of the esophagus, the remaining barium taking a cylindrical shape; (c) presence of air above the contrast medium along the entire length of the esophagus [35].

In the Hospital das Clinicas da Universidade Federal de Goiás (UFG), in the last decade, according to the Rezende's radiological classification, the distribution of stages of megaesophagus was 36%, 33%, 17%, and 14% corresponding to megaesophagus groups I to IV, respectively [19]. There has been a shift to less severe cases as compared to another series of patients seen in the same region 20 years before [31].

The esophageal transit time can also be measured by means of sequential radiographies until the esophagus is empty of the barium suspension. In chagasic megaesophagus, a correlation between esophageal transit time and degree of dilation was reported. The esophageal transit time ranged from 1 min to 36 h, with median time being 5 min in GI, 30 in GII, 2 h in GIII, and 9:15 h in GIV [36].

In endemic areas, barium esophagogram is the most useful method for the diagnosis of chagasic megaesophagus in clinical practice because of its reliability, low cost, and availability [19, 36]. However, in cases in which the clinical and radiological findings are not enough for diagnosis of the esophageal involvement, esophageal manometry should be considered.

7.2.2.1 Esophageal Manometry

Esophageal manometry has been widely performed in evaluating chagasic esophagopathy [37, 38]. In more advanced cases, the findings are similar to idiopathic achalasia [19, 39]. In conventional esophageal manometry, chagasic megaesophagus is characterized by absence of the lower esophageal sphincter relaxation (achalasia) and loss of peristalsis, with synchronous wave pressures with very low amplitude (aperistalsis) (Fig. 7.4).

In chagasic megaesophagus, the more dilated the esophagus, the lower are the esophageal body pressure waves [38, 39]. The LES basal pressure tends to be lower in chagasic megaesophagus than in idiopathic achalasia [40, 41]. In less advanced cases, without an esophageal dilation, a spectrum of different less severe abnormalities has been reported: segmental aperistalsis limited to lower esophagus, aperistalsis associated with a normal relaxing lower esophageal sphincter, low amplitude contractions [29, 39]. Those minor disorders of peristalsis can be found even in asymptomatic patients [42, 43]. An augmented response to cholinergic drug (methacholine) has been proposed as a manometric evidence of esophageal denervation in CD patients, even without esophageal dilation or symptoms [44]. The finding of manometric alterations in patients with clinical indeterminate form [41–43] suggests that only symptoms are not sufficient to indicate the presence of histological and functional abnormalities in chagasic patients.

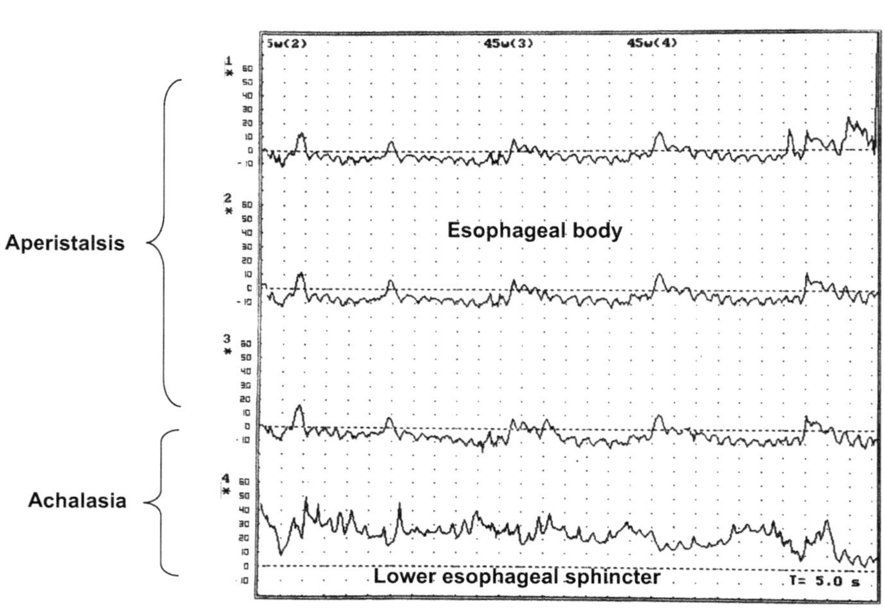

Fig. 7.4 Conventional manometry in chagasic megaesophagus. Esophageal body shows synchronous and low amplitude waves corresponding to aperistalsis (1–3). Absence of lower esophageal sphincter relaxation in response to swallows (w) corresponding to achalasia

More recently, high-resolution manometry has been applied in the assessment of esophageal involvement in Chagas disease [45, 46, 47]. In patients with chagasic megaesophagus, type I (classic achalasia) and type II (pan-pressurization) achalasia, according to Chicago Classification, has been found [48, 49]. Even in patients without esophageal dilation, achalasia type I has been reported, as well as esophageal gastric junction (EGJ) outflow obstruction (Fig. 7.5).

Minor alterations of peristalsis (fragmented peristalsis and infective esophageal motility) have been reported in CD without esophageal dilation and in asymptomatic patients, mainly in migrants living in nonendemic areas [46, 49]. However, in such cases, the esophageal manometric abnormalities described are nonspecific, regarding the presence of esophageal denervation, and could not be related to Chagas disease esophageal involvement. Therefore, performing high-resolution manometry as a method of early assessment of esophageal involvement in CD may be useful, but minor alterations could be cautiously considered, as they may represent unspecific findings and should not be overestimated.

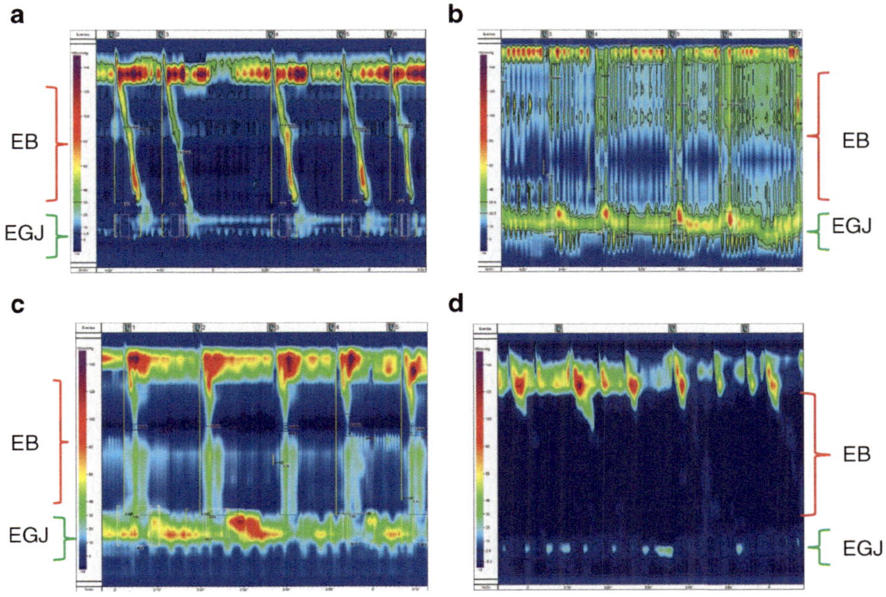

Fig. 7.5 High-resolution manometry in healthy control and in patients with chagasic esophagopathy. (**a**) Normal peristalsis and relaxation of esophageal gastric junction (EGJ) (heathy control); (**b**) Pan-pressurization and iterative motor activity in esophageal body (EB), no relaxation of esophageal gastric junction (EGJ). (**c**) Pan-pressurization (EB) and no relaxation of esophageal gastric junction (EGJ). (**d**) No pressurization (EB) and low pressure in EGJ—patient in postop. Cardioplasty (Serra-Doria operation). Courtesy of Dr. Joffre Rezende Neto

7.3 Chagasic Colopathy

Chagasic colopathy is characterized by dilation of any segment of the colon, mainly, sigmoid and rectum in a high proportion of cases [28]. The dilation of the whole colon is a rare finding [28].

Differently from congenital megacolon, in Chagas disease there is a massive loss of myenteric neurons in the dilated part of the colon [9, 50]. There is also loss of interstitial cells of Cajal [51] and glial elements degeneration [52]. In the muscular layer there is hypertrophy and irregular inflammatory infiltrate, along with a marked fibrosis [53].

The precise mechanism by which the colon denervation leads to dilation is still not clear. It was initially thought that motor incoordination of the rectum and sigmoid would promote transit difficulty leading to fecal stasis and visceral dilation [9]. There is a selective degeneration of myenteric neurons, in which VIP/NO is more preserved than cholinergic neurons [54]. This imbalance of inhibitory and excitatory neurons and fibers might contribute to the development of chagasic megacolon [54, 55]. The alterations in other cells such as glial, muscular, and interstitial cells of Cajal may also play a role in the development of colon dilation [50].

7.3.1 Clinical Manifestations

Patients with chagasic megacolon usually become symptomatic later than with megaesophagus, usually older than 40 years. As an isolated manifestation of CD, megacolon is rare finding, being associated with megaesophagus in most cases [28].

In chagasic colopathy, the main symptom is constipation. Usually, the constipation has an insidious onset and progress slowly. In the beginning, bowel movements can be improved with laxatives. With progression of the disease, colon cleansing enemas may be necessary. The interval of bowel movements can vary from a few days to weeks or even months. In a large case series of 268 chagasic megacolon seen at the first examination, Rezende and Moreira reported that 70% had gone more than 10 days and 36.6% more than 20 days without a bowel movement [56].

Approximately 25–30% cases with megacolon do not present constipation, having a normal intestinal movement frequency. Therefore, the occurrence of asymptomatic cases, or with minor or nonspecific symptoms, may contribute to underdiagnoses of chagasic megacolon [28, 57].

The occurrence of patients having normal bowel movements frequency and dilated colon suggests that other factors besides colonic motor incoordination and dilation may play an important role in the development of constipation [58]. Alterations in orocecal transit in chagasic patients with constipation, with or without megacolon, have been recently demonstrated [58]. Therefore, Bafutto et al. suggested that small bowel denervation, leading to motor and absorptive disorders due to submucosal neurons lesions, may also have a relevant role in the pathophysiology of the constipation in chagasic patients [58].

Meteorism and abdominal distension can also be found. Patients usually report abdominal discomfort and sometimes abdominal crampy pain. Dyschezia, with excessive and prolonged evacuator efforts, even with normal consistency of feces, may be reported as well [28].

The main complications of chagasic megacolon are fecaloma and sigmoid volvulus. Additionally, fecal impaction, ischemic colitis, stercoral ulcer, and visceral perforation may also occur.

A patient with fecaloma may present rectal pain and a sensation of incomplete evacuation. Depending on the size and location of the fecaloma, patients may present with paradoxical diarrhea and fecal incontinence. In cases that present with a sigmoid volvulus, severe cramping abdominal pain and bowel obstruction may be present. In cases with possible vascular ischemia of the sigmoid, signs of toxemia and peritonitis can occur [28, 56].

On physical examination, an increase in abdominal volume may be observed. The sigmoid can be easily recognized, sometimes can be localized by either percussion or palpation outside of its normal location, displaced to the right side. The presence of a fecaloma can be diagnosed by simple abdominal palpation as an elastic tumor, with presence of "Gersuny" sign [28, 56].

In case of sigmoid volvulus, there are signs suggestive of a low intestinal occlusion with sudden pain in the lower abdomen and absence of feces and gas elimination, along with a progressive increase in abdominal distension [28, 56].

7.3.2 Diagnosis

In every chagasic patient with chronic constipation, megacolon should be considered. However, it has to be considered that chronic constipation is a very common symptom in the general population and can be present in chagasic patients even without colon dilation. Moreover, other factors can aggravate the clinical symptoms, given that some studies have shown that prolonged stays at high altitudes or chewing coca leaves for long periods of time can cause dolicomegacolon or megacolon [21, 59, 60]. Additionally, patterns and behaviors observed in migrant population regarding changes in dietary habits can also contribute to constipation in Latin-American population.

Considering the symptoms and radiological findings in patients living in endemic area with high prevalence of digestive form of Chagas disease, different clinical-radiological presentation may be found, such as patients: without megacolon/megarectum and without constipation; without megacolon/megarectum and with constipation; with megacolon/megarectum and without constipation; with megacolon/megarectum and with constipation [57].

Barium enema is the most common method used for the diagnosis of chagasic megacolon. The presence of dolicocolon (elongation), megacolon, or dolicomegacolon can be found. The most frequently dilated segment is the sigmoid and the rectum [28, 56]. Dolicocolon can be one of the first radiological evidence of chagasic colopathy even before the appearance of dilation [61]. The different radiological aspect of chagasic megacolon is shown in Fig. 7.6.

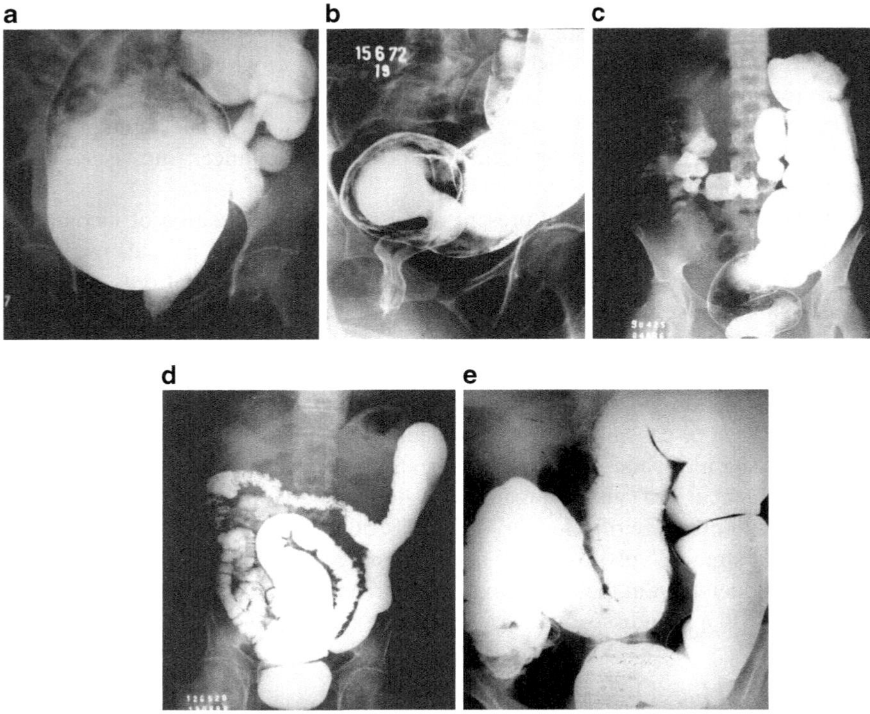

Fig. 7.6 Radiological aspects of Chagasic megacolon. (**a**) Megarectum; (**b**) megasigmoid; (**c**) megarectumsigmoid; (**d**) dolicocolon; (**e**) total megacolon

Considering that several factors, such as distension caused by pressure of the injected contrast and by air for double contrast, may influence the radiological aspect and diameter of the colon, a simplified and standardized method has been proposed, in which no colon preparation and air injection is applied [62]. Using this method, a sigmoid diameter greater than 6 cm was considered abnormal, measured in the anteroposterior film [28, 62].

In a group of 1500 patients with chagasic megacolon seen at the Hospital das Clinicas da Universidade Federal de Goias, approximately 48% present with megarectum, defined as diameter of the rectum greater than 6.5 cm in a profile X-ray of the pelvis in a barium enema examination.

One radiological classification of chagasic megacolon has been proposed based on transverse axis of the sigmoid, in which the megacolon was considered: normal (<5 cm), grade 1: 5–9 cm, grade 2:9–13 cm, and grade 3:>13 cm [63]. But until now, differently from the esophagus, there is no radiological classification widely accepted for chagasic megacolon.

The presence of fecaloma and sigmoid volvulus may be easily identified in a simple abdominal X-rays, as shown in Fig. 7.7.

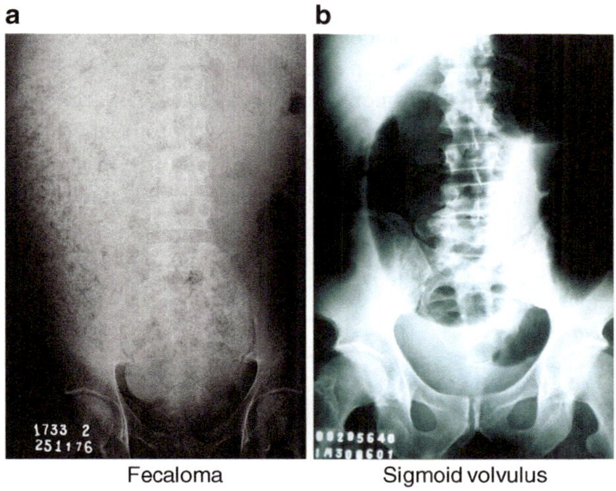

Fig. 7.7 Complications of chagasic megacolon. (**a**) Radiological aspect of a huge fecaloma. (**b**) Radiological aspect showing colon dilation with air suggestive of sigmoid volvulus

7.3.2.1 Manometry

Manometric studies of the colon of chagasic patients have demonstrated motor disturbances, mainly rectum-sigmoid motor incoordination, with synchronous contraction of the sigmoid and rectum [64]. Hypersensitivity to cholinergic stimulus by drugs such as methacholine and neostigmine resulting in a greater motor response has been described [65].

In addition, achalasia of the internal anal sphincter can be found, with absence of relaxation of this sphincter in response to rectal distension [60]. However, this finding may represent a lack of stimulation of the reflex secondary to rectal dilation [66]. Patients with megarectum, if tested with small volume of air in the balloon positioned in the rectum, may have no recto-anal inhibitory reflex elicited due to the small volume inside the balloon usually up to 50 mL. In such cases, relaxation can be demonstrated with a larger volume of the balloon distension, as much as 300 mL [66].

Chagasic patients with constipation even without colon dilation may present alterations in anorectal manometry but such findings may be similar to patients with functional constipation and could not be related to chagasic colopathy [67]. In clinical practice, anorectal manometry is not commonly performed as a clinical test for evaluation of chagasic colopathy.

7.4 Conclusion

The digestive form of CD is related to denervation of the enteric nervous system that may occur along the entire digestive tract, resulting in progressive dilation—megaformations, notably in the esophagus and colon, known as chagasic megaesophagus and megacolon. The prevalence of digestive manifestation varies considerably. Presently, in some places, as a result of cessation of vectorial transmission, a decline in prevalence has been observed. The prevalence of digestive manifestations in patients living in nonendemic areas may vary considering the original location of the patients. The main esophageal symptom is slowly progressive dysphagia, even for years. In chagasic colopathy, the main symptom is constipation. Therefore, the most common method used to evaluate chagasic esophagopathy and colopathy has been radiological examination. More recently, high-resolution manometry has been applied in the assessment of esophageal involvement in Chagas disease in less severe cases.

References

1. Langgard TJH. Diccionario de Medicina Domestica e Popular. Rio de Janeiro: Eduardo & Henrique Laemmert; 1873.
2. Chagas C. Trypanosomíase americana. Forma Aguda da Moléstia. Mem Inst Oswaldo Cruz. 1916;8:37–60.
3. Vampré E. Contribuição ao estudo do mal de engasgo. Serviço Sanitário do Estado de São Paulo. 1919;5:3–78.
4. Amorim M, Correia NA. Histopathologia e pathogenese do megaesofago e megareto. Considerações em torno de um caso de "mal do engasgo". An Fac Med Univ São Paulo. 1932;8:101–27.
5. Etzel E. Neuropathologia do megaesophago e megacolon. An Fac Med Univ São Paulo. 1934;10:383–95.
6. Freitas JLP. Contribuição para o estudo do diagnóstico da moléstia de Chagas por processos de laboratório. Tese de Doutorado em Medicina. São Paulo: FM-USP; 1947.
7. Koeberle F, Nador E. Etiologia e patogenia do megaesôfago no Brasil. Rev Paul Med. 1955;47:643–61.
8. Koeberle F. Patogenia da moléstia de Chagas. Estudo dos órgãos musculares ôcos. Rev Goiana Med. 1957;3:155–80.
9. Köeberle F. Chagas' disease and Chagas' syndromes: the pathology of American trypanosomiasis. Adv Parasitol. 1968;6:63–116.
10. Rezende JM. Megaesôfago por doença de chagas. Rev Goiana Med. 1956;2:297–314.
11. Rezende JM. Forma digestiva da moléstia de chagas. Rev Goiana Med. 1959;5:193–227.
12. Oliveira RB, Troncon LE, Dantas RO, Meneghelli UG. Gastrointestinal manifestations of Chagas' disease. Am J Gastroenterol. 1998;93(6):884–9. https://doi.org/10.1111/j.1572-0241.1998.270_r.x.
13. Vago AR, Macedo AM, Adad SJ, Reis DD, Corrêa-Oliveira R. PCR detection of Trypanosoma cruzi DNA in oesophageal tissues of patients with chronic digestive Chagas' disease. Lancet. 1996;348(9031):891–2. https://doi.org/10.1016/S0140-6736(05)64761-7.
14. Silveira AB, Arantes RM, Vago AR, Lemos EM, Adad SJ, Correa-Oliveira R, D'Avila RD. Comparative study of the presence of Trypanosoma cruzi kDNA, inflammation and denervation in chagasic patients with and without megaesophagus. Parasitology. 2005;131(Pt 5):627–34. https://doi.org/10.1017/S0031182005008061.
15. Arantes RM, Marche HH, Bahia MT, Cunha FQ, Rossi MA, Silva JS. Interferon-gamma-induced nitric oxide causes intrinsic intestinal denervation in Trypanosoma cruzi-infected mice. Am J Pathol. 2004;164(4):1361–8. https://doi.org/10.1016/s0002-9440(10)63222-1.
16. Rezende JM. The digestive tract in Chagas' disease. Mem Inst Oswaldo Cruz. 1984;79(Suppl):97–106.

17. Rezende JM. Chagasic mega syndromes and regional differences. In: New approaches in American research. Scientific publication no. 318. Washington: PAHO; 1976. p. 195–205.
18. Meneghelli UG. Evidências do declínio do megaesôfago e do megacolo chagásico: estudo epidemiológico no Hospital das Clínicas de Ribeirão Preto. Medicina (RibeirãoPreto). 1991;24:218–24.
19. Souza DH, Vaz Mda G, Fonseca CR, Luquetti A, Rezende Filho J, Oliveira EC. Current epidemiological profile of Chagasic megaesophagus in Central Brazil. Rev Soc Bras Med Trop. 2013;46(3):316–21. https://doi.org/10.1590/0037-8682-0065-2013.
20. Vizzoni AG, Varela MC, Sangenis LHC, Hasslocher-Moreno AM, do Brasil PEAA, Saraiva RM. Ageing with Chagas disease: an overview of an urban Brazilian cohort in Rio de Janeiro. Parasit Vectors. 2018;11(1):354. https://doi.org/10.1186/s13071-018-2929-y.
21. Pinto JJ, Pinazo MJ, Saravia J, Gainsborg I, Magne HR, Cuatrecasas M, Cortes-Serra N, Lozano DF, Gascon J, Torrico F. Characterization of digestive disorders of patients with chronic Chagas disease in Cochabamba, Bolivia. Heliyon. 2019;5(2):e01206. https://doi.org/10.1016/j.heliyon.2019.e01206.
22. Prata A. Clinical and epidemiological aspects of Chagas disease. Lancet Infect Dis. 2001;1(2):92–100.
23. Zingales B, Miles MA, Campbell DA, Tibayrenc M, Macedo AM, Teixeira MM, Schijman AG, Llewellyn MS, Lages-Silva E, Machado CR, Andrade SG, Sturm NR. The revised Trypanosoma cruzi subspecific nomenclature: rationale, epidemiological relevance and research applications. Infect Genet Evol. 2012;12(2):240–53. https://doi.org/10.1016/j.meegid.2011.12.009.
24. Nogueira-Paiva NC, Fonseca Kda S, Vieira PM, Diniz LF, Caldas IS, Moura SA, Veloso VM, Guedes PM, Tafuri WL, Bahia MT, Carneiro CM. Myenteric plexus is differentially affected by infection with distinct Trypanosoma cruzi strains in Beagle dogs. Mem Inst Oswaldo Cruz. 2014;109(1):51–60. https://doi.org/10.1590/0074-0276130216.
25. Pinazo MJ, Lacima G, Elizalde JI, Posada EJ, Gimeno F, Aldasoro E, Valls ME, Gascon J. Characterization of digestive involvement in patients with chronic T. cruzi infection in Barcelona, Spain. PLoS Negl Trop Dis. 2014;8(8):e3105. https://doi.org/10.1371/journal.pntd.0003105.
26. Salvador F, Treviño B, Sulleiro E, Pou D, Sánchez-Montalvá A, Cabezos J, Soriano A, Serre N, Gómez I, Prat J, Pahissa A, Molina I. Trypanosoma cruzi infection in a non-endemic country: epidemiological and clinical profile. Clin Microbiol Infect. 2014;20(7):706–12. https://doi.org/10.1111/1469-0691.12443.
27. Gobbi F, Angheben A, Anselmi M, Postiglione C, Repetto E, Buonfrate D, Marocco S, Tais S, Chiampan A, Mainardi P, Bisoffi Z. Profile of Trypanosoma cruzi infection in a tropical medicine reference center, northern Italy. PLoS Negl Trop Dis. 2014;8(12):e3361. https://doi.org/10.1371/journal.pntd.0003361.
28. Rezende JM, Moreira H. Forma digestiva da Doença de Chagas. In: Castro LP, Coelho LGV, editors. Gastroenterologia. Rio de Janeiro: Medsi Ed; 2004. p. 325–92.
29. Oliveira RB, Rezende Filho J, Dantas RO, Iazigi N. The spectrum of esophageal motor disorders in Chagas' disease. Am J Gastroenterol. 1995;90(7):1119–24.
30. Dantas RO. Comparison between idiopathic achalasia and achalasia caused by Chagas' disease: a review on the publications about the subject. Arq Gastroenterol. 2003;40(2):126–30.
31. Rezende JM, Luquetti AO. Chagasic megavisceras. In: Chagas' disease and the nervous system. Scientific publication no. 547. Washington, DC: Pan American Health Organization; 1994. p. 149–71.
32. Vaz MGM, Rezende JM, Ximenes CA, Luquetti AO. Correlação entre a sintomatologia e a evolução do megaesôfago. Rev Goiana Med. 1996;41:1–15.
33. Rezende JM, Lauar KM, Oliveira AR. Aspectos clínicos e radiológicos da aperistalsis do esôfago. Rev Bras Gastroenterol. 1960;12:247–62.
34. Rezende JM. Classificação radiológica do megaesôfago. Rev Goiana Med. 1982;28:187.
35. Lauar KM, Oliveira AR, Rezende JM. Valor do tempo de esvaziamento esofagiano no diagnóstico de esofagopatia chagásica (prova de retenção). Rev Goiana Med. 1959;5:97–102.

36. Martins P, Ferreira CS, Cunha-Melo JR. Esophageal transit time in patients with chagasic megaesophagus: lack of linear correlation between dysphagia and grade of dilatation. Medicine (Baltimore). 2018;97(10):e0084. https://doi.org/10.1097/MD.0000000000010084.
37. Pinotti HW. Contribuição para o estudo da fisiopatologia do megaesôfago. Rev Goiana Med. 1968;14:137–68.
38. Crema E, Cruvinel LA, Werneck AM, de Oliveira RM, Silva AA. Manometric and radiologic aspects of Chagas' megaesophagus: the importance to its surgical treatment. Rev Soc Bras Med Trop. 2003;36(6):665–9.
39. Rezende Filho J. Manometria na esofagopatia chagásica. In: Nakano SMS, Faintuch J, Cecconelo I, editors. Megaesôfago Chagásico: Tratamento Clinico e Cirurgico. Goiânia: Editora UCG; 2006. p. 105–13.
40. Lemme EM, Domingues GR, Pereira VL, Firman CG, Pantoja J. Lower esophageal sphincter pressure in idiopathic achalasia and Chagas disease-related achalasia. Dis Esophagus. 2001;14(3–4):232–4. https://doi.org/10.1046/j.1442-2050.2001.00190.x.
41. Dantas RO, Godoy RA, Oliveira RB, Meneghelli UG, Troncon LE. Lower esophageal sphincter pressure in Chagas' disease. Dig Dis Sci. 1990;35(4):508–12.
42. Dantas RO, Deghaide NHS, Donadi EA. Esophageal manometric and radiologic findings in asymptomatic subjects with Chagas' disease. J Clin Gastroenterol. 1999;28:245–8.
43. Sanchez-Lermen RL, Dick E, Salas JA, Fontes CJ. Upper gastrointestinal symptoms and esophageal motility disorders in indeterminate Chagas disease patients. Rev Soc Bras Med Trop. 2007;40(2):197–203.
44. Godoy RA. Estudo da esofagopatia chagásica crônica por meio do método eletromanométrico e da prova da metacolina em pacientes com e sem dilatação do esôfago. Rev Goiana Med. 1972;18:1–73.
45. Silva LC, Vicentine FP, Herbella FA. High resolution manometric findings in patients with Chagas' disease esophagopathy. Asian Pac J Trop Med. 2012;5(2):110–2. https://doi.org/10.1016/S1995-7645(12)60006-6.
46. Sánchez-Montalvá A, Moris M, Mego M, Salvador F, Accarino A, Ramírez K, Azpiroz F, Ruiz-de-Leon A, Molina I. High resolution esophageal manometry in patients with Chagas disease: a cross-sectional evaluation. PLoS Negl Trop Dis. 2016;10(2):e0004416. https://doi.org/10.1371/journal.pntd.0004416.
47. Menezes MA, Andolfi C, Herbella FA, Patti MG. High-resolution manometry findings in patients with achalasia and massive dilated megaesophagus. Dis Esophagus. 2017;30(5):1–4. https://doi.org/10.1093/dote/dow008.
48. Vicentine FP, Herbella FA, Allaix ME, Silva LC, Patti MG. High-resolution manometry classifications for idiopathic achalasia in patients with Chagas' disease esophagopathy. J Gastrointest Surg. 2014;18(2):221–4. https://doi.org/10.1007/s11605-013-2376-1.
49. Remes-Troche JM, Torres-Aguilera M, Antonio-Cruz KA, Vazquez-Jimenez G, De-La-Cruz-Patiño E. Esophageal motor disorders in subjects with incidentally discovered Chagas disease: a study using high-resolution manometry and the Chicago classification. Dis Esophagus. 2014;27(6):524–9. https://doi.org/10.1111/j.1442-2050.2012.01438.x.
50. Jabari S, de Oliveira EC, Brehmer A, da Silveira AB. Chagasic megacolon: enteric neurons and related structures. Histochem Cell Biol. 2014;142(3):235–44. https://doi.org/10.1007/s00418-014-1250-x.
51. Adad SJ, Silva GB, Jammal AA. The significantly reduced number of interstitial cells of Cajal in chagasic megacolon (CM) patients might contribute to the pathophysiology of CM. Virchows Arch. 2012;461(4):385–92. https://doi.org/10.1007/s00428-012-1299-7.
52. Silveira AB, Freitas MA, de Oliveira EC, Neto SG, Luquetti AO, Furness JB, Correa-Oliveira R, Reis D. Glial fibrillary acidic protein and S-100 colocalization in the enteroglial cells in dilated and nondilated portions of colon from chagasic patients. Hum Pathol. 2009;40(2):244–51. https://doi.org/10.1016/j.humpath.2008.04.025.
53. Silveira AB, Adad SJ, Correa-Oliveira R, Furness JB, D'Avila RD. Morphometric study of eosinophils, mast cells, macrophages and fibrosis in the colon of chronic chagasic patients with and without megacolon. Parasitology. 2007;134(Pt 6):789–96. https://doi.org/10.1017/S0031182007002296.

54. Jabari S, da Silveira AB, de Oliveira EC, Neto SG, Quint K, Neuhuber W, Brehmer A. Partial, selective survival of nitrergic neurons in chagasic megacolon. Histochem Cell Biol. 2011;135:47–57.
55. Jabari S, da Silveira AB, de Oliveira EC, Quint K, Neuhuber W, Brehmer A. Preponderance of inhibitory versus excitatory intramuscular nerve fibres in human chagasic megacolon. Int J Colorectal Dis. 2012;27:1181–9.
56. Rezende JM, Moreira H. Megacolo Chagásico. In: Porto JAF, editor. Clínica das Doenças Intestinais. Rio de Janeiro: Livraria Atheneu; 1976. p. 451–74.
57. Oliveira EC, Menezes JG, Cardoso VK, Luquetti AO, Gabriel Neto S, Garcia SB. The relationship between megacolon and constipation in Chagas' disease. Neurogastroenterol Motil. 2009;21(Suppl):5. https://doi.org/10.1111/j.1365-2982.2009.01343_1.x.
58. Bafutto M, Luquetti AO, Gabriel Neto S, Penhavel FAS, Oliveira EC. Constipation is related to small bowel disturbance rather than colonic enlargement in acquired chagasic megacolon. Gastroenterology Res. 2017;10(4):213–7. https://doi.org/10.14740/gr872w.
59. Hurtado A. Some clinical aspects of life at high altitudes. Ann Intern Med. 1960;53:247–58. https://doi.org/10.7326/0003-4819-53-2-247.
60. Frisancho OV. Dolicomegacolon Andino y Volvulos Intestinales de Altura. Rev Gastroenterol Peru. 2008;28:248–57.
61. Castro C, Hernandez EB, Rezende J, Prata A. Occurrence of dolichocolon without megacolon in chronic Chagas disease patients. Rev Soc Bras Med Trop. 2012;45:353–6.
62. Ximenes CA, Rezende JM, Moreira H, Glória M. Técnica simplificada para o diagnóstico radiológico do megacólonchagásico. Rev Soc Bras Med Trop. 1984;17:23.
63. Silva AL, Giacomin RT, Quirino VA, Miranda ES. Proposta de classificação do megacolon chagásico através do enema opaco. Rev Col Bras Cir. 2003;30:4–10.
64. Habr-Gama A, Raia A, Corrêa-Neto A. Motility of the sigmoid colon and rectum. Contribution to the physiopathology of megacolon in Chagas' disease. Dis Colon Rectum. 1971;14:291–304.
65. Vieira CB, Godoy RA, Meneghelli UG, Carril CF. Resposta do colo sigmóide não ectásico à metacolina na forma crônica da moléstia de Chagas. Arq Gastroenterol. 1966;3:21–6.
66. Cavenaghi S, Felicio OC, Ronchi LS, Cunrath GS, Melo MM, Netinho JG. Prevalence of rectoanal inhibitory reflex in chagasic megacolon. Arq Gastroenterol. 2008;45(2):128–31.
67. Salvador F, Mego M, Sánchez-Montalvá A, Morís M, Ramírez K, Accarino A, Malagelada JR, Azpiroz F, Molina I. Assessment of rectocolonic morphology and function in patients with Chagas disease in Barcelona (Spain). Am J Trop Med Hyg. 2015;92(5):898–902. https://doi.org/10.4269/ajtmh.14-0546.

Etiologic Treatment of Chagas Disease: Old Drugs, New Insights, Challenges, and Perspectives

8

Julio A. Urbina

8.1 Introduction

Chagas disease is a chronic systemic parasitosis caused by the Kinetoplastid protozoon *Trypanosoma cruzi*, which diverged from the African trypanosomes ca. 100 million years ago and evolved in the American continent into six distinct lineages (discrete typing units, DTU) [1]. This condition has afflicted humanity since its earliest presence in the New World [2] and remains the largest parasitic disease burden of the American continent, with an estimated global burden of 7.2 billion $ per year [3]. The disease is a complex condition resulting from the invasion and successful establishment of *T. cruzi*, an intracellular parasite, in key tissues of its mammalian hosts. The initial acute phase has a low (<10%) mortality and generally mild and unspecific symptoms, and evolution of a parasite-specific T_H1-biased immune response results in reduction of parasitemia to undetectable levels through the actions of parasite-specific lytic antibodies and cytotoxic cellular responses [4–6]. This acute phase is followed by a lifelong chronic condition, where the cellular immune (effector and regulatory T cells) response limits the parasite's proliferation but is unable to eradicate the infection in most hosts, leading to a sustained inflammatory response that underlies the development of one or more of the symptomatic chronic forms of the disease in 30–40% of patients, including chronic Chagas cardiomyopathy (CCC), digestive problems, and neuropathies [6–10]. The most severe of these manifestations is CCC, which typically appears decades after the initial infection, and may result in cardiac arrhythmias, ventricular aneurysm, congestive heart failure, thrombo-embolism, and sudden cardiac death; this condition is the

J. A. Urbina (✉)
Emeritus Investigator, Center for Biochemistry and Biophysics, New York, NY, USA
e-mail: jurbina@mac.com

M.-J. Pinazo Delgado, J. Gascon (eds.), *Chagas Disease*,
https://doi.org/10.1007/978-3-030-44054-1_8

first infectious cardiac disease in the world and the leading cause of cardiac disease and cardiac death in poor rural and rural-originated urban populations in Latin America [8–10].

Given its zoonotic character, this disease is not eradicable but significant advances have taken place in the control of the vectorial and transfusional transmission of the disease in parts of the American continent, particularly by the Southern Cone initiative, which led to a significant drop in the prevalence and the population at risk [11–14]. Nevertheless, the disease is far from being controlled, due to the uneven extent and quality of control programs in other parts of the continent and limitations of both diagnostic methods and currently available specific treatments [13, 15, 16]. Unfortunately, an effective prophylactic vaccine has yet to be developed, although some recent results look promising [17–19].

8.2 Etiological Treatment and Its Relevance in the Chronic Stage

The first treatments in animal models and humans were developed empirically mainly through the efforts of academic investigators such as the late Professor Zigman Brener by repurposing broad-spectrum antimicrobial agents, an effort that led to the identification of two nitroheterocyclic compounds, nifurtimox (NFX), a 5-nitrofuran, and benznidazole (BNZ), a 2-nitromidazole, clinically developed empirically ca. 50 years ago [20, 21]. High efficacy of NFX and BNZ in the acute stage of the disease was found in both experimental animals and humans, and later the drugs were introduced pragmatically for the treatment of the chronic condition. It has been consistently verified that these drugs have significant activity in congenital and adult acute *T. cruzi* infections (>95 and 60–80% of parasitological cures, as defined by negativization of all parasitological and conventional serological tests), as well as in early chronic infections, with 60–70% radical parasitological cures observed among children up to 14 years old in Brazil and Argentina after several years of follow-up (reviewed in [15]). However, an important limitation of such compounds is their limited and variable curative activity in the established chronic stage of the disease, the most prevalent clinical presentation [22–26]. Furthermore, the efficacy varies according to the selection of endpoints for evaluation, duration of follow-up [22–26], as well as geographical area, possibly due to differences in drug susceptibility among the different *T. cruzi* DTUs [1, 27]. Of important note, opportunities exist for the control and prevention of congenital transmission, as demonstrated by recent studies on benznidazole treatment of women of childbearing potential [28–32].

In addition, both drugs have adverse side effects that can lead to treatment discontinuation in 10–25% of patients and are related to their mechanism of action (generation of nitro-reduction intermediates that lead to oxidative stress for nifurtimox or reductive stress for benznidazole, see [33]).

Despite the limitations of currently available etiological treatments, it should be pointed out that the main reason that discouraged for decades the treatment of chronic Chagas disease patients was a protracted controversy over the pathogenesis of the disease. Although the role of *T. cruzi* in the pathology of the acute phase of Chagas disease and the importance of etiological treatment in that stage have always been widely accepted, the participation of the parasite in the pathogenesis of chronic Chagas disease was the subject of decades of controversies [6–8, 34]. Several studies implicated autoimmune phenomena as a primary factor leading to the persistent inflammation associated with chronic Chagas disease pathological manifestations, including CCC [35, 36]. According to such hypothesis, after the initial infection the parasite triggers an autoimmune response in the host and its persistence should not play a pivotal role in the pathogenesis of the disease; thus, even a successful antiparasitic treatment may not lead to an improvement of the clinical outcome of the patients. In fact, the autoimmune hypothesis of chronic Chagas disease pathogenesis stalled for decades the development of new specific chemotherapeutic approaches for this disease, as antiparasitic treatment in the chronic stage was considered irrelevant [34, 37]. This notion, together with the limited available information on long-term efficacy of currently available drugs in chronic infections, is one of the main factors responsible for the abysmally low treatment coverage of this condition (<1%) [38].

However, the results of many studies carried out in the last two decades have consistently concluded that the persistence of parasites, coupled with an unbalanced immune response in some individuals that may include autoimmune reactions, is a necessary and sufficient condition for the sustained inflammatory responses that underlie the characteristic lesions of chronic Chagas disease [7, 8, 39–43]. This new paradigm indicates, in contrast with previous notions, that eradication of *T. cruzi* may be a prerequisite to arrest the evolution of chronic Chagas disease and avert its irreversible long-term consequences and that this condition must be treated primarily as an infectious, not autoimmune, condition, which implies that etiological treatment should be offered to all seropositive patients [6, 34, 37]. Consistent with the parasite persistent hypothesis for the pathogenesis of the chronic disease are the results of many observational clinical studies, which have shown that most chronic patients subjected to antiparasitic treatment with benznidazole, although not parasitologically cured (see above), had a significant reduction in the occurrence of electrocardiographic changes and a lower frequency of deterioration of their clinical condition [44–47] as well as high efficacy in the prevention of congenital transmission of the parasite by seropositive women [28–32]. Seen from this new perspective, the positive effect of currently available drugs on the patients' clinical evolution and congenital transmission despite their inability to eradicate the parasite can be explained by a drug-induced reduction of their parasite load, although there is limited long-term data to confirm their ability to induce sterile cure [48].

Based on this paradigm shift [34, 37], in the last decade a series of clinical studies has been launched to rigorously evaluate the safety and efficacy of currently available and novel anti-*T. cruzi* drugs in chronic patients, whose results will be discussed in the next section.

8.3 Recent Randomized Clinical Trials of Etiological Treatments of Chronic Chagas Disease

8.3.1 Benznidazole Evaluation for Interrupting Trypanosomiasis (BENEFIT) Trial

BENEFIT ([49, 50]; ClinicalTrials.gov Identifier: NCT00123916) was the first randomized, double-blind, placebo-controlled phase III trial on the effects of benznidazole (300 mg/d, 40–80 days, depending on weight) on a composite endpoint of death, rescued cardiac arrest, sustained ventricular tachycardia, implant of a pacemaker or implantable cardiac defibrillator, heart failure, stroke or systemic embolism and heart transplant in chronic Chagas disease patients with cardiac compromise (NYHA Class I–III and up to 75 years of age) when compared with placebo. Follow-up period was 4–8 years (mean: 6 years). The study recruited and randomized 2850 patients in Brazil, Argentina, Bolivia, Colombia, and El Salvador. The BENEFIT Pilot Study was a randomized, double-blind trial on the effects of benznidazole (300 mg/d, 40–80 days, depending on weight) on the blood *T. cruzi* load of chronic Chagas disease patients with cardiac compromise, evaluated by PCR. 1896 patients were randomized at baseline and 1618 patients evaluated at the end of treatment, 1530 patients at 2 years, and 1487 patients at the final follow-up visit. Sixty percent of the originally randomized patients had positive basal PCR.

The main authors' conclusion was "*Trypanocidal therapy with benznidazole in patients with established Chagas cardiomyopathy significantly reduced serum parasite detection but did not significantly reduce cardiac clinical deterioration through 5 years of follow-up*."

Other specific findings and important caveats are listed below:

1. The rate of drug interruption because of an adverse event was significantly higher in the benznidazole group than in the placebo group (23.9% vs. 9.5%, $P < 0.001$).
2. The reported overall relative parasitological efficacy of etiological treatment in treated patients, as measured by PCR conversion rates, declined over time as among those patients with positive results at baseline and treated with benznidazole. PCR conversion rates were 66.2% at the end of treatment vs. 33.5% in the placebo group, 55.4% vs. 35.3% at 2 years, and 46.7% vs. 33.1% at 5 years or more ($p < 0.001$ for all comparisons). Such findings may indicate relapse of parasitemia among treated patients, which could explain the lack of clinical efficacy; however, the data was not collected in the same subset of patients.
3. Important caveat: Analysis of patients' clinical evolution stratified by sustained parasitological efficacy of benznidazole treatment [51] was not be carried out as the circulating parasite load in treated patients was not measured at all time points of the follow-up interval.
4. The effect of treatment on PCR conversion varied according to geographic region: in Brazil, 86.3% of patients in the benznidazole group vs. 24.3% in the placebo group at the end of treatment [odds ratio (OR) 7.20, 95% CI

4.53–11.4], in Argentina and Bolivia 73.0% vs. 28.6% (OR 3.32, 95% CI, 2.43–4.54), and in Colombia and El Salvador 45.6% vs. 43.9% (OR 1.15, 95% CI 0.81–1.62).

5. Ulterior analyses [10] of patients' clinical evolution stratified by geographic region revealed significant regional differences in the efficacy of benznidazole treatment: hazard ratio for clinical events at the end of follow-up [95% CI]: Brazil, 0.85 [0.71 to 1.02]; Bolivia, 0.88 [0.51–1.50]; Argentina, 1.33 [0.94–1.88]; Colombia, 0.91 [0.64–1.29]; El Salvador, 1.01 [0.42 to 2.43]. It is important to note that there was a correlation of these results with the effects of treatment on PCR conversion rates (see above), a fact consistent with the parasite persistence hypothesis of chronic Chagas disease pathogenesis.
6. There is evidence that a substantial proportion of recruited patients presented late-stage chronic cardiac disease [10].

8.3.2 Etiologic Treatment with Benznidazole in Adult Patients with Chronic Chagas Disease. A Randomized Clinical Trial (TRAENA)

TRAENA [52], ClinicalTrials.gov Identifier: NCT02386358, was a randomized, double-blind, placebo-controlled phase 3 trial on the effects of benznidazole (5 mg/kg d, 60 days) on conventional (c-ELISA) and non-conventional (ELISA F29) serology, blood *T. cruzi* load (qPCR), and clinical evolution of adult chronic Chagas disease patients with or without cardiac compromise. The study randomized 763 patients (552 completed treatment PP, 272 with benznidazole and 280 with placebo) and lasted from 1999 to 2012. 74.7% of patients were in the indeterminate stage, 22% in NYHA Class I, and 3.1% in NYHA Class II and III. The mean follow-up period was 7 years. Preliminary results of the study were presented at international meetings. This initial analysis showed that at the end of the follow-up 29.1% of patients of the benznidazole arm were negative by c-ELISA and 39.9% were negative by ELISA F29, while for the placebo arm 12.2% were negative by c-ELISA and 10.7% were negative by ELISA F29. In the benznidazole-treated arm, the median blood parasite load at time 0, 60 days post-treatment, and 12–14 months post-treatment measured by qPCR were 6.59 [95% CI, 5.11 to 10.50], 4.10 [95% CI, 0.10 to 7.15], and 0.00 parasite equivalents/mL, respectively, with $p = 0.32$ between 60 days and T0 and $p < 0.0001$ between 12–14 months and T0. No further evaluations of circulating parasite levels were reported. Fifty eight clinical events occurred in both the benznidazole and placebo arms, including overall deaths, deaths attributable to Chagas disease, pacemaker/cardioverter defibrillator, arrhythmia with hemodynamic decompensation and heart failure. ITT and PP analysis showed no differences between the two arms; however, due to the number of events, the study was not powered for this assessment.

The authors' main conclusion is: *The franc parasiticide effect [observed] by serology and qPCR was not associated with differences in clinical events.*

Other specific findings and important caveats are listed below:

1. Among benznidazole-treated patients, 77% reported adverse effects and 22% abandoned treatment due to intolerance to the drug.
2. The reported results for the parasitological events cover only 12–14 months after the end of treatment, compared with a mean follow-up of 7 years, and do not allow to conclude that the original strong suppression of the circulating parasite load was sustained over the follow-up time.
3. Important caveat: The study sample size ($N = 750$) was calculated for an 80% power to detect a 50% reduction in the incidence of clinical events in the benznidazole arm. Such large estimated reduction among a study population predominantly in the indeterminate stage of the disease was probably an overestimation. In addition, there is a small number of clinical events in patients without target organ involvement. In the opinion of several independent experts, such limtations rendered the study underpowered to detect possible differences between the clinical outcomes of the experimental groups.

8.3.3 Ergosterol Biosynthesis Inhibitors as Potential Agents for the Etiologic Treatment of Chronic Chagas Disease

Ergosterol biosynthesis inhibitors (EBI), particularly those targeting cytochrome P-450 sterol C14α-demethylase (CYP51), are the mainstay therapy for the prevention and treatment of invasive fungal infections in human beings worldwide. Compounds of this class, such as posaconazole and ravuconazole, also have potent and highly selective anti-*T. cruzi* activity in vitro and animal models, including strains that are intrinsically resistant to benznidazole and nifurtimox, and even if the hosts are immunosuppressed (reviewed in [53–56]; see also [57, 58]). Recent clinical studies have evaluated the safety and parasitological efficacy of such compounds vis-á-vis benznidazole in chronic Chagas disease patients, and the results are described in the next three sections. Due to the relatively short follow-up times of these studies (10–12 months), compared with the slow progression of the chronic disease, the clinical efficacy of these treatments could not be ascertained, nor the sustainability of the observed reduction of the circulating parasite load beyond 1 year.

8.3.3.1 The CHAGASAZOL Trial

CHAGASAZOL ([59]; ClinicalTrials.gov Identifier: NCT01162967) was a randomized, open-label clinical trial, at three centers that were participating in the International Health Program of the Catalonian Institute of Health, Barcelona, Spain. Seventy eight chronic Chagas disease patients who met the enrollment criteria were randomly assigned, in a 1:1:1 ratio, to receive oral benznidazole at a dose of 150 mg twice daily, oral posaconazole at a dose of 100 mg twice daily (low-dose posaconazole group), or oral posaconazole at a dose of 400 mg twice daily (high-dose posaconazole group) for 60 days. The country of origin of 75 of the 78 patients was Bolivia. The mean (±SD) age of the patients was 39 ± 9 years. A total of 51 of the patients (65%) were classified as having an indeterminate form of Chagas disease, 17 (22%) as having cardiac involvement, 5 (6%) as having gastrointestinal involvement, and 5 (6%) as having involvement of more than one organ system. Patients were followed for 40 weeks after the end of the treatment period. A qPCR assay for *T. cruzi* DNA was performed at 8, 16, 24, and 40 weeks after the end of the treatment period. The mean posaconazole serum concentrations on day 14 of treatment (when the steady-state level of the drug is reached) was 0.909 ± 0.384 μg/mL in the low-dose posaconazole group and 1.666 ± 0.935 μg in the high-dose group, from which average AUC_{0-24} values were estimated to be 20.656 ± 8.809 μg per milliliter per hour and 42.202 ± 22.840 μg per milliliter per hour, respectively.

The key result was that in the intention-to-treat analysis, 92% of the patients in the low-dose posaconazole group and 81% in the high-dose posaconazole group, as compared with 38% in the benznidazole group, tested positive on qPCR during the follow-up period, while in the per-protocol analysis the values were 90%, 80%, and 5.9%, respectively. However, among posaconazole-treated patients there was a clear dose–response effect on the time to parasitemia relapse.

The authors' main conclusion was: "*In patients with chronic Chagas disease, treatment with low-dose or high-dose posaconazole resulted in a significantly larger percentage of treatment failures than did treatment with benznidazole, which is the current standard of care.*"

Other specific findings and caveats are listed below:

1. In the benznidazole group, five patients (19.2%) had to discontinue treatment due to adverse events (allergic dermatitis), whereas no serious adverse events were reported in the posaconazole groups.
2. Caveat: The most plausible explanation for the failure of posaconazole in the study in contrast with its remarkable anti-*T. cruzi* activity in vitro and in animal models [53–56] is that even the high dose (400 mg twice daily) which is the optimal for humans in its liquid suspension formulation [60] leads to systemic exposures that are 10–20% of those measured in mice at the curative anti-*T. cruzi* dose of 20 mg/kg d [61]. This indicates that the patients in this trial, as well as in the subsequent STOP CHAGAS study (see below), were underdosed with posaconazole as anti-*T. cruzi* therapy [15, 62]. The recent development of a delayed-release tablet formulation of the drug, with up to fourfold higher oral bioavailability than the liquid suspension formula, and no adverse side effects up to 400 mg/day [63–65], provides a way to assess the true efficacy of posaconazole for the treatment of human *T. cruzi* infections.

8.3.3.2 The STOP CHAGAS Trial

The STOP CHAGAS trial ([66]; ClinicalTrials.gov Identifier: NCT01377480) was prospective, multicenter randomized placebo-controlled study conducted in 120 subjects from Latin America and Spain (from eight clinical centers in Argentina, one in Chile, three in Colombia, three in Guatemala, one in Mexico, and three in Spain) who were randomized to four groups: posaconazole 400 mg b.i.d.; benznidazole 200 mg + placebo b.i.d.; benznidazole 200 mg b.i.d. +

posaconazole 400 mg b.i.d.; or placebo 10 mg b.i.d. The mean (±SD) age of the patients was 38.3 ± 0.5 years. All patients were followed at days 15, 30, 45, and 60 (end of treatment) and at days 90, 120, 150, 180, and 360 after randomization. *T. cruzi* DNA was detected by qPCR at 30, 60, 90, 120, 150, 180, and 360 days. The primary efficacy outcome is the proportion of subjects with persistent negative qPCR by day 180; the secondary outcome was negative qPCR at 360 days. The key results were that 13.3% of those receiving posaconazole and 10% receiving placebo achieved the primary outcome, compared with 80% receiving benznidazole + posaconazole and 86.7% receiving benznidazole monotherapy ($p < 0.0001$ vs. posaconazole/placebo); the secondary outcome was reached in 96% of patients for both benznidazole monotherapy benznidazole + posaconazole versus placebo (17%) and posaconazole (16%, $p < 0.0001$). Posaconazole monotherapy and posaconazole combined with benznidazole achieved high qPCR conversion rates during treatment (30 days, 93.3% and 88.9%; and 60 days, 90% and 92.3%) that were similar to benznidazole (89.7% and 89.3%); all were superior to placebo (10% and 16.7%, $p < 0.0001$).

The authors' main conclusion was: "*Posaconazole demonstrated trypanostatic activity during treatment, but it is ineffective long-term in asymptomatic T. cruzi carriers. Benznidazole monotherapy is superior to posaconazole, with high RT-PCR conversion rates sustained at one year. No advantages were observed with combined therapy versus benznidazole monotherapy.*"

Other specific findings and caveats are listed below:

1. Serious adverse events were rare (six patients) and only observed in the benznidazole-treated patients; however, permanent discontinuation was reported in 19 patients (31.7%) receiving either benznidazole monotherapy or combined with posaconazole. No serious adverse effects or discontinuation of treatment were observed among patients receiving posaconazole monotherapy.
2. Caveat: As in the CHAGASAZOL trial (see above), the sub-optimal posaconazole exposure for anti-*T. cruzi* therapy with the standard antifungal dose of the liquid suspension formulation (400 mg b.i.d.) used in this study is one of the possible factors for its documented therapeutic failure, as monotherapy and in combination with benznidazole [15, 62].

8.3.3.3 The E1224 Trial

The E1224 trial ([67]; ClinicalTrials.gov Identifier: NCT01489228) was a proof-of-concept, double-blind, randomized phase 2 clinical trial sponsored by the Drugs for Neglected Diseases initiative (DNDi), to investigate the safety and efficacy of three oral E1224 (fosravuconazole, a water-soluble ravuconazole prodrug) regimens and benznidazole versus placebo in adult chronic indeterminate Chagas disease. Two hundred thirty one patients were enrolled and randomly assigned to five oral treatment groups: high-dose E1224 (duration 8 weeks, total dose 4000 mg, $n = 45$), low-dose E1224 (8 weeks, 2000 mg, $n = 48$), short-dose E1224 (4 weeks + 4 weeks placebo, 2400 mg, $n = 46$), benznidazole (60 days, 2.5 mg/kg b.i.d., $n = 45$), or placebo (8 weeks, E1224-matched tablets, $n = 47$). The primary efficacy endpoint was parasitological response to all treatments at the end of treatment (eot), assessed by qPCR. The secondary efficacy endpoints were sustainability of parasitological response up to 12 months after eot; parasite clearance and changes in parasite load; incidence of conversion to negative response in conventional and non-conventional (antigen trypomastigote chemiluminescent ELISA [AT CL-ELISA]) serological response.

The key results were that parasite clearance was observed with all E1224 treatments during the treatment phase, but no sustained response was seen with low-dose and short-dose regimens, whereas 13 patients (29%, 95% CI 16.4–44.3) had sustained response with the high-dose regimen compared with four (9%, 2.4–20.4) in the placebo group ($p < 0.0001$). Benznidazole had a rapid and sustained effect on parasite clearance, with 37 patients (82%, 67.9–92.0) having sustained response at 12-month follow-up. Reversible, dose-dependent liver enzyme increases were seen with E1224 and benznidazole. One hundred eighty seven (81%) participants developed treatment-emergent adverse events and six (3%) developed treatment-emergent serious adverse events.

The authors' main conclusion was: "*E1224 displayed a transient, suppressive effect on parasite clearance, whereas benznidazole showed early and sustained efficacy until 12 months of follow-up.*"

Other specific results and caveats are listed below:

1. Ravuconazole systemic exposures are substantially higher with E1224 prodrug administration, resulting from increased bioavailability of the prodrug and the long plasma terminal half-life of free ravuconazole in humans (4–8 days); the steady state of ravuconazole is achieved within a week and allows for once weekly dosing for maintenance of target plasma concentrations.
2. The parasite load in the high-dose E1224 group remained significantly lower than in the placebo group, *with no difference from the benznidazole group on adjusted post-hoc comparison (p = 0.97, adjusted).*
3. Pharmacokinetic/pharmacodynamic (PK/PD) models indicated that an increase in high-dose E1224 treatment duration would significantly reduce the probability of relapse from 61% (95% CI 48–73) with 8 weeks to 44% (95% CI 24–67) with 12 weeks.
4. The authors concluded that, given the absence of a sustained response to E1224 high dose at 12 months in the majority of patients, and the potential requirement for extended treatments, the drug will not be further investigated as monotherapy However, "*given its favorable safety profile, combinations of E1224 with existing drugs, such as benznidazole or nifurtimox, should be considered, especially given that animal model studies show combination therapy has the potential to improve treatment response and shorten treatment duration.*"

Following these findings, DNDi sponsored a new study, the BENDITA trial (ClinicalTrials.gov Identifier: NCT0337866), to investigate the safety and efficacy of combinations of E1224 (fosravuconazole) and benznidazole in shorter treatments and the initial results were released in March 2019 (https://www.dndi.org/2019/

media-centre/press-releases/study-shows-dramatically-shorter-treatment-chagas-effective-and-safer/).

The study incorporated six experimental groups:

1. The standard 8-week treatment, with a daily dose of 300 mg/day of benznidazole in monotherapy
2. A 4-week treatment with a daily dose of 300 mg/day of benznidazole in monotherapy
3. A 2-week treatment with a daily dose of 300 mg/day of benznidazole in monotherapy
4. A 4-week treatment with a lower daily dose of 150 mg/day of benznidazole in monotherapy
5. A 4-week treatment with a lower daily dose of 150 mg/day of benznidazole, in combination with 300 mg/week of fosravuconazole
6. An 8-week treatment, with a lower weekly dose of 300 mg of benznidazole, in combination with 300 mg/week of fosravuconazole

Efficacy was measured through sustained parasitological response at 6 months assessed by qPCR, with a final assessment at 12 months after the end of the treatment.

Preliminary results have been presented in an international meeting. More than ≥80% of patients in all treatment groups responded to treatment (i.e., had no detectable parasite in blood tests after completing treatment or at 6 or 12 months after eot), compared to 3.3% in the placebo group; at these follow-up times, there were no significant differences in efficacy among the groups receiving combination treatments (#5 and 6) compared with benznidazole monotherapy (#1 and 2). As expected, treatment-related side effects leading to treatment discontinuation decreased markedly in the shorter/lower dose treatment groups, going from 20% for the standard treatment (#1) to 0–10% among patients in groups #2–6.

Interpretation and caveat: These interesting results clearly show that benznidazole treatments with lower doses or shorter duration, alone or in combination with fosravuconazole, can lead to a marked and sustained reduction of the parasite load of patients up to 12 months post-treatment, indistinguishable from that of the standard treatment, and with a very significant reduction of toxicity. However, a longer follow-up of the patients would be required to establish the true therapeutic efficacy of the new treatment schemes.

8.4 Recent Insights, Challenges, and New Perspectives on the Etiological Treatment of Chronic Chagas Disease

The results of the recent randomized clinical trials on the parasitological efficacy of etiological treatment in chronic Chagas disease patients, when most of the parasite load consists of intracellular organisms (amastigotes) in deep tissues, have provided some important insights that can be summarized as follows: (a) The parasitological efficacy of the standard treatment with benznidazole (2.5 mg/kg b.i.d, 60 days) reported in the randomized studies discussed above is consistently high at the end of treatment and at 1 year post-treatment, but not necessarily sustained as seen in the BENEFIT trial [50], previous observational studies [22–26], and recent study in a dog model of chronic Chagas disease [68]; (b) The antiparasitic efficacy of benznidazole may vary among the endemic regions, probably resulting from different in vivo drug susceptibilities of the local circulating parasite populations [1, 27], and the clinical efficacy seems also to depend of the stage of the disease [10, 15, 20]; (c) Although EBIs such as posaconazole and ravuconazole are active against experimental animals and human *T. cruzi* infections, effective dose, formulation, and duration of treatment for humans, in monotherapy or combination with benznidazole or nifurtimox, have not been established.

The parasite persistence hypothesis for the pathogenesis of chronic Chagas disease indicates that to slow or stop the clinical evolution of the disease an anti-*T. cruzi* treatment must be able to induce a profound and sustained reduction of the patient's parasite load (although not necessarily sterile cure, [48]). The results summarized above suggest that to achieve this objective higher doses and/or longer treatments with the currently available drugs, new drug candidates or combinations may be required. In the case of nitroheterocyclic drugs (benznidazole and nifurtimox), higher doses or longer treatments at current doses are not an option due to the well-known toxicity of such compounds for the dose/duration of the current standard treatments. However, novel findings suggest alternative treatment schemes of higher efficacy and lower toxicity:

(a) A crucial insight on this issue was provided by Tarleton's team [58], who showed in a murine model of acute Chagas disease that reducing the dosing frequency of benznidazole or nifurtimox from daily (continuous) to every 5 days (intermittent) provided the same parasitological efficacy using a much lower total dose of the drug; such findings indicate that the drugs act on the parasite through a critical peak concentration (Cmax effect), rather than a continuous exposure (AUC effect). Based on this insight, Álvarez et al. [69] recently showed that an intermittent treatment scheme with benznidazole (2.5 mg/kg d twice a day every 5 days, for a total of 60 days) had the same antiparasitic efficacy at the end of treatment as the standard scheme (2.5 mg/kg b.i.d, 60 days) of patients with chronic *T. cruzi* infections, with a marked reduction of adverse effects, as expected.
(b) A recent population pharmacokinetic study of benznidazole in chronic Chagas disease patients [70] found that using half of the dosing frequency of the standard treatment (from 2.5 mg/kg b.i.d. to 2.5 mg/kg q.d.) the levels of the drug can be sustained in its therapeutic range.
(c) Based in these results, a clinical trial to evaluate the safety and efficacy of a novel treatment scheme for benznidazole and nifurtimox in chronic Chagas disease patients, *using half of the currently accepted dosing frequency but extending the treatment up to 90 days* was approved by NIH in 2019 and is currently underway (TESEO trial, ClinicalTrials.gov Identifier: NCT0398152). The para-

sitological efficacy of these treatments will be evaluated in a 3-year follow-up using qPCR and a series of potential biomarkers of early response to the drug treatment.

(d) Alternatively, fexinidazole, a 2-substituted 5-nitroimidazole originally discovered by Hoechst AG in the 1980s as a broad-spectrum antimicrobial agent, and rediscovered by DNDi, has a potent and selective anti-*T. brucei* agent in murine models of stage I and II HAT [71] and successfully completed a Phase 2/3 clinical trial for late-stage human African *Trypanosoma brucei gambiense* trypanosomiasis in Central Africa (DRC and CAR; [72]). The drug and its two main metabolites (sulfoxide and sulfone) are also active against *T. cruzi* in in vitro and animal models, being well tolerated and more effective than benznidazole and posaconazole, even against drug-resistant strains [73–75]. Unfortunately, a DNDi-sponsored Phase II trial in Bolivia aimed at investigating safety and efficacy of the drug in 140 adult indeterminate Chagas disease patients was interrupted after only 47 patients had enrolled due to safety issues, including neutropenia at the high doses; a new proof-of-concept (Phase 2) study has been designed and is being run at five sites in Spain. The target conclusion date is mid-2019 (https://www.dndi.org/diseases-projects/portfolio/fexinidazole-chagas/).

(e) Concerning EBIs, to evaluate true potential of this class of compounds for the treatment of human *T. cruzi* infections rigorous PK/PD analyses are essential to translate the results of the preclinical studies in vitro and animal models to the clinical application, as recently reviewed for *Leishmania* and other intracellular pathogens ([76]; see also [16, 62]). In the specific case of posaconazole, the recently developed delayed-release tablet formulation of the drug [63–65] will be required to achieve the exposure levels associated with curative activity in acute and chronic animal models of human Chagas disease, but also an increase in the treatment duration should be considered, as a chronic Chagas disease patient with systemic lupus erythematosus and under immunosuppressive treatment was parasitologically cured (as assessed by repeated qPCR tests) by posaconazole given at the standard antifungal dose of its liquid suspension (400 mg b.i.d.) for 90 days [77]. For ravuconazole, administered as its prodrug E1224 (fosravuconazole), the results of the E1224 trial confirmed the prediction of PK/PD analyses that the required exposure levels of the free drug for the anti-*T. cruzi* activity can be achieved using the high-dose treatment of the prodrug (duration 8 weeks, total dose 4000 mg) with good tolerability, but the same models indicate that higher efficacy could be achieved by increasing the duration of treatment from eight to 12 weeks or with shorter combination treatments with benznidazole [67]. However, it must be pointed out that the potential of EBIs for the etiological treatment of Chagas disease is not restricted to particular compounds or a single compound class, as they comprise a large chemical space that includes inhibitors of different steps of the de novo biosynthesis pathway of ergosterol and other 24-alkyl sterols, which are essential for survival and virulence of many fungal and protozoan pathogens as these parasites cannot use the abundant supply of cholesterol present in their vertebrate hosts. Although posaconazole is one of the more potent and specific CYP51 inhibitors against invasive fungal pathogens and Kinetoplastid protozoa [56, 78, 79], the eventual confirmation of its safety and efficacy for the treatment of human *T. cruzi* infections would only be a proof of concept for this class of compounds as specific treatments of Chagas disease, as this drug is a very expensive proprietary compound (Merck) and thus inaccessible for the treatment of the very poor populations afflicted by Chagas disease. Even so, the remarkable results of the preclinical studies with this drug and analogs against *T. cruzi* in vitro and in vivo sparked several research programs that led to the determination of the 3D structure of *T. cruzi* CYP51 at atomic resolution [80, 81] and a structure-based searches for *T. cruzi*-specific CYP51 inhibitors that led to the identification of novel compounds (azoles and non-azoles) with in vitro and in vivo potency and specificity comparable or superior to posaconazole and ravuconazole (reviewed in [79, 82]; see also: [83–87]). Such advances, mostly from non-profit academic institutions, may lead the way towards the clinical development of *T. cruzi*-specific CYP51 drug candidates.

(f) Given the high requirements for the etiological treatment of chronic Chagas disease, combination chemotherapies are a logical choice, as they can result in higher efficacy with lower toxicity as well as to limit the development of drug resistance by the pathogen [55, 88, 89]. This approach is currently the standard treatment for resilient infections caused by intracellular pathogens, such as HIV-AIDS, tuberculosis, leprosy, malaria, and visceral leishmaniasis, but has not been clinically applied for the treatment of human Chagas disease, except in the STOP CHAGAS and BENDITA trials. Preclinical studies have shown potent in vivo synergism of combinations of benznidazole and EBIs such as posaconazole, ravuconazole, and itraconazole [57, 58, 90–92]. As such synergy is not observed in vitro [91], the in vivo synergism most probably results from drug–drug interactions in animals treated with the combinations. This concept was demonstrated by a PK study in a murine model treated with benznidazole, itraconazole and their combination, which found that co-administration of itraconazole with benznidazole led to a 1.5-fold reduction of the plasma Cmax of the latter, with a concomitant 2.66-fold increase of its volume of distribution and a 7.5-fold increase of the elimination half-life [93]; such modification of the PK profile of benznidazole indicates that in the combination treatment the drug has higher penetration and permanence in the internal organs where *T. cruzi* reproduces, most probably as a result of inhibition by itraconazole of P450 cytochromes (CYP), which are the main enzymes involved in benznidazole breakdown [94]. The BENDITA study, designed to evaluate the safety and efficacy of combination treatments with benznidazole and E1224 (fosravuconazole), was preceded by a Phase I drug–drug interaction study to assess the safety and pharmacokinetic interactions of the two drugs administered separately and in combination, and no major clinically relevant safety or tolerability issues were identified (https://www.dndi.org/diseases-projects/portfolio/new-benz-regimens/); based on these antecedents, the final results on parasitological efficacy of these treatments are widely expected.

A final challenge in the quest for safer and more potent anti-*T. cruzi* agents is the lack of validated biomarkers of early response to etiological treatment in patients with chronic *T. cruzi* infections. In these patients: (a) Conventional serology responds very slowly to parasite clearance (in many years to decades); (b) Even the most sensitive blood PCR tests are negative in a significant proportion of chronic patients; thus, PCR should be used to confirm therapeutic failure, although quantitative PCR can effectively evaluate modifications of the circulating parasite's burden above the limit of detection; (c) Evaluation of drug response in terms of clinical evolution would require several years of follow-up. The lack of such biomarkers has been a major stumbling block for the development of new anti-*T. cruzi* drugs, as well as for the evaluation of the true activity of currently available drugs in patients with chronic disease. Recent approaches, such as non-conventional serology [95, 96], aptamer-based methods to detect parasite's secreted antigens [97, 98], and proteomic biomarkers of active *T. cruzi* infection [99–101], are providing promising results [102, 103]; see Chap. 9.

8.5 Conclusions

The search for a safe and effective etiological treatment for chronic Chagas disease has been a long and tortuous process, resulting from the biological complexity of the disease and its socio-economical determinants, including the shortage of political will by governments of endemic countries and economic interest by pharmaceutical companies [16]. The efficacy of clinically available drugs, NFX and BNZ, in the acute stage of the disease has been repeatedly verified in both experimental animals and humans; they were introduced pragmatically for the treatment of the chronic condition but its efficacy was soon found to be significantly lower than in the acute phase by the original criteria of cure (seroconversion of the patient). Furthermore, the predominance of the autoimmune hypothesis on pathogenesis of the chronic stage essentially stopped for decades the search of novel anti-*T. cruzi* agents for this prevalent stage of the disease, as they were considered irrelevant [34]. However, in the last two decades the paradigm shift for the pathogenesis of the chronic condition towards the persistence of the parasite [37] rekindled the interest in etiological treatment, as it is now accepted that this condition must be treated as an infectious, not autoimmune, disease (see Sect. 8.2).

Key insights from recent randomized clinical studies with benznidazole in chronic patients are that the parasitological efficacy of the current standard treatment (benznidazole at 2.5 mg/kg b.i.d. for 60 days) is high by the end of treatment 12 months after eot, but rigurous confirmation that this activity will be sustained is still lacking; furthermore, the parasitological and clinical efficacy varies markedly depending on the criteria used for assessment of response, geographical location of the study, and the stage of the chronic disease (Sect. 8.3). Additional data on alternative regimens of existing drugs in monotherapy and combination should be performed in different geographical regions and using standardized methods for evaluation of treatment response.

Such variable efficacy of current and novel treatment in the management of chronic *T. cruzi* infections could result from several reasons: (a) The presence of a load of intracellular parasites in deep tissues that is not eliminated by the standard treatments due to limited drug accessibility, a well-known fact in experimental animal and human cancers; (b) In vivo drug resistance by many parasite strains [27], which is not observed in vitro [104–106] and could be related to the virulence and histotropism of the infecting parasite population, as well as to the intensity and quality of the immunological response to it; (c) The existence of recently reported "dormant" [viable but not actively replicative] intracellular amastigote stages that are resistant to sustained antiparasitic drug pressure [107]; however, the significance of this finding is controversial as there have been many studies that have demonstrated sterile cure of *T. cruzi* infections in animal models with different drugs and combinations, using immunosuppression after prolonged periods post-treatment [57, 58]. Taken together, these facts suggest that to attain a profound and sustained reduction of chronic patients' parasite load higher doses and/or longer treatments or combination therapies will probably be required. The recent demonstration in animal models of Chagas disease that intermittent treatment with benznidazole and nifurtimox is as effective as continuous (daily) dosing, which indicates that these drugs act through a Cmax effect, has led to the design and testing of clinical trials where the frequency of dosing is reduced but the treatment duration is increased (Sect. 8.4).

Finally, it is fundamental to consider that, resulting from the *T. cruzi* persistence in infected tissues, Chagas disease is characterized by a chronic inflammatory status in the parasite's target organs, particularly heart, GI tract, and CNS, which is the proximal cause of the physiopathological manifestations in the chronic stage [108, 109]. There is a growing interest to understand these inter-related issues, which could lead to combinations of etiological treatment with immunomodulators and anti-inflammatory agents [110–113], promoters of inflammation resolution [114–117], and immunotherapy [19], as such combinations may have synergistic effects in controlling the progression of the disease; see Chaps. 9 and 10.

References

1. Zingales B, Miles MA, Campbell DA, Tibayrenc M, Macedo AM, Teixeira MM, et al. The revised Trypanosoma cruzi subspecific nomenclature: rationale, epidemiological relevance and research applications. Infect Genet Evol. 2012;12(2):240–53. https://doi.org/10.1016/j.meegid.2011.12.009.
2. Aufderheide AC, Salo W, Madden M, Streitz J, Buikstra J, Guhl F, et al. A 9,000-year record of Chagas' disease. Proc Natl Acad Sci U S A. 2004;101(7):2034–9.
3. Lee BY, Bacon KM, Bottazzi ME, Hotez PJ. Global economic burden of chagas disease: a computational simulation model. Lancet Infect Dis. 2013;13(4):342–8. https://doi.org/10.1016/S1473-3099(13)70002-1.
4. Brener Z, Gazzinelli RT. Immunological control of trypanosoma cruzi infection and pathogenesis of Chagas' disease. Int Arch Allergy Immunol. 1997;114(2):103–10.
5. Padilla AM, Bustamante JM, Tarleton RL. CD8+ T cells in trypanosoma cruzi infection. Curr Opin Immunol. 2009;21(4):385–90. https://doi.org/10.1016/j.coi.2009.07.00.

6. Machado FS, Dutra WO, Esper L, Gollob KJ, Teixeira MM, Factor SM, et al. Current understanding of immunity to Trypanosoma cruzi infection and pathogenesis of Chagas disease. Semin Immunopathol. 2012;34(6):753–70. https://doi.org/10.1007/s00281-012-0351-7.
7. Gutierrez FR, Guedes PM, Gazzinelli RT, Silva JS. The role of parasite persistence in pathogenesis of Chagas heart disease. Parasite Immunol. 2009;31(11):673–85. https://doi.org/10.1111/j.1365-3024.2009.01108.
8. Marin-Neto JA, Cunha-Neto E, Maciel BC, Simoes MV. Pathogenesis of chronic Chagas heart disease. Circulation. 2007;115(9):1109–23.
9. Rassi A Jr, Rassi A, Marin-Neto JA. Chagas heart disease: pathophysiologic mechanisms, prognostic factors and risk stratification. Mem Inst Oswaldo Cruz. 2009;104(Suppl 1):152–8. S0074-02762009000900021 [pii].
10. Rassi A, Marin JA, Rassi A. Chronic Chagas cardiomyopathy: a review of the main pathogenic mechanisms and the efficacy of aetiological treatment following the benznidazole evaluation for interrupting trypanosomiasis (BENEFIT) trial. Mem Inst Oswaldo Cruz. 2017;112(3):224–35. https://doi.org/10.1590/0074-02760160334.
11. Dias JC, Silveira AC, Schofield CJ. The impact of Chagas disease control in Latin America: a review. Mem Inst Oswaldo Cruz. 2002;97(5):603–12.
12. Vazquez-Prokopec GM, Spillmann C, Zaidenberg M, Kitron U, Gurtler RE. Cost-effectiveness of Chagas disease vector control strategies in northwestern Argentina. PLoS Negl Trop Dis. 2009;3(1):e363. https://doi.org/10.1371/journal.pntd.000036.
13. Reithinger R, Tarleton RL, Urbina JA, Kitron U, Gurtler RE. Eliminating Chagas disease: challenges and a roadmap. BMJ. 2009;338:b1283.
14. Horstick O, Runge-Ranzinger S. Protection of the house against Chagas disease, dengue, leishmaniasis, and lymphatic filariasis: a systematic review. Lancet Infect Dis. 2018;18(5):e147–58. https://doi.org/10.1016/S1473-3099(17)30422-X.
15. Urbina JA. Recent clinical trials for the etiological treatment of chronic Chagas disease: advances, challenges and perspectives. J Eukaryot Microbiol. 2015;62(1):149–56. https://doi.org/10.1111/jeu.1218.
16. Urbina JA. The long road towards a safe and effective treatment of chronic Chagas disease. Lancet Infect Dis. 2018;18(4):363–5. https://doi.org/10.1016/S1473-3099(17)30535-2.
17. Serna C, Lara JA, Rodrigues SP, Marques AF, Almeida IC, Maldonado RA. A synthetic peptide from Trypanosoma cruzi mucin-like associated surface protein as candidate for a vaccine against Chagas disease. Vaccine. 2014;32(28):3525–32. https://doi.org/10.1016/j.vaccine.2014.04.026.
18. Hotez PJ, Bottazzi ME, Strych U. New vaccines for the world's poorest people. Annu Rev Med. 2016;67:405–17. https://doi.org/10.1146/annurev-med-051214-024241.
19. Jones K, Versteeg L, Damania A, Keegan B, Kendricks A, Pollet J, et al. Vaccine-Linked chemotherapy improves benznidazole efficacy for acute Chagas disease. Infect Immun. 2018;86(4):e00876–17. https://doi.org/10.1128/IAI.00876-17.
20. Rodriques Coura J, de Castro SL. A critical review on Chagas disease chemotherapy. Mem Inst Oswaldo Cruz. 2002;97(1):3–24.
21. Steverding D. The history of Chagas disease. Parasit Vectors. 2014;7:317.
22. Britto C, Silveira C, Cardoso MA, Marques P, Luquetti A, Macedo V, Fernandes O. Parasite persistence in treated chagasic patients revealed by xenodiagnosis and polymerase chain reaction. Mem Inst Oswaldo Cruz. 2001;96(6):823–6.
23. Aguiar C, Batista AM, Pavan TB, Almeida EA, Guariento ME, Wanderley JS, Costa SC. Serological profiles and evaluation of parasitemia by PCR and blood culture in individuals chronically infected by Trypanosoma cruzi treated with benznidazole. Trop Med Int Health. 2012;17(3):368–73. https://doi.org/10.1111/j.1365-3156.2011.02936.x.
24. Fernandes CD, Tiecher FM, Balbinot MM, Liarte DB, Scholl D, Steindel M, Romanha A. Efficacy of benznidazol treatment for asymptomatic chagasic patients from state of Rio grande do Sul evaluated during a three years follow-up. Mem Inst Oswaldo Cruz. 2009;104(1):27–32. S0074-02762009000100004 [pii].

25. Machado-de-Assis GF, Silva AR, Do Bem VA, Bahia MT, Martins-Filho OA, Dias JC, et al. Post-therapeutic cure criteria in Chagas' disease: conventional serology followed by supplementary serological, parasitological, and molecular tests. Clin Vaccine Immunol. 2012;19(8):1283–91. https://doi.org/10.1128/CVI.00274-12.
26. Anez N, Carrasco H, Parada H, Crisante G, Rojas A, Fuenmayor C, et al. Myocardial parasite persistence in chronic chagasic patients. Am J Trop Med Hyg. 1999;60(5):726–32.
27. Filardi LS, Brener Z. Susceptibility and natural resistance of Trypanosoma cruzi strains to drugs used clinically in chagas disease. Trans R Soc Trop Med Hyg. 1987;81(5):755–9.
28. Sosa-Estani S, Cura E, Velazquez E, Yampotis C, Segura EL. Etiological treatment of young women infected with Trypanosoma cruzi, and prevention of congenital transmission. Rev Soc Bras Med Trop. 2009;42(5):484–7.
29. Fabbro DL, Danesi E, Olivera V, Codebó MO, Denner S, Heredia C, et al. Trypanocide treatment of women infected with Trypanosoma cruzi and its effect on preventing congenital chagas. PLoS Negl Trop Dis. 2014;8(11):e3312. https://doi.org/10.1371/journal.pntd.0003312.
30. Moscatelli G, Moroni S, García-Bournissen F, Ballering G, Bisio M, Freilij H, Altcheh J. Prevention of congenital Chagas through treatment of girls and women of childbearing age. Mem Inst Oswaldo Cruz. 2015;110(4):507–9.
31. Murcia L, Simón M, Carrilero B, Roig M, Segovia M. Treatment of infected women of childbearing age prevents congenital Trypanosoma cruzi infection by eliminating the parasitemia detected by PCR. J Infect Dis. 2017;215(9):1452–8. https://doi.org/10.1093/infdis/jix087.
32. Álvarez MG, Vigliano C, Lococo B, Bertocchi G, Viotti R. Prevention of congenital Chagas disease by benznidazole pre-treatment in reproductive-age women. An observational study. Acta Trop. 2017;174:149–52. https://doi.org/10.1016/j.actatropica.2017.07.004.
33. Docampo R. Recent developments in the chemotherapy of Chagas' disease. Curr Pharm Design. 2001;7:1157–64.
34. Machado FS, Tyler KM, Brant F, Esper L, Teixeira MM, Tanowitz HB. Pathogenesis of Chagas disease: time to move on. Front Biosci (Elite Ed). 2012;4:1743–58.
35. Kalil J, Cunha-Neto E. Autoimmunity in Chagas disease cardiomyopathy: fulfilling the criteria at last? Parasitol Today. 1996;12(10):396–9. https://doi.org/10.1016/0169-4758(96)10058-2.
36. Bonney KM, Luthringer DJ, Kim SA, Garg NJ, Engman DM. Pathology and pathogenesis of Chagas heart disease. Annu Rev Pathol. 2018;14:421–47. https://doi.org/10.1146/annurev-pathol-020117-043711.
37. Viotti R, Alarcón de Noya B, Araujo-Jorge T, Grijalva MJ, Guhl F, López MC, et al. Towards a paradigm shift in the treatment of chronic Chagas disease. Antimicrob Agents Chemother. 2014;58(2):635–9. https://doi.org/10.1128/AAC.01662-13.
38. Ribeiro I, Sevcsik AM, Alves F, Diap G, Don R, Harhay MO, et al. New, improved treatments for Chagas disease: from the R&D pipeline to the patients. PLoS Negl Trop Dis. 2009;3(7):e484.
39. Levin MJ. In chronic Chagas heart disease, don't forget the parasite. Parasitol Today. 1996;12(11):415–6. https://doi.org/10.1016/0169-4758(96)20051-1.
40. Brandariz S, Schijman A, Vigliano C, Viotti R, Levin MJ. Role of parasites in the pathogenesis of Chagas' cardiomyopathy. Lancet. 1996;347:914–0.
41. Tarleton RL, Zhang L. Chagas disease etiology: autoimmunity or parasite persistence? Parasitol Today. 1999;15(3):94–9.
42. Tarleton RL. Parasite persistence in the aetiology of Chagas disease. Int J Parasitol. 2001;31(5–6):550–4.
43. Hyland KV, Leon JS, Daniels MD, Giafis N, Woods LM, Bahk TJ, et al. Modulation of autoimmunity by treatment of an infectious disease. Infect Immun. 2007;75(7):3641–50. https://doi.org/10.1128/IAI.00423-0.
44. Viotti R, Vigliano C. Etiological treatment of chronic Chagas disease: neglected 'evidence' by evidence-based medicine. Expert Rev Anti Infect Ther. 2007;5(4):717–26.
45. Fabbro DL, Streiger ML, Arias ED, Bizai ML, del Barco M, Amicone NA. Trypanocide treatment among adults with chronic Chagas disease living in Santa Fe city (Argentina), over a

mean follow-up of 21 years: parasitological, serological and clinical evolution. Rev Soc Bras Med Trop. 2007;40(1):1–10. https://doi.org/10.1590/s0037-86822007000100001.
46. Fragata-Filho AA, França FF, Fragata Cda S, Lourenço AM, Faccini CC, Costa CA Evaluation of parasiticide treatment with benznidazol in the electrocardiographic, clinical, and serological evolution of Chagas disease. PLoS Negl Trop Dis. 2016;10(3):e0004508. https://doi.org/10.1371/journal.pntd.0004508.
47. Cardoso CS, Ribeiro ALP, Oliveira CDL, Oliveira LC, Ferreira AM, Bierrenbach AL, et al. Beneficial effects of benznidazole in Chagas disease: NIH sami-trop cohort study. PLoS Negl Trop Dis. 2018;12(11):e0006814. https://doi.org/10.1371/journal.pntd.000681.
48. Urbina JA, McKerrow JH. Drug susceptibility of genetically engineered Trypanosoma cruzi strains and sterile cure in animal models as a criterion for potential clinical efficacy of anti-T. cruzi drugs. Antimicrob Agents Chemother. 2015;59(12):7923–4. https://doi.org/10.1128/AAC.01714-15.
49. Marin-Neto JA, Rassi AJ, Morillo CA, Avezum A, Connolly SJ, Sosa-Estani S, et al. Rationale and design of a randomized placebo-controlled trial assessing the effects of etiologic treatment in Chagas' cardiomyopathy: the benznidazole evaluation for interrupting trypanosomiasis (BENEFIT). Am Heart J. 2008;156(1):37–43. https://doi.org/10.1016/j.ahj.2008.04.001.
50. Morillo CA, Marin-Neto JA, Avezum A, Sosa-Estani S, Rassi A, Rosas F, et al. Randomized trial of benznidazole for chronic Chagas' cardiomyopathy. N Engl J Med. 2015;373(14):1295–306. https://doi.org/10.1056/NEJMoa1507574.
51. Urbina JA, Gascon J, Ribeiro I. Benznidazole for chronic Chagas' cardiomyopathy. N Engl J Med. 2016;374(2):189. https://doi.org/10.1056/NEJMc1514453.
52. Riarte A. TRAENA: Placebo-controlled evaluation of impact of benznidazole treatment on long-term disease progression in adults with chronic Chagas disease. 62nd Annual Meeting of the American 460 Society of Tropical Medicine and Hygiene, November 13–17, 2013, Washington, DC.
53. Urbina JA. Ergosterol biosynthesis and drug development for Chagas disease. Mem Inst Oswaldo Cruz. 2009;104(Suppl 1):311–8. https://doi.org/10.1590/s0074-02762009000900041.
54. Urbina JA. Specific chemotherapy of Chagas disease: relevance, current limitations and new approaches. Acta Trop. 2010;115(1–2):55–68. https://doi.org/10.1016/j.actatropica.2009.10.023.
55. Urbina JA. New insights in Chagas' disease treatment. Drugs Future. 2010;35(5):409–19. https://doi.org/10.1358/dof.2010.35.5.1484391.
56. Buckner FS, Urbina JA. Recent developments in sterol 14-demethylase inhibitors for Chagas disease. Int J Parasitol Drugs Drug Resist. 2012;2:236–42. https://doi.org/10.1016/j.ijpddr.2011.12.002.
57. Diniz Lde F, Urbina JA, de Andrade IM, Mazzeti AL, Martins TA, Caldas IS, et al. Benznidazole and posaconazole in experimental Chagas disease: positive interaction in concomitant and sequential treatments. PLoS Negl Trop Dis. 2013;7(8):e2367. https://doi.org/10.1371/journal.pntd.0002367.
58. Bustamante JM, Craft JM, Crowe BD, Ketchie SA, Tarleton RL. New, combined, and reduced dosing treatment protocols cure Trypanosoma cruzi infection in mice. J Infect Dis. 2014;209(1):150–62. https://doi.org/10.1093/infdis/jit420.
59. Molina I, Gómez i, Prat J, Salvador F, Treviño B, Sulleiro E, Serre N, et al. Randomized trial of posaconazole and benznidazole for chronic Chagas' disease. N Engl J Med. 2014;370(20):1899–908. https://doi.org/10.1056/NEJMoa1313122.
60. Ullmann AJ, Cornely OA, Burchardt A, Hachem R, Kontoyiannis DP, Topelt K, et al. Pharmacokinetics, safety, and efficacy of posaconazole in patients with persistent febrile neutropenia or refractory invasive fungal infection. Antimicrob Agents Chemother. 2006;50(2):658–66.
61. Nomeir AA, Kumari P, Hilbert MJ, Gupta S, Loebenberg D, Cacciapuoti A, et al. Pharmacokinetics of SCH 56592, a new azole broad-spectrum antifungal agent, in mice, rats, rabbits, dogs, and cynomolgus monkeys. Antimicrob Agents Chemother. 2000;44(3):727–31.

62. Urbina JA. Pharmacodynamics and follow-up period in the treatment of human Trypanosoma cruzi infections with posaconazole. J Am Coll Cardiol. 2017;70(2):299–300. https://doi.org/10.1016/j.jacc.2017.03.611.
63. Guarascio AJ, Slain D. Review of the new delayed-release oral tablet and intravenous dosage forms of posaconazole. Pharmacotherapy. 2015;35(2):208–19. https://doi.org/10.1002/phar.1533.
64. Cumpston A, Caddell R, Shillingburg A, Lu X, Wen S, Hamadani M, et al. Superior serum concentrations with posaconazole delayed-release tablets compared to suspension formulation in hematological malignancies. Antimicrob Agents Chemother. 2015;59(8):4424–8. https://doi.org/10.1128/AAC.00581-15.
65. Durani U, Tosh PK, Barreto JN, Estes LL, Jannetto PJ, Tande AJ. Retrospective comparison of posaconazole levels in patients taking the delayed-release tablet versus the oral suspension. Antimicrob Agents Chemother. 2015;59(8):4914–8. https://doi.org/10.1128/AAC.00496-15.
66. Morillo CA, Waskin H, Sosa-Estani S, Del Carmen Bangher M, Cuneo C, Milesi R, et al. Benznidazole and posaconazole in eliminating parasites in asymptomatic T. cruzi carriers: the STOP-CHAGAS trial. J Am Coll Cardiol. 2017;69(8):939–47. https://doi.org/10.1016/S1473-3099(17)30538-8.
67. Torrico F, Gascon J, Ortiz L, Alonso-Vega C, Pinazo MJ, Schijman A, et al. Treatment of adult chronic indeterminate Chagas disease with benznidazole and three E1224 dosing regimens: a proof-of-concept, randomised, placebo-controlled trial. Lancet Infect Dis. 2018;18(4):419–30. https://doi.org/10.1016/S1473-3099(17)30538-8.
68. Santos FM, Mazzeti AL, Caldas S, Gonçalves KR, Lima WG, Torres RM, Bahia MT. Chagas cardiomyopathy: the potential effect of benznidazole treatment on diastolic dysfunction and cardiac damage in dogs chronically infected with Trypanosoma cruzi. Acta Trop. 2016;161:44–54. https://doi.org/10.1016/j.actatropica.2016.05.007.
69. Álvarez MG, Hernández Y, Bertocchi G, Fernández M, Lococo B, Ramírez JC, et al. New scheme of intermittent benznidazole administration in patients chronically infected with Trypanosoma cruzi: a pilot short-term follow-up study with adult patients. Antimicrob Agents Chemother. 2016;60(2):833–7. https://doi.org/10.1128/AAC.00745-15.
70. Soy D, Aldasoro E, Guerrero L, Posada E, Serret N, Mejía T, et al. Population pharmacokinetics of benznidazole in adult patients with Chagas disease. Antimicrob Agents Chemother. 2015;59(6):3342–9. https://doi.org/10.1128/AAC.05018-14.
71. Torreele E, Bourdin Trunz B, Tweats D, Kaiser M, Brun R, Mazue G, et al. Fexinidazole—a new oral nitroimidazole drug candidate entering clinical development for the treatment of sleeping sickness. PLoS Negl Trop Dis. 2010;4(12):e923. https://doi.org/10.1371/journal.pntd.0000923.
72. Mesu VKBK, Kalonji WM, Bardonneau C, Mordt OV, Blesson S, Simon F, et al. Oral fexinidazole for late-stage African Trypanosoma brucei gambiense trypanosomiasis: a pivotal multicentre, randomised, non-inferiority trial. Lancet. 2018;391(10116):144–54. https://doi.org/10.1016/S0140-6736(17)32758-7.
73. Bahia MT, de Andrade IM, Martins TA, do Nascimento AF, Diniz Lde F, Caldas IS, et al. Fexinidazole: a potential new drug candidate for Chagas disease. PLoS Negl Trop Dis. 2012;6(11):e1870. https://doi.org/10.1371/journal.pntd.0001870.
74. Bahia MT, Nascimento AF, Mazzeti AL, Marques LF, Gonçalves KR, Mota LW, et al. Antitrypanosomal activity of fexinidazole metabolites, potential new drug candidates for Chagas disease. Antimicrob Agents Chemother. 2014;58(8):4362–70. https://doi.org/10.1128/AAC.02754-13.
75. Caldas S, Caldas IS, Cecílio AB, Diniz LD, Talvani A, Ribeiro I, Bahia MT. Therapeutic responses to different anti-Trypanosoma cruzi drugs in experimental infection by benznidazole-resistant parasite stock. Parasitology. 2014;21:1–10. https://doi.org/10.1017/S0031182014000882.
76. Croft SL. Leishmania and other intracellular pathogens: selectivity, drug distribution and PK-PD. Parasitology. 2017;6:1–11. https://doi.org/10.1017/S0031182017001664.

77. Pinazo MJ, Espinosa G, Gallego M, Lopez-Chejade PL, Urbina JA, Gascon J. Successful treatment with posaconazole of a patient with chronic chagas disease and systemic lupus erythematosus. Am J Trop Med Hyg. 2010;82(4):583–7. https://doi.org/10.4269/ajtmh.2010.09-0620.
78. Perfect JR. The antifungal pipeline: a reality check. Nat Rev Drug Discov. 2017;16(9):603–16. https://doi.org/10.1038/nrd.2017.46.
79. Lepesheva GI, Friggeri L, Waterman MR. CYP51 as drug targets for fungi and protozoan parasites: past, present and future. Parasitology. 2018;12:1–17. https://doi.org/10.1017/S0031182018000562.
80. Lepesheva GI, Hargrove TY, Anderson S, Kleshchenko Y, Furtak V, Wawrzak Z, et al. Structural insights into inhibition of sterol 14alpha-demethylase in the human pathogen Trypanosoma cruzi. J Biol Chem. 2010;285(33):25582–90. https://doi.org/10.1074/jbc.M110.133215.
81. Chen CK, Leung SS, Guilbert C, Jacobson MP, McKerrow JH, Podust LM. Structural characterization of CYP51 from Trypanosoma cruzi and Trypanosoma brucei bound to the antifungal drugs posaconazole and fluconazole. PLoS Negl Trop Dis. 2010;4(4):e651. https://doi.org/10.1371/journal.pntd.0000651.
82. Choi JY, Podust LM, Roush WR. Drug strategies targeting CYP51 in neglected tropical diseases. Chem Rev. 2014;114(22):11242–71. https://doi.org/10.1021/cr5003134.
83. Hoekstra WJ, Hargrove TY, Wawrzak Z, da Gama Jaen Batista D, da Silva CF, Nefertiti AS, et al. Clinical candidate VT-1161's antiparasitic effect in vitro, activity in a murine model of Chagas disease, and structural characterization in complex with the target enzyme CYP51 from Trypanosoma cruzi. Antimicrob Agents Chemother. 2016;60(2):1058–66. https://doi.org/10.1128/AAC.02287-15.
84. Calvet CM, Vieira DF, Choi JY, Kellar D, Cameron MD, Siqueira-Neto JL, et al. 4-Aminopyridyl-based CYP51 inhibitors as anti-Trypanosomas cruzi drug leads with improved pharmacokinetic profile and in vivo potency. J Med Chem. 2014;57(16):6989–7005. https://doi.org/10.1021/jm500448u.
85. Calvet CM, Choi JY, Thomas D, Suzuki B, Hirata K, Lostracco-Johnson S, et al. 4-aminopyridyl-based lead compounds targeting CYP51 prevent spontaneous parasite relapse in a chronic model and improve cardiac pathology in an acute model of Trypanosoma cruzi infection. PLoS Negl Trop Dis. 2017;11(12):e0006132. https://doi.org/10.1371/journal.pntd.0006132.
86. Ottilie S, Goldgof GM, Calvet CM, Jennings GK, LaMonte G, Schenken J, et al. Rapid Chagas disease drug target discovery using directed evolution in drug-sensitive yeast. ACS Chem Biol. 2017;12(2):422–34. https://doi.org/10.1021/acschembio.6b01037.
87. Guedes-da-Silva FH, Batista DG, Da Silva CF, De Araújo JS, Pavão BP, Simões-Silva MR, et al. Antitrypanosomal activity of sterol 14α-demethylase (CYP51) inhibitors VNI and VFV in the swiss mouse models of Chagas disease induced by the Trypanosoma cruzi Y strain. Antimicrob Agents Chemother. 2017;61(4):e02098–16. https://doi.org/10.1128/AAC.02098-16.
88. Kappagoda S, Singh U, Blackburn BG. Antiparasitic therapy. Mayo Clin Proc. 2011;86(6):561–83. https://doi.org/10.4065/mcp.2011.0203.
89. Fügi MA, Kaiser M, Tanner M, Schneiter R, Mäser P, Guan XL. Match-making for posaconazole through systems thinking. Trends Parasitol. 2014;31(2):46–51. https://doi.org/10.1016/j.pt.2014.11.004.
90. Assíria Fontes Martins T, de Figueiredo Diniz L, Mazzeti AL, da Silva do Nascimento ÁF, Caldas S, Caldas IS, et al. Benznidazole/itraconazole combination treatment enhances anti-Trypanosoma cruzi activity in experimental Chagas disease. PLoS One. 2015;10(6):e0128707. https://doi.org/10.1371/journal.pone.0128707.
91. Diniz LF, Mazzeti AL, Caldas IS, Ribeiro I, Bahia MT. Outcome of E1224/benznidazole combination treatment upon infection with multi-drug resistant Trypanosoma cruzi strain in mice. Antimicrob Agents Chemother. 2018;62(6):e00401–18. https://doi.org/10.1128/AAC.00401-18.

92. Guedes da Silva FH, Batista DDGJ, Da Silva CF, Pavão BP, Batista MM, Moreira ODC, et al. Successful aspects of the co-administration of sterol 14α-demethylase inhibitor VFV and benznidazole in experimental mouse models of chagas disease caused by the drug-resistant strain of Trypanosoma cruzi. ACS Infect Dis. 2019;5:365–71. https://doi.org/10.1021/acsinfecdis.8b0025.
93. Moreira da Silva R, Oliveira LT, Silva Barcellos NM, de Souza J, de Lana M. Preclinical monitoring of drug association in experimental chemotherapy of Chagas' disease by a new HPLC-UV method. Antimicrob Agents Chemother. 2012;56(6):3344–8. https://doi.org/10.1128/AAC.05785-11.
94. Raether W, Hänel H. Nitroheterocyclic drugs with broad spectrum activity. Parasitol Res. 2003;90(Supp 1):S19–39. https://doi.org/10.1007/s00436-002-0754-9.
95. Almeida IC, Ferguson MA, Schenkman S, Travassos LR. Lytic anti-alpha-galactosyl antibodies from patients with chronic Chagas' disease recognize novel o-linked oligosaccharides on mucin-like glycosyl-phosphatidylinositol-anchored glycoproteins of Trypanosoma cruzi. Biochem J. 1994;304(Pt 3):793–802.
96. Fernandez-Villegas A, Pinazo MJ, Maranon C, Thomas MC, Posada E, Carrilero B, et al. Short-term follow-up of chagasic patients after benznidazole treatment using multiple serological markers. BMC Infect Dis. 2011;11:206. https://doi.org/10.1186/1471-2334-11-206.
97. Nagarkatti R, de Araujo FF, Gupta C, Debrabant A. Aptamer based, non-pcr, non-serological detection of Chagas disease biomarkers in Trypanosoma cruzi infected mice. PLoS Negl Trop Dis. 2014;8(1):e2650. https://doi.org/10.1371/journal.pntd.0002650.
98. de Araujo FF, Nagarkatti R, Gupta C, Marino AP, Debrabant A. Aptamer-based detection of disease biomarkers in mouse models for Chagas drug discovery. PLoS Negl Trop Dis. 2015;9(1):e3451. https://doi.org/10.1371/journal.pntd.0003451.
99. Miao Q, Santamaria C, Bailey D, Genest J, Ward BJ, Ndao M. Apolipoprotein A-I truncations in Chagas disease are caused by cruzipain, the major cysteine protease of Trypanosoma cruzi. Am J Pathol. 2014;184(4):976–84. https://doi.org/10.1016/j.ajpath.2013.12.018.
100. Santamaria C, Chatelain E, Jackson Y, Miao Q, Ward BJ, Chappuis F, Ndao M. Serum biomarkers predictive of cure in Chagas disease patients after nifurtimox treatment. BMC Infect Dis. 2014;14:302.
101. Ruiz-Lancheros E, Rasoolizadeh A, Chatelain E, Garcia-Bournissen F, Moroni S, Moscatelli G, et al. Validation of apolipoprotein A-1 and fibronectin fragments as markers of parasitological cure for congenital Chagas disease in children treated with benznidazole. Open Forum Infect Dis. 2018;5(11):ofy236. https://doi.org/10.1093/ofid/ofy236.
102. Pinazo MJ, Thomas MC, Bua J, Perrone A, Schijman AG, Viotti RJ, et al. Biological markers for evaluating therapeutic efficacy in Chagas disease: a systematic review. Expert Rev Anti Infect Ther. 2014;12(4):479–96. https://doi.org/10.1586/14787210.2014.899150.
103. Pinazo MJ, Pinto J, Ortiz L, Sánchez J, García W, Saravia R, et al. A strategy for scaling up access to comprehensive care in adults with chagas disease in endemic countries: the Bolivian Chagas platform. PLoS Negl Trop Dis. 2017;11(8):e0005770. https://doi.org/10.1371/journal.pntd.0005770.
104. Neal RA, van Bueren J. Comparative studies of drug susceptibility of five strains of Trypanosoma cruzi in vivo and in vitro. Trans R Soc Trop Med Hyg. 1988;82(5):709–14. https://doi.org/10.1371/journal.pntd.0000740.
105. Canavaci AM, Bustamante JM, Padilla AM, Perez Brandan CM, Simpson LJ, Xu D, et al. In vitro and in vivo high-throughput assays for the testing of anti-trypanosoma cruzi compounds. PLoS Negl Trop Dis. 2010;4(7):e740.
106. Moreno M, D'ávila DA, Silva MN, Galvão LM, Macedo AM, Chiari E, et al. Trypanosoma cruzi benznidazole susceptibility in vitro does not predict the therapeutic outcome of human Chagas disease. Mem Inst Oswaldo Cruz. 2010;105(7):918–24.
107. Sánchez-Valdéz FJ, Padilla A, Wang W, Orr D, Tarleton RL. Spontaneous dormancy protects Trypanosoma cruzi during extended drug exposure. Elife. 2018;7:e34039. https://doi.org/10.7554/eLife.34039.

108. Trachtenberg BH, Hare JM. Inflammatory cardiomyopathic syndromes. Circ Res. 2017;121(7):803–18. https://doi.org/10.1161/CIRCRESAHA.117.310221.
109. Shapiro H, Meymandi S, Shivkumar K, Bradfield JS. Cardiac inflammation and ventricular tachycardia in Chagas disease. Heart Rhythm Case Rep. 2017;3(8):392–5. https://doi.org/10.1016/j.hrcr.2017.05.007.
110. Pereira IR, Vilar-Pereira G, Silva AA, Moreira OC, Britto C, Sarmento ED, Lannes-Vieira J. Tumor necrosis factor is a therapeutic target for immunological unbalance and cardiac abnormalities in chronic experimental Chagas' heart disease. Mediators Inflamm. 2014;2014:798078. https://doi.org/10.1155/2014/798078.
111. Pereira IR, Vilar-Pereira G, Moreira OC, Ramos IP, Gibaldi D, Britto C, et al. Pentoxifylline reverses chronic experimental chagasic cardiomyopathy in association with repositioning of abnormal CD8+ t-cell response. PLoS Negl Trop Dis. 2015;9(3):e0003659. https://doi.org/10.1371/journal.pntd.0003659.
112. Vilar-Pereira G, Carneiro VC, Mata-Santos H, Vicentino AR, Ramos IP, Giarola NL, et al. Resveratrol reverses functional Chagas heart disease in mice. PLoS Pathog. 2016;12(10):e1005947. https://doi.org/10.1371/journal.ppat.1005947.
113. Cruz JS, Machado FS, Ropert C, Roman-Campos D. Molecular mechanisms of cardiac electromechanical remodeling during Chagas disease: role of TNF and TGF-β. Trends Cardiovasc Med. 2017;27(2):81–91.
114. Tabas I, Glass CK. Anti-inflammatory therapy in chronic disease: challenges and opportunities. Science. 2013;339(6116):166–72. https://doi.org/10.1126/science.1230720.
115. Serhan CN. Pro-resolving lipid mediators are leads for resolution physiology. Nature. 2014;510(7503):92–101. https://doi.org/10.1038/nature13479.
116. González-Herrera F, Cramer A, Pimentel P, Castillo C, Liempi A, Kemmerling U, et al. Simvastatin attenuates endothelial activation through 15-epi-lipoxin A4 production in murine chronic Chagas cardiomyopathy. Antimicrob Agents Chemother. 2017;61(3):e02137–16. https://doi.org/10.1128/AAC.02137-16.
117. López-Muñoz RA, Molina-Berríos A, Campos-Estrada C, Abarca-Sanhueza P, Urrutia-Llancaqueo L, Peña-Espinoza M, Maya JD. Inflammatory and pro-resolving lipids in trypanosomatid infections: a key to understanding parasite control. Front Microbiol. 2018;9:1961. https://doi.org/10.3389/fmicb.2018.01961.

9 Management of Chronic Chagasic Cardiomyopathy in Endemic and Non-endemic Countries: Challenges and Limitations

Roberto M. Saraiva and Sheba Meymandi

9.1 Routine Assessment of Patients with Chagas Disease

Routine follow-up of patients with chronic Chagas disease is driven by the screening of the development of the determined forms of the disease. The most important is to determine if the patient has developed the cardiac form of the disease and to assess the risk of cardiac complications and clinical events. Most patients with chronic Chagas disease present the indeterminate form of the disease without clinical evidence of cardiac or digestive disease. However, 2 to 3% of these patients per year will evolve into the cardiac form of the disease. In fact, up to 30% of the patients will present with the cardiac form of the disease during their life time. The cardiac form has different classifications according to different guidelines (see Chap. 6) [1–3].

Although some researchers claim that patients who are in the indeterminate form of the disease for more than 50 years have a very low probability of presenting disease progression, there is no safe age limit to stop the follow-up of patients with the indeterminate form. All those patients must undergo an electrocardiogram (ECG) every year to detect if the patient has progressed to the cardiac form as changes in the ECG are used to define if a person with Chagas disease present the cardiac form (Table 9.1) [2, 3]. Some of these changes demand immediate treatment and others have prognostic significance [2, 3].

In addition to history, physical exam, and baseline blood work, patients with the indeterminate form of Chagas disease should have ECG, Holter, and echocardiogram performed when they are first diagnosed. Echocardiogram is a key tool for the

R. M. Saraiva (✉)
Instituto Nacional de Infectologia Evandro Chagas, Fundação Oswaldo Cruz,
Rio de Janeiro, Brazil
e-mail: roberto.saraiva@ini.fiocruz.br

S. Meymandi
David Geffen School of Medicine, Olive View-UCLA Medical Center, Sylmar, CA, USA

M.-J. Pinazo Delgado, J. Gascon (eds.), *Chagas Disease*,
https://doi.org/10.1007/978-3-030-44054-1_9

Table 9.1 Electrocardiographic changes in Chagas disease

ECG changes used to define progression to CCC	ECG changes found in Chagas disease but not diagnostic of CCC
Complete right bundle branch block	Uncomplete right bundle branch block
Left bundle branch block	Left anterior fascicular block
Second- or third-degree atrioventricular block	First-degree atrioventricular block
Sinus bradycardia with less than 40 beats/min	Sinus bradycardia from 41 to 59 beats/min
Primary T-wave abnormalities	Low voltage
Polymorphic or repetitive ventricular extrasystoles	
Sinus node dysfunction	
Electric inactive areas	
Atrial fibrillation	

CCC chronic Chagasic cardiomyopathy

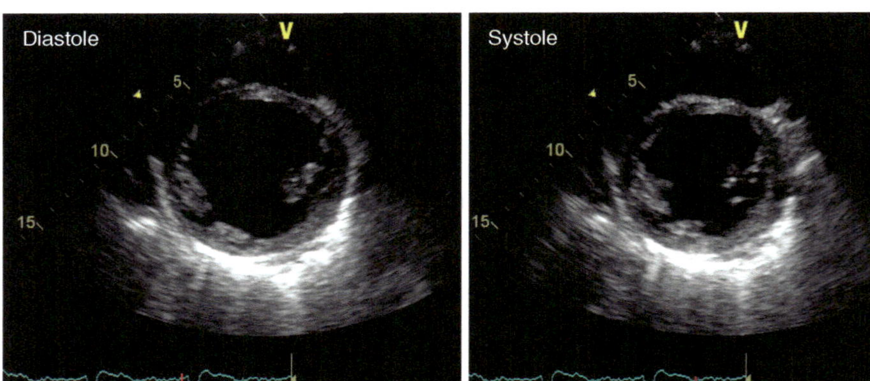

Fig. 9.1 LV short-axis images at the end of the diastole (left panel) and at the end of the systole (right panel). Note the diffuse change in wall motion contractility in this patient with LV systolic dysfunction

classification of the stage of the cardiac form, identification of complications, follow-up, and risk assessment of patients with Chagas disease. The echocardiogram identifies chamber size, global and regional left ventricle (LV) contractility, LV aneurysms, LV diastolic dysfunction, left atrium (LA) size and function, and right ventricle (RV) systolic dysfunction [4]. The LV segments that most commonly present wall motion abnormalities are the inferior and the lateral-inferior walls, and the apex [5]. Those changes include hypokinesis, akinesis, and dyskinesis [5] (Fig. 9.1). Even patients with the indeterminate form by ECG criteria may present wall motion abnormalities in 13% of the echocardiograms [5] and a low prevalence of apical aneurysms (around 2% [4]).

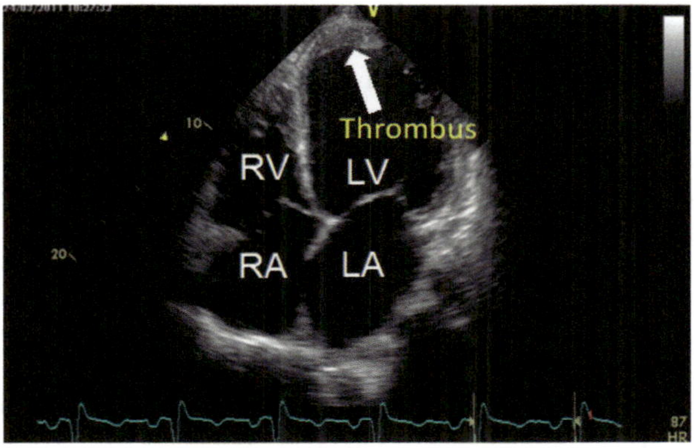

Fig. 9.2 LV 4-chamber view with color Doppler: Note LA and LV enlargements complicated by thrombus localized at the LV apex. *LA* left atrium, *LV* left ventricle, *RA* right atrium, *RV* right ventricle

LV aneurysms prevalence in patients with Chagas disease is around 14%. In patients with the indeterminate form, LV aneurysms are found in only 2% of the patients while in patients with moderate or severe LV systolic dysfunction LV aneurysms prevalence increases to 55% (ranging from 47% to 64%) [4]. Most aneurysms are apical but they may also be found in lateral-inferior wall [5]. Apical aneurysms are more associated with intraventricular thrombi and stroke risk (Fig. 9.2), while lateral-inferior aneurysms are more associated with arrhythmia risk.

RV systolic dysfunction can also be detected by the echocardiogram and usually is a late finding in patients with CCC and HF. It is usually attributed to the burden generated by chronic pulmonary hypertension secondary to LV systolic dysfunction [6] but others also argue that direct damage to the RV myocardium due to Chagas disease may play a role in the development of RV dysfunction [7].

LV diastolic dysfunction may also contribute to symptoms and present prognostic value in CCC. Diastolic dysfunction is present in patients with the indeterminate form but is more prevalent in patients with the cardiac form. Within patients with CCC, diastolic dysfunction prevalence and severity increases as CCC progresses from the initial stage to the more advanced stages [8].

Other findings in the echocardiogram of patients with CCC are mitral and tricuspid regurgitation. Mitral regurgitation is due to the distortion of the mitral annulus and subvalvular apparatus created by the LV remodeling and fibrosis of the inferolateral wall. When moderate to severe, mitral regurgitation may worsen HF symptoms and pulmonary congestion, and carries an ominous prognosis in HF [9]. Tricuspid regurgitation is due to dilation of the tricuspid annulus, pulmonary hypertension, and/or the presence of pacemaker lead through the tricuspid valve. Tricuspid regurgitation may worsen right-sided HF and increase the need for diuretics in order to compensate HF.

In poor areas without easy access to echocardiography, a chest X-ray in patients with CCC may reveal an increase in cardiothoracic ratio with or without pulmonary congestion [2] and determine who should be referred to secondary or tertiary clinical centers. The exam should be performed in both posterior-anterior and lateral projections with a barium swallow, in order to assess atrial and ventricular size, and esophagus enlargement. In the acute phase of Chagas disease, chest X-ray can identify increased cardiothoracic ratio suggestive of pericardial effusion. Those patients should be promptly referred to further diagnostic and treatment [10].

Holter monitoring is indicated in the initial evaluation of all patients with CCC in order to evaluate the presence and density of complex ventricular arrhythmias, sinus node disease, and atrioventricular conduction [2, 3, 11].

After the initial evaluation, all patients with the cardiac form must also undergo annual ECG and echocardiograms. Additional echocardiograms may be obtained for patients with worsening functional class, developing new electrocardiographic changes, having aborted sudden cardiac death, and undergoing assessment for pacemaker or Implantable Cardiac Defibrillator (ICD) devices or after their implantation. Additional ECGs should be obtained for those with symptoms or signs of cardiac arrhythmias. Additional Holter exams must be done whenever patients develop symptoms compatible with arrhythmias that were not diagnosed by an ECG. The most important symptom is presyncope or syncope. Occult nonsustained ventricular tachycardia (VT) is more common in CCC than in HF due to other etiologies and increases in frequency from stage B (40%) to stages C and D of the cardiac form (90%) [12] and is an independent predictor of death [13]. Electrophysiological study should be indicated if patient persist with symptoms and no cause is identified.

The frequency the patient with cardiac form must return to the medical office will be determined by their clinical status, complications, and treatment. Pharmacological treatment in patients with CCC will be described below.

9.2 Arrhythmias

Arrhythmias in CCC can be either bradyarrhythmias or tachyarrhythmias and are an important cause of pacemaker implantation in Latin America [14]. The most important points when evaluating bradyarrhythmia in a patient with CCC are if there are any reversible factor aggravating it or when to indicate a pacemaker implantation. Patients may present with presyncope, syncope, fatigue, atypical chest pain, or exertional dyspnea, even with preserved LV systolic function. In case of sick sinus syndrome, 24-h Holter monitoring can reveal bradycardia, sinus pauses, sinoatrial block, and bradycardia-tachycardia syndrome [15, 16], and electrophysiological study can reveal abnormal sinus node recovery time and sinoatrial conduction [17]. Cardiac stress test is also useful to evaluate the chronotropic response to exercise, which can be diminished as a result of autonomic impairment [18, 19], and to uncover complex ventricular arrhythmias [20, 21].

All medications taken by the patient must be analyzed and those capable of worsening heart conduction should be withheld. Advanced atrioventricular block and symptomatic sinus sick syndrome are the main indications for pacemaker implantation in CCC. Recommendations for pacemaker implantation in CCC disease follow the same guidelines for other conditions. However, some aspects should receive special consideration. The RV electrode should be implanted at the midseptal position due to the possibility of excessive fibrosis at the apical position [22, 23]. Moreover, patients with LV systolic dysfunction may evolve with worsening of their LV systolic function and HF after pacemaker implantation as the LV pacing may induce LV systolic dyssynchrony. In patients with left bundle branch block and HF who need a pacemaker implantation, a device with resynchronization therapy should be the choice, while in patients with CCC and right bundle branch block there is scarce evidence to support resynchronization in case pacemaker implantation is indicated.

The most common arrhythmia in CCC is isolated premature ventricular complexes which do not need treatment unless they are symptomatic. On the other hand, malignant ventricular tachyarrhythmias are the main cause of sudden death in Chagas disease. Asymptomatic nonsustained VT also does not need treatment in patients with preserved LV systolic function. On the other hand, pharmacological treatment of patients with symptomatic nonsustained VT or asymptomatic nonsustained VT in patients with LV systolic dysfunction is controversial [24–26]. Although amiodarone, the drug of choice in patients with CCC, improves symptoms and reduces the density of ventricular arrhythmia [27, 28], it has high toxicity (e.g., dermatitis, pulmonary fibrosis, and thyroid dysfunction) and there is no evidence that amiodarone decreases mortality among patients with CCC [28, 29]. Nevertheless, amiodarone should be used in high-risk patients with LV systolic dysfunction and nonsustained VT associated with symptoms, especially dizziness and syncope, and in patients with delayed enhancement by Cardiac Magnetic Resonance (CMR) [30], late potentials in the signal-averaged ECG [31], T-wave variability [32], and T-wave microalternans [33]. In addition, amiodarone should be considered in patients with a high percentage of ventricular ectopic beats and nonsustained VT by 24-h Holter monitoring because these can result in tachycardiomyopathy [34]. Amiodarone has been shown to have intrinsic anti-*Trypanosoma cruzi* activity both in vitro and in vivo which may have an added benefit [35].

Another treatment and prophylaxis against malignant ventricular arrhythmias in patients with CCC is the use of ICD. In fact, patients with CCC have a higher risk of malignant ventricular arrhythmia and sudden death [36] than patients with HF due to other etiologies, even if matched by ventricular dysfunction [37]. ICDs are indicated in patients with HF and LV ejection fraction under 35% with or without a previous history of VT [38]. However, the studies that supported this recommendation included few patients with CCC, if any. The CHAGASICS trial (Amiodarone Against ICD Therapy in Chagas Cardiomyopathy for Primary Prevention of Death) is underway to test whether ICD therapy is useful for the primary prevention of death in patients with CCC [39]. At the moment, ICDs are recommended in CCC for secondary prevention after documented VT, ventricular fibrillation, or aborted sudden death, patients with left ventricle ejection fraction (LVEF) <35% and documented syncope secondary to VT, patients with LVEF >35% who have experienced syncope secondary to VT, and in patients with CCC with syncope and inducible sustained VT during electrophysiological study [3]. In a single study, patients with CCC and LVEF <40% with documented prior life-threatening arrhythmia had a better survival under ICDs than patients using only amiodarone [40]. Amiodarone should be prescribed even after ICDs placement in CCC in order to decrease the number of shocks as patients with CCC can have frequent shocks due to intense ventricular arrhythmic activity [37], and an excessive number of shocks may be deleterious [41], causing myocardial necrosis and exacerbating ventricular dysfunction [42].

It is important to predict patients with increased risk of VT as sudden death can be the first manifestation of a malignant arrhythmia. Studies have demonstrated that the extension of myocardial fibrosis as assessed by delayed enhancement on CMR imaging can identify patients with high risk of VT [30, 43], even in the absence of global LV systolic dysfunction [30, 44]. As the cost of CMR is high and not available in many endemic areas, a scoring based on ECG, Selvester QRS scoring system, was proposed to estimate scar size by changes in Q-, R-, and S-wave duration, amplitude, and morphology. This score showed a good correlation with CMR and with a previous history of VT [43]. De Souza et al. developed a score based on clinical, echocardiographic, and ECG data (QT dispersion, syncope, premature ventricular contractions, and LV function) to predict sudden death and classifies patients into low (0–2 points), intermediate (3–4 points), and high (>5 points) risk of sudden death [45]. Another phenomenon associated with risk of VT in Chagas disease is myocardial sympathetic denervation [46] that can be identified by myocardial scintigraphy by iodine-123 metaiodobenzylguanidine testing. The detection of areas of cardiac fibrosis by single-photon emission computed tomography and areas of myocardial sympathetic denervation indicate patients at risk of developing malignant ventricular arrhythmia [7, 47]. Nevertheless, ICD implantation for primary VT prophylaxis is still not indicated in every day clinical practice based on such findings in complementary exams.

In case of medication failure, intolerable side effects, or repetitive appropriate shocks, ablation therapy of VT (catheter based) is an option to treat recurrent VT in patients with CCC [48]. As VT in CCC is typically reentrant [49], it is possible to map the circuits and disrupt them by ablation. However, mapping should be extensive as multiple circuits with complex anatomy are usually present. Most common sites of origin of reentrant circuits are the same ventricular segments most affected by Chagas disease as seen in echocardiograms or CMR [50, 51]. One important aspect is that fibrosis is not necessarily subendocardial or transmural in CCC as in ischemic cardiomyopathy. CMR identifies areas of cardiac fibrosis that is gradually more prevalent from patients with the indeterminate form (around 20% [44, 52]) and stage A of the cardiac form (43.7% [44]) to the more advanced stages of the cardiac form (89% [52] to 100% [30]). The walls most frequently involved were the inferolateral and apex [30, 44, 52, 53]. Cardiac fibrosis pattern is mostly midwall and subepicardial [53] when studying patients with the indeterminate or at the initial stages of the cardiac form while the predominant pattern is transmural [30, 52] among patients with LV systolic dysfunction. In fact, one-third of reentrant circuits are located on the epicardial surface. Thus careful electrophysiological mapping is necessary to achieve successful ablation [54]. VT ablation is recommended in CCC based on its indication for other clinical conditions [55]: symptomatic sustained monomorphic VT, including VT terminated by ICD, that recurs despite drug therapy or when antiarrhythmic drugs are not tolerated or not desired and when there is a suspected trigger that can be targeted for ablation; control of incessant sustained monomorphic VT or VT storm that is not the result of a transient reversible cause; and bundle branch reentrant or interfascicular VT. Some patients with refractory VT or VT storm may require bilateral cardiac sympathetic denervation to reduce the burden of ventricular arrhythmias in addition to endocardial and epicardial ablation [56].

9.3 Stroke

Chagas disease is a major cause of stroke in Latin America. In endemic areas, up to 20% of patients that suffer a stroke have Chagas disease [57]. The main cause of stroke in Chagas disease is cardioembolic due to intracardiac thrombi arising mainly from apical aneurysms. Stroke can also be due to atrial fibrillation in CCC [58, 59]. Risk factors for stroke include apical aneurysm, LV thrombus, severe atrial dilation, LV systolic dysfunction, older age, and atrial fibrillation [60–63]. Other causes for stroke in Chagas disease, such as small vessel disease and large vessel atherosclerosis, are being increasingly noticed [64] and are related to concomitant risk factors (hypertension, hyperlipidemia, smoking) [64], proinflammatory and prothrombotic disease state [64–66], or endothelial dysfunction [67].

Most patients with stroke present signs and symptoms of ischemia of the brain in the distribution of the anterior or medium cerebral arteries, as expected by its cardioembolic nature [64, 68]. Stroke may also contribute to cognitive impairment and dementia in endemic areas [69, 70]. It is important to bear in mind that stroke can be the first and only manifestation of Chagas disease [71–74] and that the incidence of cardioembolic events in patients with HF due to CCC is higher than in patients with HF due to other etiologies.

Transthoracic echocardiography is indicated in all patients with Chagas disease and thromboembolic events in order to rule out mural thrombi in LV, especially in LV apical aneurysms. In case of documented or suspected atrial fibrillation, transesophageal echocardiograms must be added to the diagnostic workup as it is more sensitive for LA thrombi than transthoracic echocardiogram [74]. Among patients with Chagas disease and stroke, 23% had LV thrombus, 47% had LV aneurysm, and only 5% had thrombi in the LA appendage [74]. In case the source of cardioembolic event is still unclear, Holter monitoring is indicated to investigate occult paroxysmal atrial fibrillation [75]. Eventually, CMR may detect intracardiac thrombi, but there is no data to support its routine use in the diagnostic workup of patients with Chagas disease and stroke [76].

Stroke prediction models in Chagas disease are scarce. A study from our group based on a cohort of >1000 patients identified four variables: LV systolic dysfunction, apical aneurysm, primary ST changes on the ECG, and age above 48 years [62]. A score was created with 2 points attributed for LV systolic dysfunction and 1 point for each of the other variables. We recommended that anticoagulation should be indicated in patients with a score of 4 to 5, as the annual risk of stroke was 4.4%. For patients with a score of 2 to 3, the risk of stroke is lower and can be similar to the risk of bleeding, and we recommended either anticoagulation or aspirin. Patients with a score of 1 had a low incidence of ischemic events, and we recommend aspirin or no treatment [62]. However, there is no randomized, controlled study that evaluated anticoagulation in patients with CCC and an external validation is needed for this prediction model. Therefore, anticoagulation is recommended based on retrospective studies and specialist's consensus. In case of paroxysmal or permanent atrial fibrillation, primary stroke prophylaxis with warfarin is indicated in patients with CHA2DS2-VASc score $\geq$ 1–2 or with LV systolic dysfunction [2]. Additional recommendations for warfarin include intracardiac thrombi, previous stroke, or transient ischemic attack, especially in the presence of apical aneurysm.

In case of acute stroke, the experience with thrombolytic is limited in CCC but short-term treatment with thrombolytics seems to have similar success compared to non-Chagas stroke [77, 78]. The use of antiplatelet agents for secondary prophylaxis in patients with Chagas disease and stroke considered to be noncardioembolic is recommended based on studies with non-Chagas groups of patients.

9.4 Heart Failure

Heart failure evolution in CCC is insidious and in general will manifest on average only 20 or more years after infection. The clinical manifestations are characterized by a predominance of signs and symptoms of right-sided heart failure.

An important aspect of management of patients with CCC is risk stratification to identify patients with increased risk of death. The single most powerful predictor is LV systolic function, measured by LV ejection fraction. In fact, LV systolic dysfunction is the commonest and more consistent independent mortality predictor [79]. Other known predictors of poor outcome in CCC include RV dysfunction [80], LA volume [60] and function [8], and LV diastolic function [8, 63]. LV diastolic dysfunction, evaluated by the ratio of the early diastolic mitral inflow over the early myocardial diastolic velocity, functional class, LV ejection fraction, RV Tei index, and maximum LA volume correlated independently with adverse outcomes in multivariate analysis [63].

A risk score that combines several aspects of CCC is the Rassi score which includes six independent prognostic factors: New York Heart Association class III or IV (5 points), increased cardiothoracic ratio on chest X-ray (5 points), LV systolic dysfunction on echocardiography (3 points), nonsustained VT on 24-h Holter monitoring (3 points), low QRS voltage on EKG (2 points), and male sex (2 points) [13]. Patients were classified into three risk groups: low risk (0 to 6 points), intermediate risk (7 to 11 points), and high risk (12 to 20 points). The 10-year mortality rates for these three groups were 10%, 44%, and 84%, respectively [13].

Biomarkers have also been extensively studied and although some biomarkers differ between patients with the indeterminate and cardiac form [81], few studies showed prognostic value of these biomarkers and usually did not evaluate them against clinical and echocardiographic parameters. Nonetheless, brain natriuretic peptide, transforming growth factor β1, and metalloproteinase were described to be able to predict adverse outcomes in CCC [82–85].

Medical treatment of patients with CC and HF is largely based on scientific evidence obtained for treatment of HF due to other etiologies. Most trials that established the clinical benefit of the drugs currently used in HF did not include a significant number of patients with CCC or did not include them at all. Therefore, recommendations are based on guidelines for HF [2, 3]. Patients with asymptomatic LV systolic dysfunction should be started on angiotensin-converting enzyme (ACE) inhibitors and β-blockers. Patients with HF should also be treated with a combination of β-blockers and ACE inhibitors. Loop diuretics must also be added to treat fluid retention and alleviate symptoms. The maximum dose of β-blockers may not be achieved in patients with CCC as Chagas disease may also affect the conduction system. In fact, a more gradual increase in β-blockers dose is recommended, but many patients use the maximum recommended dose of carvedilol without complications. Moreover, adrenergic blockade could have an additional and specific benefit in CCC because patients may have autonomic imbalance (due to predominant parasympathetic dysautonomia and areas of denervated myocardium) and a potential for the development of ventricular arrhythmia. Therefore, β-blockers may have a potential benefit in decreasing the risk of ventricular arrhythmia and sudden death, further suggesting their usefulness in CCC. Some studies have shown potential additional benefits with treatment using the beta blocker carvedilol as it decreased the parasite load through inhibition of proliferation of epi- and amastigotes and reduced the inflammatory infiltration in cardiac cells [86, 87].

In patients who persist in NYHA class II to IV and LV ejection fraction ≤35%, aldosterone receptor antagonist (spironolactone) should be added. In patients who do not tolerate ACE inhibitors due to adverse drug reactions, angiotensin receptor blockers (ARB) may be used instead. In patients who persist symptomatic and with reduced LV ejection fraction, digoxin and nitrate/hydralazine combination may be added to the prescription. Digoxin is specially recommended in patients with difficult to control atrial fibrillation with normal renal function. However, digoxin must be used with caution in CCC due to the risk of developing atrioventricular nodal dysfunction and should be avoided in patients with conduction abnormalities at or below the atrioventricular node as it prolongs the effective refractory period of atrioventricular nodal tissue and can worsen heart block. Finally, in refractory cases, sacubitril/valsartan may be used and alternative therapies may be considered. Patients with fluid retention refractory to loop diuretics and spironolactone may also use thiazides diuretics [2, 3].

Beyond the use of drugs to control HF, non-pharmacological management and patient education in patients with CCC and HF should be undertaken as recommended in international guidelines for HF. It is very important that patients with HF are followed by a multidisciplinary team with integral care in order for the non-pharmacological management and patient education to be successful. The following strategies and recommendations can be adopted in patients with CCC and HF:

pharmaceutical care [88], identification and treatment of reversible precipitating factors (poor compliance with prescribed therapy, use of medications known to worsen HF, infections, arrhythmias, among others), advice on controlling the amount of salt in the diet and fluid restriction, encouragement to carry out daily physical and leisure-time activities that do not induce symptoms, exercise training programs in stable patients in NYHA class II–III [89], patient education on how to identify HF deterioration (e.g., weigh on a regular basis, identify HF symptoms), avoid smoking, and immunization against pneumococcal and influenza [2, 3].

Small or non-controlled studies tested drugs usually prescribed for HF in patients with CCC. A small, single-blind clinical trial of patients with CCC demonstrated that captopril had a beneficial effect on neurohormones with a reduction in heart rate and in the incidence of ventricular arrhythmias [90]. ACE inhibitors and mineralocorticoid receptor antagonists were shown to improve functional class and to lower BNP [91]. β-blockers may improve the patients' quality of life, decrease BNP, and increase LVEF [91], but their effect on survival is still controversial. While a retrospective study demonstrates a lower mortality among patients in use of carvedilol [92], a meta-analysis of two small randomized trials did not have enough power to confirm or reject the hypothesis of beneficial effect of carvedilol in CCC [93].

When overt and refractory HF ensues, alternative therapies are still possible in CCC. Cardiac resynchronization therapy (CRT) is based on an implantation of a cardiac device with electrodes placed on the right atrium, RV, and LV. CRT is indicated to patients with NYHA functional class III and IV who remain symptomatic despite stable, optimal HF medical therapy and have a LVEF $\leq 35\%$ and a prolonged QRS duration [38]. However, most experience with CRT therapy comes from studies that included patients with left bundle branch block who presented a clear desynchrony in ventricular contraction. Patients with CCC and left bundle branch block or with pacemaker may benefit from a CRT device if their clinical condition meets the conditions outlined above. However, most patients with CCC present right bundle branch block and the benefit of a CRT implant in this case is not as clear or not indicated [2, 3].

Other alternative therapy for CCC with HF refractory to optimal medical therapy is heart transplant. Chagas disease is not a contraindication for heart transplant and selection criteria do not differ from the general transplantation evaluation except that megaesophagus or megacolon is a relative contraindication [94]. Moreover, survival after heart transplant after 1 month (83%), 1 year (71%), and 10 years (46%) is even better than in the general heart transplant population [95]. Cardiopulmonary test is important to evaluate indication for heart transplant and also to guide cardiac rehabilitation in patients with advanced HF [3].

Heart transplant in CCC has a specific issue that is the risk of reactivation of the acute phase of Chagas disease with myocarditis due to immunosuppression. However, antitrypanosomal therapy prior to heart transplant is controversial (see Chap. 11). Patients must be followed and quantitative *T cruzi* PCR obtained whenever there is a suspicion of Chagas disease reactivation. Usually the test is sensible enough to detect reactivation before allograft dysfunction develops [96].

Benznidazole is the drug of choice for treatment of *T cruzi* reactivation [2]; however, treatment will not eliminate the infection and close lifelong surveillance is warranted to treat new episodes of reactivation.

Other important issue when dealing with patients with CCC is that chronic patients are ageing and prevalence of comorbidities is increasing [97]. Therefore, patients with chest pain or heart failure should have their comorbidities evaluated and treated as well. It is not uncommon for coronary artery disease to be the cause or aggravate HF symptoms. However, the investigation of coronary artery disease in Chagas disease is challenging. First, physicians must bear in mind that ST changes in exercise stress test require careful interpretation due to the background ST- and T-wave abnormalities present in CCC [98]. Secondly, nuclear medicine tests also need careful interpretation as single-photon emission computed tomography reveals areas of fixed perfusion defects and wall motion abnormality which correspond to fibrosis due to the destruction of the myocardium by Chagas disease and not to coronary artery disease [99, 100]. Even areas of "ischemia" (mild, multiple reversible perfusion defects not related to coronary anatomy) may be due to microvascular disease and not to epicardial coronary stenosis [101]. In fact, microvascular disease is associated with Chagas chest pain syndrome and may be a potential cause of myocardial damage in Chagas disease [50, 101]. Coronary artery disease should be suspected in presence of transmural reversible defects that follow coronary anatomy. Therefore, direct visualization of the coronary anatomy is frequently needed. In fact, many patients with angina-like symptoms are referred to cardiac catheterization and angiography frequently reveals normal epicardial coronary arteries. On the other hand, coronary artery disease is being increasingly found in association with CCC. In our opinion, whenever symptoms of angina are typical and risk factors are present, coronary disease stratification should follow the current guidelines for patients with the indeterminate form. On the other hand, non-invasive exams of patients with CCC are more difficult to interpret and cardiac catheterization should be performed when clinical suspicion is high. Cardiac catheterization with contrast ventriculography can also reveal global and regional LV systolic function and apical aneurysms, even small ones [3, 46].

Another issue in CCC is the use of new imaging technologies such as CMR and new echocardiographic techniques including real-time three-dimensional echocardiography (RT3DE) and speckle tracking imaging (STI) with analysis of two-dimensional strain (ε). Although research papers have found different potential utilities for those tests, the cost of these exams and the unavailability in many endemic areas have limited the adoption of these techniques. Guidelines on the use of new imaging modalities can be found elsewhere [46] and are beyond the scope of this chapter. Briefly, RT3DE and CMR can improve the evaluation of chambers' volume and function over bidimensional echocardiography as they do not depend on geometrical models and mathematical formulas. Other CMR applications have already been described previously in this chapter. LA function can be analyzed by both RT3DE and two-dimensional strain. LA function has three different phases: reservoir (LA filling phase), conductive (during the early filling phase of the LV diastole), and contractile (LA systole). In Chagas disease, using RT3DE and two-dimensional strain we found that the conductive and reservoir LA function are depressed in all stages of the cardiac form, while LA contractile function is depressed only in the group of patients with HF (stages C and D) [8]. We recently described that a low total LA emptying fraction and higher age are independent predictors of new-onset AF in patients with Chagas disease [102]. LA contractile function measured by LA ε together with end-systolic LV diameter, and E' velocity were independent predictors of the combined end-point of all-cause mortality, stroke, heart transplant, atrial fibrillation, or admission due to worsening HF or cardiac arrhythmias [8]. Analysis of the LV ε may also yield new prognostic index in Chagas disease. We described that LV longitudinal, circumferential, and radial ε were decreased in patients at stage B, and further decreased in patients with HF (stages C and D), while LV torsion was progressively decreased in all stages of the cardiac form [8]. In fact, among patients with HF due to other etiologies, peak LV longitudinal ε has prognostic value independent from LVEF [103, 104].

9.5 Final Remarks

In Chagas disease, it is crucial to diagnose patients while in the acute phase or during the first years of the chronic phase. Successful Chagas disease etiologic treatment will decrease the progression to the cardiac form of the disease. Nonetheless, screening and diagnosis of patients at the chronic phase with cardiac form is also critical as early treatment of cardiac complications can help prevent/delay the occurrence of cardiac events and death.

References

1. Nunes MCP, Beaton A, Acquatella H, Bern C, Bolger AF, Echeverria LE, et al. Chagas cardiomyopathy: an update of current clinical knowledge and management: a scientific statement from the American Heart Association. Circulation. 2018;138:e169–209.
2. Dias JC, Ramos AN Jr, Gontijo ED, Luquetti A, Shikanai-Yasuda MA, Coura JR, et al. 2 nd Brazilian consensus on Chagas disease, 2015. Rev Soc Bras Med Trop. 2016;49(Suppl 1):3–60.
3. Andrade JP, Marin-Neto JA, Paola AA, Vilas-Boas F, Oliveira GM, Bacal F, et al. I Latin American guidelines for the diagnosis and treatment of Chagas cardiomyopathy. Arq Bras Cardiol. 2011;97:1–48.
4. Acquatella H. Echocardiography in Chagas heart disease. Circulation. 2007;115:1124–31.
5. Viotti RJ, Vigliano C, Laucella S, Lococo B, Petti M, Bertocchi G, et al. Value of echocardiography for diagnosis and prognosis of chronic Chagas disease cardiomyopathy without heart failure. Heart. 2004;90:655–60.
6. Nunes MC, Barbosa MM, Brum VA, Rocha MO. Morphofunctional characteristics of the right ventricle in Chagas' dilated cardiomyopathy. Int J Cardiol. 2004;94:79–85.
7. Marin-Neto JA, Bromberg-Marin G, Pazin-Filho A, Simoes MV, Maciel BC. Cardiac autonomic impairment and early myocardial damage involving the right ventricle are independent phenomena in Chagas' disease. Int J Cardiol. 1998;65:261–9.

8. Nascimento CA, Gomes VA, Silva SK, Santos CR, Chambela MC, Madeira FS, et al. Left atrial and left ventricular diastolic function in chronic chagas disease. J Am Soc Echocardiogr. 2013;26:1424–33.
9. Koelling TM, Aaronson KD, Cody RJ, Bach DS, Armstrong WF. Prognostic significance of mitral regurgitation and tricuspid regurgitation in patients with left ventricular systolic dysfunction. Am Heart J. 2002;144:524–9.
10. Pinto AY, Valente SA, Valente VC, Ferreira Junior AG, Coura JR. Acute phase of Chagas disease in the Brazilian Amazon region: study of 233 cases from Para, Amapa and Maranhao observed between 1988 and 2005. Rev Soc Bras Med Trop. 2008;41:602–14.
11. Rassi JA, Rassi A, Marin-Neto JA. Chagas heart disease: pathophysiologic mechanisms, prognostic factors and risk stratification. Mem Inst Oswaldo Cruz. 2009;104(Suppl 1):152–8.
12. Rassi JA, Gabriel RA, Gabriel RS, Rassi JL, Rassi A. Ventricular arrhythmia in Chagas disease. Diagnostic, prognostic, and therapeutic features. Arq Bras Cardiol. 1995;65:377–87.
13. Rassi A Jr, Rassi A, Little WC, Xavier SS, Rassi SG, Rassi AG, et al. Development and validation of a risk score for predicting death in Chagas' heart disease. N Engl J Med. 2006;355:799–808.
14. Rincon LG, Rocha MO, Pires MT, Oliveira BG, Barros VC, Barros MV, et al. Clinical profile of Chagas and non-Chagas' disease patients with cardiac pacemaker. Rev Soc Bras Med Trop. 2006;39:245–9.
15. Ribeiro AL, Marcolino MS, Prineas RJ, Lima-Costa MF. Electrocardiographic abnormalities in elderly chagas disease patients: 10-year follow-up of the Bambui cohort study of aging. J Am Heart Assoc. 2014;3:e000632.
16. Marcolino MS, Palhares DM, Ferreira LR, Ribeiro AL. Electrocardiogram and Chagas disease: a large population database of primary care patients. Glob Heart. 2015;10:167–72.
17. Cedraz SS, Silva PC, Minowa RK, Aragao JF, Silva DV, Morillo C, et al. Electrophysiological characteristics of Chagas disease. Einstein (Sao Paulo). 2013;11:291–5.
18. Gallo L Jr, Neto JA, Manco JC, Rassi A, Amorim DS. Abnormal heart rate responses during exercise in patients with Chagas' disease. Cardiology. 1975;60:147–62.
19. Pereira MH, Brito FS, Ambrose JA, Pereira CB, Levi GC, Neto VA, et al. Exercise testing in the latent phase of Chagas' disease. Clin Cardiol. 1984;7:261–5.
20. Pedrosa RC, Campos MC. Exercise testing and 24 hours Holter monitoring in the detection of complex ventricular arrhythmias in different stages of chronic Chagas' heart disease. Rev Soc Bras Med Trop. 2004;37:376–83.
21. De Paola AA, Gomes JA, Terzian AB, Miyamoto MH, Martinez Fo EE. Ventricular tachycardia during exercise testing as a predictor of sudden death in patients with chronic chagasic cardiomyopathy and ventricular arrhythmias. Br Heart J. 1995;74:293–5.
22. de Souza SF, Nascimento BR, Nunes MC, da Silva JL, de Carvalho VT, Beaton AZ, et al. Effect of pacemaker site on B-type natriuretic peptide levels and left ventricular function in a population with high prevalence of Chagas disease. Int J Cardiol. 2015;190:315–8.
23. da Silva JO, Borges MC, de Melo CS, Nascente GA, Correia D. Alternative sites for right ventricular pacing in Chagas disease: a comparative study of the mid-septum and inflow tract. Pacing Clin Electrophysiol. 2014;37:1166–73.
24. Rassi A Jr, Rassi SG, Rassi A. Sudden death in Chagas' disease. Arq Bras Cardiol. 2001;76:75–96.
25. Ribeiro AL, Nunes MP, Teixeira MM, Rocha MO. Diagnosis and management of Chagas disease and cardiomyopathy. Nat Rev Cardiol. 2012;9:576–89.
26. Bestetti RB, Cardinalli-Neto A. Sudden cardiac death in Chagas' heart disease in the contemporary era. Int J Cardiol. 2008;131:9–17.
27. Miranda CH, Figueiredo AB, Maciel BC, Marin-Neto JA, Simoes MV. Sustained ventricular tachycardia is associated with regional myocardial sympathetic denervation assessed with 123I-metaiodobenzylguanidine in chronic Chagas cardiomyopathy. J Nucl Med. 2011;52:504–10.

28. Stein C, Migliavaca CB, Colpani V, Rosa PRD, Sganzerla D, Giordani NE, et al. Amiodarone for arrhythmia in patients with Chagas disease: a systematic review and individual patient data meta-analysis. PLoS Negl Trop Dis. 2018;12:e0006742.
29. Piccini JP, Berger JS, O'Connor CM. Amiodarone for the prevention of sudden cardiac death: a meta-analysis of randomized controlled trials. Eur Heart J. 2009;30:1245–53.
30. Mello RP, Szarf G, Schvartzman PR, Nakano EM, Espinosa MM, Szejnfeld D, et al. Delayed enhancement cardiac magnetic resonance imaging can identify the risk for ventricular tachycardia in chronic Chagas' heart disease. Arq Bras Cardiol. 2012;98:421–30.
31. Ribeiro AL, Cavalvanti PS, Lombardi F, Nunes MC, Barros MV, Rocha MO. Prognostic value of signal-averaged electrocardiogram in Chagas disease. J Cardiovasc Electrophysiol. 2008;19:502–9.
32. Ribeiro AL, Rocha MO, Terranova P, Cesarano M, Nunes MD, Lombardi F. T-wave amplitude variability and the risk of death in Chagas disease. J Cardiovasc Electrophysiol. 2011;22:799–805.
33. Barbosa MPT, da Costa Rocha MO, Neto ES, Brandao FV, Lombardi F, Ribeiro ALP. Usefulness of microvolt T-wave alternans for predicting outcome in patients with Chagas disease with implantable cardioverter defibrillators. Int J Cardiol. 2016;222:80–5.
34. Brugada P. Chagas' disease and tachycardiomyopathy. Arq Bras Cardiol. 1991;56:5–7.
35. Benaim G, Sanders JM, Garcia-Marchan Y, Colina C, Lira R, Caldera AR, et al. Amiodarone has intrinsic anti-Trypanosoma cruzi activity and acts synergistically with posaconazole. J Med Chem. 2006;49:892–9.
36. Barbosa MP, Carmo AA, Rocha MO, Ribeiro AL. Ventricular arrhythmias in Chagas disease. Rev Soc Bras Med Trop. 2015;48:4–10.
37. Barbosa MP, da Costa Rocha MO, de Oliveira AB, Lombardi F, Ribeiro AL. Efficacy and safety of implantable cardioverter-defibrillators in patients with Chagas disease. Europace. 2013;15:957–62.
38. Ponikowski P, Voors AA, Anker SD, Bueno H, Cleland JG, Coats AJ, et al. 2016 ESC guidelines for the diagnosis and treatment of acute and chronic heart failure: the task force for the diagnosis and treatment of acute and chronic heart failure of the European Society of Cardiology (ESC). Developed with the special contribution of the Heart Failure Association (HFA) of the ESC. Eur J Heart Fail. 2016;18:891–975.
39. Martinelli M, Rassi A Jr, Marin-Neto JA, De Paola AA, Berwanger O, Scanavacca MI, et al. CHronic use of Amiodarone aGAinSt Implantable cardioverter-defibrillator therapy for primary prevention of death in patients with Chagas cardiomyopathy study: rationale and design of a randomized clinical trial. Am Heart J. 2013;166:976–82.
40. Gali WL, Sarabanda AV, Baggio JM, Ferreira LG, Gomes GG, Marin-Neto JA, et al. Implantable cardioverter-defibrillators for treatment of sustained ventricular arrhythmias in patients with Chagas' heart disease: comparison with a control group treated with amiodarone alone. Europace. 2014;16:674–80.
41. Cardinalli-Neto A, Bestetti RB, Cordeiro JA, Rodrigues VC. Predictors of all-cause mortality for patients with chronic Chagas's heart disease receiving implantable cardioverter defibrillator therapy. J Cardiovasc Electrophysiol. 2007;18:1236–40.
42. Poole JE, Johnson GW, Hellkamp AS, Anderson J, Callans DJ, Raitt MH, et al. Prognostic importance of defibrillator shocks in patients with heart failure. N Engl J Med. 2008;359:1009–17.
43. Strauss DG, Cardoso S, Lima JA, Rochitte CE, Wu KC. ECG scar quantification correlates with cardiac magnetic resonance scar size and prognostic factors in Chagas' disease. Heart. 2011;97:357–61.
44. Tassi EM, Continentino MA, Nascimento EM, Pereira BB, Pedrosa RC. Relationship between fibrosis and ventricular arrhythmias in Chagas heart disease without ventricular dysfunction. Arq Bras Cardiol. 2014;102:456–64.
45. de Souza AC, Salles G, Hasslocher-Moreno AM, de Sousa AS, Alvarenga Americano do Brasil PE, Saraiva RM, et al. Development of a risk score to predict sudden death in patients with Chaga's heart disease. Int J Cardiol. 2015;187:700–4.

46. Nunes MCP, Badano LP, Marin-Neto JA, Edvardsen T, Fernandez-Golfin C, Bucciarelli-Ducci C, et al. Multimodality imaging evaluation of Chagas disease: an expert consensus of Brazilian Cardiovascular Imaging Department (DIC) and the European Association of Cardiovascular Imaging (EACVI). Eur Heart J Cardiovasc Imaging. 2018;19:459–60.
47. Ribeiro AL, Moraes RS, Ribeiro JP, Ferlin EL, Torres RM, Oliveira E, et al. Parasympathetic dysautonomia precedes left ventricular systolic dysfunction in Chagas disease. Am Heart J. 2001;141:260–5.
48. Santangeli P, Muser D, Maeda S, Filtz A, Zado ES, Frankel DS, et al. Comparative effectiveness of antiarrhythmic drugs and catheter ablation for the prevention of recurrent ventricular tachycardia in patients with implantable cardioverter-defibrillators: a systematic review and meta-analysis of randomized controlled trials. Heart Rhythm. 2016;13:1552–9.
49. De Paola AA, Horowitz LN, Miyamoto MH, Pinheiro R, Ferreira DF, Terzian AB, et al. Angiographic and electrophysiologic substrates of ventricular tachycardia in chronic Chagasic myocarditis. Am J Cardiol. 1990;65:360–3.
50. Sarabanda AV, Sosa E, Simoes MV, Figueiredo GL, Pintya AO, Marin-Neto JA. Ventricular tachycardia in Chagas' disease: a comparison of clinical, angiographic, electrophysiologic and myocardial perfusion disturbances between patients presenting with either sustained or nonsustained forms. Int J Cardiol. 2005;102:9–19.
51. Sosa E, Scanavacca M, D'Avila A, Bellotti G, Pilleggi F. Radiofrequency catheter ablation of ventricular tachycardia guided by nonsurgical epicardial mapping in chronic Chagasic heart disease. Pacing Clin Electrophysiol. 1999;22:128–30.
52. Rochitte CE, Oliveira PF, Andrade JM, Ianni BM, Parga JR, Avila LF, et al. Myocardial delayed enhancement by magnetic resonance imaging in patients with Chagas' disease: a marker of disease severity. J Am Coll Cardiol. 2005;46:1553–8.
53. Gomes VA, Alves GF, Hadlich M, Azevedo CF, Pereira IM, Santos CR, et al. Analysis of regional left ventricular strain in patients with Chagas disease and normal left ventricular systolic function. J Am Soc Echocardiogr. 2016;29:679–88.
54. Sosa E, Scanavacca M, D'Avila A, Pilleggi F. A new technique to perform epicardial mapping in the electrophysiology laboratory. J Cardiovasc Electrophysiol. 1996;7:531–6.
55. Aliot EM, Stevenson WG, Almendral-Garrote JM, Bogun F, Calkins CH, Delacretaz E, et al. EHRA/HRS expert consensus on catheter ablation of ventricular arrhythmias: developed in a partnership with the European heart rhythm association (EHRA), a registered branch of the European Society of Cardiology (ESC), and the Heart Rhythm Society (HRS); in collaboration with the American College of Cardiology (ACC) and the American Heart Association (AHA). Heart Rhythm. 2009;6:886–933.
56. Vaseghi M, Gima J, Kanaan C, Ajijola OA, Marmureanu A, Mahajan A, et al. Cardiac sympathetic denervation in patients with refractory ventricular arrhythmias or electrical storm: intermediate and long-term follow-up. Heart Rhythm. 2014;11:360–6.
57. Cardoso RN, Macedo FY, Garcia MN, Garcia DC, Benjo AM, Aguilar D, et al. Chagas cardiomyopathy is associated with higher incidence of stroke: a meta-analysis of observational studies. J Card Fail. 2014;20:931–8.
58. Samuel J, Oliveira M, Correa De Araujo RR, Navarro MA, Muccillo G. Cardiac thrombosis and thromboembolism in chronic Chagas' heart disease. Am J Cardiol. 1983;52: 147–51.
59. Aras R, da Matta JA, Mota G, Gomes I, Melo A. Cerebral infarction in autopsies of chagasic patients with heart failure. Arq Bras Cardiol. 2003;81:414–3.
60. Nunes MC, Barbosa MM, Ribeiro AL, Colosimo EA, Rocha MO. Left atrial volume provides independent prognostic value in patients with Chagas cardiomyopathy. J Am Soc Echocardiogr. 2009;22:82–8.
61. Nunes MC, Kreuser LJ, Ribeiro AL, Sousa GR, Costa HS, Botoni FA, et al. Prevalence and risk factors of embolic cerebrovascular events associated with Chagas heart disease. Glob Heart. 2015;10:151–7.
62. Sousa AS, Xavier SS, Freitas GR, Hasslocher-Moreno A. Prevention strategies of cardioembolic ischemic stroke in Chagas' disease. Arq Bras Cardiol. 2008;91:306–10.

63. Rocha MO, Nunes MC, Ribeiro AL. Morbidity and prognostic factors in chronic chagasic cardiopathy. Mem Inst Oswaldo Cruz. 2009;104(Suppl 1):159–66.
64. Carod-Artal FJ, Vargas AP, Horan TA, Nunes LG. Chagasic cardiomyopathy is independently associated with ischemic stroke in Chagas disease. Stroke. 2005;36:965–70.
65. Herrera RN, Diaz de Amaya EI, Perez Aguilar RC, Joo TC, Maranon R, Berman SG, et al. Inflammatory and prothrombotic activation with conserved endothelial function in patients with chronic, asymptomatic Chagas disease. Clin Appl Thromb Hemost. 2011;17:502–7.
66. Pinazo MJ, Posada EJ, Izquierdo L, Tassies D, Marques AF, de Lazzari E, et al. Altered hypercoagulability factors in patients with chronic chagas disease. Potential biomarkers of therapeutic response. PLoS Negl Trop Dis. 2016;10:e0004269.
67. Nisimura LM, Estato V, de Souza EM, Reis PA, Lessa MA, Castro-Faria-Neto HC, et al. Acute Chagas disease induces cerebral microvasculopathy in mice. PLoS Negl Trop Dis. 2014;8:e2998.
68. Dias Junior JO, da Costa Rocha MO, de Souza AC, Kreuser LJ, de Souza Dias LA, Tan TC, et al. Assessment of the source of ischemic cerebrovascular events in patients with Chagas disease. Int J Cardiol. 2014;176:1352–4.
69. Lima-Costa MF, Castro-Costa E, Uchoa E, Firmo J, Ribeiro AL, Ferri CP, et al. A population-based study of the association between Trypanosoma cruzi infection and cognitive impairment in old age (the Bambui study). Neuroepidemiology. 2009;32:122–8.
70. Mangone CA, Sica RE, Pereyra S, Genovese O, Segura E, Riarte A, et al. Cognitive impairment in human chronic Chagas' disease. Arq Neuropsiquiatr. 1994;52:200–3.
71. Paixao LC, Ribeiro AL, Valacio RA, Teixeira AL. Chagas disease: independent risk factor for stroke. Stroke. 2009;40:3691–4.
72. Carod-Artal FJ, Vargas AP, Falcao T. Stroke in asymptomatic Trypanosoma cruzi-infected patients. Cerebrovasc Dis. 2011;31:24–8.
73. Oliveira-Filho J, Viana LC, Vieira-de-Melo RM, Faical F, Torreao JA, Villar FA, et al. Chagas disease is an independent risk factor for stroke: baseline characteristics of a Chagas disease cohort. Stroke. 2005;36:2015–7.
74. Nunes MC, Barbosa MM, Rocha MO. Peculiar aspects of cardiogenic embolism in patients with Chagas' cardiomyopathy: a transthoracic and transesophageal echocardiographic study. J Am Soc Echocardiogr. 2005;18:761–7.
75. Cardoso R, Garcia D, Fernandes G, He LI, Lichtenberger P, Viles-Gonzalez J, et al. The prevalence of atrial fibrillation and conduction abnormalities in Chagas' disease: a meta-analysis. J Cardiovasc Electrophysiol. 2016;27:161–9.
76. Lee-Felker SA, Thomas M, Felker ER, Traina M, Salih M, Hernandez S, et al. Value of cardiac MRI for evaluation of chronic Chagas disease cardiomyopathy. Clin Radiol. 2016;71:618–7.
77. Trabuco CC, Pereira de Jesus PA, Bacellar AS, Oliveira-Filho J. Successful thrombolysis in cardioembolic stroke from Chagas disease. Neurology. 2005;64:170–1.
78. Cougo-Pinto PT, Dos Santos BL, Dias FA, Camilo MR, Alessio-Alves FF, Barreira CM, et al. Safety of IV thrombolysis in acute ischemic stroke related to Chagas disease. Neurology. 2013;81:1773–5.
79. Rassi A Jr, Rassi A, Rassi SG. Predictors of mortality in chronic Chagas disease: a systematic review of observational studies. Circulation. 2007;115:1101–8.
80. Nunes MC, Rocha MO, Ribeiro AL, Colosimo EA, Rezende RA, Carmo GA, et al. Right ventricular dysfunction is an independent predictor of survival in patients with dilated chronic Chagas' cardiomyopathy. Int J Cardiol. 2008;127:372–9.
81. Dutra WO, Menezes CA, Magalhaes LM, Gollob KJ. Immunoregulatory networks in human Chagas disease. Parasite Immunol. 2014;36:377–87.
82. Saraiva RM, Waghabi MC, Vilela MF, Madeira FS, da Silva GM, Xavier SS, et al. Predictive value of transforming growth factor-beta1in Chagas disease: towards a biomarker surrogate of clinical outcome. Trans R Soc Trop Med Hyg. 2013;107:518–25.
83. Heringer-Walther S, Moreira MC, Wessel N, Saliba JL, Silvia-Barra J, Pena JL, et al. Brain natriuretic peptide predicts survival in Chagas' disease more effectively than atrial natriuretic peptide. Heart. 2005;91:385–7.

84. Lima-Costa MF, Cesar CC, Peixoto SV, Ribeiro AL. Plasma B-type natriuretic peptide as a predictor of mortality in community-dwelling older adults with Chagas disease: 10-year follow-up of the Bambui cohort study of aging. Am J Epidemiol. 2010;172:190–6.
85. Sherbuk JE, Okamoto EE, Marks MA, Fortuny E, Clark EH, Galdos-Cardenas G, et al. Biomarkers and mortality in severe Chagas cardiomyopathy. Glob Heart. 2015;10:173–80.
86. Rivero CV, Vanrell C, Cueto J, Romano P. Carvedilol (Crv) inhibits the*Trypanosoma cruzi* autophagic pathway affecting parasite replication and survival. ID Week Conference 2016. Abstract.
87. Horta AL, Leite AL, Paula CG, Figueiredo VP, Talvani A. Potential role of Carvedilol in the cardiac immune response induced by experimental infection with Trypanosoma cruzi. Biomed Res Int. 2017;2017:9205062.
88. Sperandio da Silva GM, Chambela MC, Sousa AS, Sangenis LH, Xavier SS, Costa AR, et al. Impact of pharmaceutical care on the quality of life of patients with Chagas disease and heart failure: randomized clinical trial. Trials. 2012;13:244.
89. Mendes FS, Sousa AS, Souza FC, Pinto VL, Silva PS, Saraiva RM, et al. Effect of physical exercise training in patients with Chagas heart disease: study protocol for a randomized controlled trial (PEACH study). Trials. 2016;17:433.
90. Roberti RR, Martinez EE, Andrade JL, Araujo VL, Brito FS, Portugal OP, et al. Chagas cardiomyopathy and captopril. Eur Heart J. 1992;13:966–70.
91. Botoni FA, Poole-Wilson PA, Ribeiro AL, Okonko DO, Oliveira BM, Pinto AS, et al. A randomized trial of carvedilol after renin-angiotensin system inhibition in chronic Chagas cardiomyopathy. Am Heart J. 2007;153:544–8.
92. Bestetti RB, Otaviano AP, Cardinalli-Neto A, da Rocha BF, Theodoropoulos TA, Cordeiro JA. Effects of B-blockers on outcome of patients with Chagas' cardiomyopathy with chronic heart failure. Int J Cardiol. 2011;151:205–8.
93. Marti-Carvajal AJ, Kwong JS. Pharmacological interventions for treating heart failure in patients with Chagas cardiomyopathy. Cochrane Database Syst Rev. 2016;7:CD009077.
94. Bacal F, Silva CP, Pires PV, Mangini S, Fiorelli AI, Stolf NG, et al. Transplantation for Chagas' disease: an overview of immunosuppression and reactivation in the last two decades. Clin Transpl. 2010;24:E29–34.
95. Bestetti RB, Theodoropoulos TA. A systematic review of studies on heart transplantation for patients with end-stage Chagas' heart disease. J Card Fail. 2009;15:249–55.
96. Diez M, Favaloro L, Bertolotti A, Burgos JM, Vigliano C, Lastra MP, et al. Usefulness of PCR strategies for early diagnosis of Chagas' disease reactivation and treatment follow-up in heart transplantation. Am J Transplant. 2007;7:1633–40.
97. Vizzoni AG, Varela MC, Sangenis LHC, Hasslocher-Moreno AM, do Brasil PEAA, Saraiva RM. Ageing with Chagas disease: an overview of an urban Brazilian cohort in Rio de Janeiro. Parasit Vectors. 2018;11:354.
98. Simoes MV, Ayres EM, Santos JL, Schmidt A, Pintya AO, Maciel BC, et al. Detection of myocardial ischemia in chronic Chagas disease patients with atypic precordial pain by exercise and Holter tests. Arq Bras Cardiol. 1993;60:315–9.
99. Peix A, Garcia R, Sanchez J, Cabrera LO, Padron K, Vedia O, et al. Myocardial perfusion imaging and cardiac involvement in the indeterminate phase of Chagas disease. Arq Bras Cardiol. 2013;100:114–7.
100. Abuhid IM, Pedroso ER, Rezende NA. Scintigraphy for the detection of myocardial damage in the indeterminate form of Chagas disease. Arq Bras Cardiol. 2010;95:30–4.
101. Marin-Neto JA, Marzullo P, Marcassa C, Gallo JL, Maciel BC, Bellina CR, et al. Myocardial perfusion abnormalities in chronic Chagas' disease as detected by thallium-201 scintigraphy. Am J Cardiol. 1992;69:780–4.
102. Nascimento CAS, Pacheco NP, Alves GF, Madeira FS, Costa AR, Holanda MT, et al. Total left atrial emptying fraction determined by real time 3-dimensional echocardiography is an independent predictor of new-onset atrial fibrillation in patients with Chagas disease. Heart Failure 2017 meeting. Eur J Heart Fail. 2017;19:S21. Abstract.

103. Nahum J, Bensaid A, Dussault C, Macron L, Clemence D, Bouhemad B, et al. Impact of longitudinal myocardial deformation on the prognosis of chronic heart failure patients. Circ Cardiovasc Imaging. 2010;3:249–56.
104. Cho GY, Marwick TH, Kim HS, Kim MK, Hong KS, Oh DJ. Global 2-dimensional strain as a new prognosticator in patients with heart failure. J Am Coll Cardiol. 2009;54:618–24.

Management of Chronic Digestive Involvement in Patients with Chagas Disease in Endemic and Non-endemic Countries: Challenges and Limitations

10

Joffre Rezende Filho and Enio Chaves de Oliveira

10.1 Introduction

Patients with Chagas disease (CD) with digestive involvement commonly present with dysphagia or constipation [1, 2] or otherwise do not have any digestive symptoms but have a positive serology for CD and are referred for further evaluation. The management of those patients will depend upon the clinical manifestations and will be discussed separately.

10.2 Chagasic Oesophagopathy

The main symptom of chagasic oesophagopathy is dysphagia. Therefore, this symptom should be further characterized: duration, intensity, types of meals, slow ingestion. There is not linear correlation between intensity of dysphagia, duration of the disease and degree of oesophageal dilation. Some patients with huge megaoesophagus do not complain of dysphagia, in contrast with others with severe dysphagia and less dilation [1, 3]. Patients may present mild dysphagia for a long time without

J. Rezende Filho (✉)
Division of Gastroenterology, Hospital das Clinicas, Universidade Federal de Goias, Goiânia, Goiás, Brazil

Departamento de Clinica Medica, Faculdade de Medicina, Universidade Federal de Goiás, Goiânia, Goiás, Brazil

Instituto de Gastroenterologia de Goiania, Goiânia, Goiás, Brazil
e.mail: joffrerf@terra.com.br

E. Chaves de Oliveira
Chagas Disease Study Center, Hospital das Clinicas, Universidade Federal de Goias, Goiânia, Goiás, Brazil

Departamento de Cirurgia, Faculdade de Medicina, Universidade Federal de Goiás, Goiânia, Goiás, Brazil

M.-J. Pinazo Delgado, J. Gascón (eds.), *Chagas Disease*,
https://doi.org/10.1007/978-3-030-44054-1_10

getting worse and in others it may progress in short period of time [1]. Regurgitation is another oesophageal symptom that needs further consideration. In general, persistent supine passive regurgitation correlates with more advanced megaoesophagus and is associated with malnutrition [1, 4]. It should be emphasized that chagasic patients may present with other oesophageal disorders causing dysphagia. Therefore, the differential diagnosis should be made with oesophageal/gastric cancer, gastro-oesophageal reflux disease, eosinophilic oesophagitis and others [1].

Other digestive symptoms such as chest pain, dyspepsia and constipation should also be explored as well as cardiac symptoms such as dyspnoea and palpitations [1]. The presence of chronic chagasic cardiopathy and other associated clinical conditions such as hypertension, diabetes and chronic pulmonary obstructive disease should be taken in consideration as they can influence the treatment [1, 5, 6].

Radiological examination is the first test to be done in order to make the diagnosis of such patients. The barium oesophagogram with fluoroscopy shows not only the degree of oesophageal dilation but also functional alterations such as abnormal or absence of peristaltic contractions and the rate of oesophageal emptying [7, 8]. A timed barium oesophagram can also be made [3]. In chagasic oesophagopathy, radiological findings are oesophageal diameter increase, slow emptying, affiliated distal portion ("bird-beak" appearance), variable oesophageal dilation and/or elongation, and decrease of gastric air-bubble [3, 7]. The degree of oesophageal dilation does not always correspond to the duration of the disease [1, 3, 7]. There are cases that stay for a long time with the same degree and others that quickly progress to great dilation [1, 7].

For therapeutic purposes, it is appropriate to divide the chagasic megaoesophagus into different stages or groups. The Rezende's classification into four groups is widely used [5–8]. In Groups III and IV in which great oesophageal dilation is found, no further tests need to be performed to diagnose the oesophageal involvement in Chagas disease. In other hand, in those cases in which there is mild (group II) or no oesophageal dilation (group I), performing oesophageal manometry could confirm the diagnosis.

Conventional or high-resolution oesophageal manometry demonstrates alterations in the oesophageal peristalsis and in the lower oesophageal sphincter relaxation (achalasia) [9, 10]. The manometric features in chagasic megaoesophagus also have influence in choosing the most appropriate treatment [11].

Upper digestive endoscopy should also be performed in all patients with megaoesophagus in order to evaluate the oesophageal mucosa, to exclude other possible diagnosis such as pseudoachalasia and also to evaluate gastric and duodenal possible abnormalities [12].

No treatment of chagasic megaoesophagus restores normal oesophageal physiology. The goal of the treatment is mainly to relieve dysphagia and to facilitate oesophageal emptying. The treatment of chagasic megaoesophagus can be done by means of dietary measures, clinical treatment, endoscopic balloon dilation, surgery, and per oral endoscopic myotomy (POEM), as will be discussed below. The

Table 10.1 Therapeutic measures available for treatment of chagasic megaoesophagus

Treatment of chagasic megaoesophagus
Dietary modifications
Medications: nitrates, nifedipine
Endoscopic balloon dilation
Botulinum toxin intramural injection
Laparoscopy cardiomyotomy
Per oral endoscopic myotomy
Cardioplasty
Oesophagectomy

treatment will depend upon the stage of the disease, the patient profile and the availability of the method (Table 10.1) [1, 5, 6].

The patient should eat slowly, in a quiet place, chew the food as thoroughly as possible, swallow small bolus, adjusting the food consistency to the dysphagia intensity, giving preference to softer foods. Irritant foods to the oesophageal mucosa and very cold foods, that aggravate the dysphagia, should be avoided. Drinking more water during the meals may help oesophageal emptying. Ingestion of food before going to bed at night could lead to supine regurgitation and possible aspiration, and therefore should be avoided. Those measures should be recommended even after other treatment has been done [1, 5].

The use of medications that relax the lower oesophageal sphincter pressure, isosorbide dinitrate 2.5–5 mg, sublingual, 15 min before meals or nifedipine 10 mg sublingual 30 min before meals may lead to some dysphagia relief and accelerate oesophageal emptying in patients with chagasic megaoesophagus [13, 14]. Those medications can be prescribed as a clinical alternative option in cases that is not possible to perform balloon dilation or surgery. However, one has to consider that they have short duration of drug action, so it is necessary to use in every meal. In addition, they are associated with relevant side effects such as headache and hypotension. Patients with cardiac involvement should not take those medications.

Botulinum toxin is a potent inhibitor of acetylcholine release from nerve endings. When applied in oesophageal wall in cases of achalasia, it counteracts the unopposed lower esophageal sphincter (LES) contraction that is mediated by cholinergic nerves, thereby lowering LES pressure [15]. In one study evaluating patients with chagasic megaoesophagus, stage II and III, the intramural application, through endoscopy, of 100 unit of botulinum toxin (BOTOX) in four quadrants of the LES was associated with dysphagia relief in some (14/21) patients [15]. The application of botulinum toxin in the treatment of chagasic megaoesophagus can be considered a palliative and an alternative option in selected non-advanced cases. It has the advantages of being less invasive and that can be repeated many times [1, 7].

Forced balloon dilation of the cardia has been performed for a long time in the treatment of chagasic megaoesophagus [1, 16]. The principle of the procedure is to weaken the lower esophageal sphincter by tearing its muscle fibers by generating radial force. Nowadays, pneumatic dilation (PD) has been done as an endoscope-guided procedure without fluoroscopic control, under sedation, using a low-compliance balloon such as Rigiflex dilator [17]. It has the advantage of determining mucosal injury during the dilation process, so that a repeat endoscopy is not needed to assess the mucosal tearing. In the experience at the Hospital das Clinicas da UFG, better outcomes are observed in patients with mega Groups II and III. In cases with most advanced dilation with dolicomegaoesophagus (group IV), the procedure does not provide benefit and is not recommended [1, 18]. Patients without oesophageal dilation (group I) with mild or intermittent dysphagia may not need balloon dilation, but can be performed if dysphagia is severe. Alternatively, it can be done oesophageal bouginage with Maloney bougie [1, 19]. Some patients will need more than one dilation session in order to obtain improvement of symptoms. The main complication of this procedure is oesophageal perforation that has been reported in less than 2% [1, 19].

The surgical treatment of chagasic megaoesophagus has evolved and is now being considered the gold standard procedure [5]. The surgical treatment provides better outcomes as compared to balloon dilation in long-term follow-up. The surgical procedure may vary according to the stage of the disease. Laparoscopic cardiomyotomy associated with a partial antireflux valve (Heller–Pinotti operation) [20, 21] is the most performed operation in the treatment of cases of chagasic megaoesophagus groups I, II and some patients of group III [1, 5, 11, 20, 21]. This operation decreases oesophageal sphincter pressure, facilitates oesophageal emptying and improves dysphagia in most patients [21]. However, after some time, patients may complain of dysphagia again, needing another procedure.

In more advanced cases, considered as those showing atonic and massive dilation in which no contractions is seen in oesophageal manometry [22] or in cases with dolicomegaoesophagus (group IV), oesophagectomy is indicated as primary therapy [22]. The oesophageal resection is also indicated as rescue therapy for those with persistent achalasia after failure of previous less invasive treatment options. Considering the risk of oesophagectomy, other procedures have been suggested including: cardioplasty plus truncal vagotomy and Roux-en-Y partial gastrectomy (Serra-Doria operation) [23], and less commonly Thal–Hatafuku operation [24].

In the last decade, per oral endoscopic myotomy (POEM) was introduced for treatment of idiopathic achalasia [25]. It has progressively been performed in the treatment of chagasic megaoesophagus as well [26]. POEM is a feasible procedure, performed through a submucosal tunnel, which allows myotomy of the distal oesophagus and of the lower esophageal sphincter [25]. This procedure requires a great endoscopic expertise and is presently not widely available. The procedure is similar to a laparoscopic Heller myotomy without a fundoplication. Short-term studies have shown that POEM is very effective in relieving dysphagia and regurgitation [25]. It can be done even after balloon dilation or Heller myotomy + fundoplication failure [27]. As compared to laparoscopic myotomy and balloon dilation, results in short-term follow-up seem to be similar [27]. However, POEM is associated with a higher incidence of gastro-oesophageal reflux disease compared to alternative therapies (Heller myotomy with fundoplication or PD) [27]. In the near future, POEM probably will be more widely available and possibly will be considered the first option for treatment [27].

Table 10.2 Treatment of chagasic megaoesophagus according to radiological classification [7, 8]

Radiological classification (stage)	Preferred Treatment of megaoesophagus
Group I (no dilation)	1. Dietary orientation 2. Oesophageal bouginage 3. Endoscopic balloon dilation
Group II–III	1. Laparoscopic Heller cardiomyotomy (LHC) + fundoplication 2. Endoscopic balloon dilation 3. POEM
Group IV (dolicomegaoesophagus)	1. Oesophagectomy 2. Serra-Doria operation (cardioplasty + hemigastrectomy) 3. POEM
LHC failure	1. Oesophagectomy 2. Serra-Doria operation (cardioplasty + hemigastrectomy) 3. POEM

The great challenge in treating chagasic patients with megaoesophagus is that some of those patients are presenting at old ages, with advanced or with failure of previous treatments, having comorbidities, malnutrition, and sometimes with associated chronic chagasic cardiopathy, limiting the therapeutic options. At the Hospital das Clinicas da UFG, in the last decade, the median age of patients with chagasic megaoesophagus was 55 years old [28]. Our approach is to individualize the treatment, based upon the clinical conditions and the radiological classification as discussed above [1]. The treatment options presently applied at our hospital is shown in Table 10.2.

10.3 Chagasic Colopathy

The main symptom of chagasic colopathy is constipation [1] Therefore, this symptom should be further characterized: duration, mode of evolution, presence of dyssynergic defecation, previous use of laxatives, history of faecaloma, excessive straining. Abdominal pain, sensation of bloating or abdominal distention, meteorism and flatulence may also need to be explored. In chagasic megacolon, constipation usually is of slow and insidious onset and of progressive evolution, becoming persistent with time, leading to frequent use and abuse of laxatives. On the other hand, about 25% of patients with chagasic megacolon does not complain of constipation, reporting normal bowel movement frequency [29].

Any patient with megaoesophagus or chronic chagasic cardiopathy or coming from an endemic area for Chagas disease presenting the above symptoms should be considered as possibly a case of chagasic colopathy and should be investigated. Physical examination may reveal a dilated sigmoid loop or a presence of a faecaloma. Proctological examination should be performed routinely. Rectal palpation permits the detection of megarectum as well as of faecaloma.

Radiological examination is next step towards confirming the diagnosis. A faecaloma can be observed in a simple plain abdominal X-ray. In order to demonstrate colon dilation, barium enema can be performed using a simplified method without previous preparation. Rectal and sigmoid enlargement can be easily demonstrated [30].

The differential diagnosis of chagasic megacolon should be made with secondary dilatations to organic obstacles such as stenosis, tumours, adherences, extrinsic compression and rectosigmoid endometriosis [1, 31]. In case of a chagasic patient with constipation but without colon dilatation, the differential diagnosis between functional constipation occurring in a chagasic patient and constipation secondary to the presence of an early stage of chagasic colopathy without dilation is still hard to define [32, 33].

The treatment of the constipation associated with chagasic colopathy can be done with dietary and habit modifications, use of laxatives and surgical treatment [1, 5, 31]. In those patients presenting with colon enlargement but with normal bowel movements, no further treatment is necessary. Patients with two/three bowel movements/week without presence of fecaloma should be oriented to ingest more liquids, fruits and vegetables [5, 31]. However, a high insoluble fibre diet is controversial, considering that such intervention may induce faecaloma formation and aggravate symptoms [1, 5, 31]. Abundant hydric ingestion should be emphasized, at least 2 L/day [5, 31]. The use of laxatives is recommended to patients not responding properly to dietary interventions. The osmotic laxatives—polyethylene glycol, 20% mannitol, lactulose—are the most appropriate. Lubricant laxative as mineral oil may also be useful [5]. However, with progression of the disease, many patients will need more and progressive doses of laxatives to obtain a bowel movement, finally turning to use of water cleansing enemas [1, 31].

The treatment of faecaloma depends on its location and faecal consistency. Those localized in the rectum and with not so hard faeces or those localized higher in the rectum and sigmoid can be removed by using water and glycerine enemas or rectal slow infusion of saline, and those with hard stools can be firstly manually removed, under anaesthesia, followed by repeated glycerine enemas. After the removal, those enemas can be repeated two or three times/week [1, 5, 31].

The surgical treatment of chagasic megacolon is indicated for those cases that present with great sigmoid dilation, prolonged faeces retention, recurrence of faecalomas and repeated sigmoid volvulus. The goal of the surgical treatment is to improve or normalize bowel movements. The most utilized surgical technique is anterior rectosigmoidectomy with immediate posterior colorectal end-to-side stapled anastomosis [34, 35] and Duhamel–Haddad operation [36, 37]. Most of the patients improve the constipation after surgery. More recently, both operations are becoming progressively performed by laparoscopic approach [37, 38]. Emergency operations may need to be done in cases of sigmoid volvulus that does not resolve with other decompressive measures such as endoscopy, in cases of ulceration, obstruction or colonic perforation [5, 39]. In those situations, Hartman's operation and sigmoid fixation or sigmoidostomy are most frequent techniques utilized [40, 41].

10.4 Conclusion

Patients with Chagas disease with digestive involvement commonly present with dysphagia or constipation need therapeutic intervention. The treatment of chagasic megaoesophagus can be done by means of dietary measures, clinical treatment, endoscopic balloon dilation, surgery and per oral endoscopic myotomy. The treatment will depend upon the stage of the disease, the patient profile and the availability of the method. The treatment of the constipation associated with Chagas colopathy can be done with dietary and habits modifications, use of laxatives and surgical treatment. The surgical treatment of chagasic megacolon is indicated for those cases that present with great sigmoid dilation, prolonged faeces retention, recurrence of faecalomas and repeated sigmoid volvulus.

References

1. Rezende JM, Moreira H. Forma digestiva da Doença de Chagas. In: Castro LP, Coelho LGV, editors. Gastroenterologia. Rio de Janeiro: Medsi Ed; 2004. p. 325–92.
2. Oliveira RB, Troncon LEA, Dantas RO, Meneghelli UG. Gastrointestinal manifestations of Chagas disease. Am J Gastroenterol. 1998;93(6):884–9. https://doi.org/10.1111/j.1572-0241.1998.270_r.x.
3. Martins P, Ferreira CS, Cunha-Melo JR. Esophageal transit time in patients with chagasic megaesophagus: lack of linear correlation between dysphagia and grade of dilatation. Medicine (Baltimore). 2018;97(10):e0084. https://doi.org/10.1097/MD.0000000000010084.
4. Vaz MGM, Rezende JM, Ximenes CA, Luquetti AO. Correlação entre a sintomatologia e a evolução do megaesôfago. Rev Goiana Med. 1996;41:1–15.
5. Dias JCP, Ramos AN Jr, Gontijo ED, Luquetti A, Shikanai-Yasudama MA, Coura JR, et al. II Consenso Brasileiro em Doença de Chagas, 2015. Epidemiol Serv Saúde. 2016;25(Suppl): 7–86.
6. Pinazo MJ, Canas E, Elizalde JI, Garcia M, Gascon J, Gimeno F, et al. Diagnosis, management and treatment of chronic Chagas' gastrointestinal disease in areas where Trypanosoma cruzi infection is not endemic. Gastroenterol Hepatol. 2010;33(3):191–200. https://doi.org/10.1016/j.gastrohep.2009.07.009.
7. Rezende JM, Lauar KM, Oliveira AR. Aspectos clínicos e radiológicos da aperistalsis do esôfago. Rev Bras Gastroenterol. 1960;12:247–62.
8. Rezende JM. Classificação radiológica do megaesôfago. Rev Goiana Med. 1982;28:187.
9. Rezende Filho J. Manometria na esofagopatia chagásica. In: Nakano-Stefani MS, Faintuch J, Cecconelo I, editors. Megaesôfago Chagásico: Tratamento Clinico e Cirurgico. Goiânia: Editora UCG; 2006. p. 105–13.
10. Vicentine FP, Herbella FA, Allaix ME, Silva LC, Patti MG. High-resolution manometry classifications for idiopathic achalasia in patients with Chagas' disease esophagopathy. J Gastrointest Surg. 2014;18(2):221–4. https://doi.org/10.1007/s11605-013-2376-1.
11. Crema E, Cruvinel LA, Werneck AM, de Oliveira RM, Silva AA. Manometric and radiologic aspects of Chagas' megaesophagus: the importance to its surgical treatment. Rev Soc Bras Med Trop. 2003;36(6):665–9.
12. Rezende JM, Rosa H, Vaz MG, Andrade-Sá N, Porto JD, Neves Neto J, Ximenes JA. Endoscopy in megaesophagus. Prospective study of 600 cases. Arq Gastroenterol. 1985;22(2): 53–62.
13. Rezende Filho J, de Oliveira RB, Dantas RO, Iazigi N. The effect of isosorbide dinitrate on esophageal emptying in chagasic megaesophagus. Arq Gastroenterol. 1990;27(3):115–9.

14. Borges Migliavaca C, Stein C, Colpani V, René Pinto de Sousa Miguel S, Nascimento Cruz L, Oliveira Dantas R, Falavigna M. Isosorbide and nifedipine for Chagas' megaesophagus: a systematic review and meta-analysis. PLoS Negl Trop Dis. 2018;12(9):e0006836. https://doi.org/10.1371/journal.pntd.0006836.
15. Brant C, Moraes-Filho JP, Siqueira E, Nasi A, Libera E, Morais M, Rohr M, Macedo EP, Alonso G, Ferrari AP. Intrasphincteric botulinum toxin injection in the treatment of chagasic achalasia. Dis Esophagus. 2003;16(1):33–8. https://doi.org/10.1046/j.1442-2050.2003.00287.x.
16. Esper FE, Mineiro V, Santos EF, Moraes DM, Andrade NB. Dilatação da cárdia no tratamento da disfagia de pacientes com megaesôfago chagásico. Arq Gastroenterol. 1988;25:69–74.
17. Moura EGH, Maluf Filho F, Sakai P, Ishioka S. Dilatação pneumática da cárdia em portadores de megaesôfago chagásico. GED. 1991;10:83–6.
18. Caldeira LM, Rezende Filho J, Rezende JM, Ximenes JAA, Daher R. Endoscopic pneumatic dilation in Chagasic Achalasia: a clinical, radiological and manometric study with one year follow-up. Neurogastroenterol Motil. 2006;18(8):714.
19. Rezende JM. Dilatação de acalásia da cárdia. In Castro LP, Savassi-Rocha PR, Lima DCA, Tanure, JC, editors. Tópicos em Gastroenterologia. Diagnóstico e Tratamento. Rio de Janeiro: MEDSI, 1998. p. 25-34.
20. Pinotti HW, Rodrigues JJG, Ellenbogen G, Raia A. Nova técnica no tratamento cirúrgico do megaesôfago. Esofagocardiomiotomiaassociada com esofago-fundogastropexia. Rev Goiana Med. 1974;20:1–13.
21. Pinotti HW, Habr-Gama A, Cecconello I, Felix VN, Zilberstein B. The surgical treatment of megaesophagus and megacolon. Dig Dis. 1993;11(4-5):206–15. https://doi.org/10.1159/000171413.
22. Crema E, Ribeiro LB, Terra JA Jr, Silva AA. Laparoscopic transhiatal subtotal esophagectomy for the treatment of advanced megaesophagus. Ann Thorac Surg. 2005;80(4):1196–201. https://doi.org/10.1016/j.athoracsur.2004.10.059.
23. Ponciano H, Cecconello I, Alves L, Ferreira BD, Gama-Rodrigues J. Cardiaplasty and Roux-en-Y partial gastrectomy (Serra-Dória procedure) for reoperation of achalasia. Arq Gastroenterol. 2004;41(3):155–61.. S0004-28032004000300004 [pii]
24. Alves AP, Oliveira PG, Oliveira JM, Mesquita DM, Santos JH. Long-term results of the modified Thal procedure in patients with chagasic megaesophagus. World J Surg. 2014;38:1425–30. https://doi.org/10.1007/s00268-013-2445-3.
25. Inoue H, Tianle KM, Ikeda H, Hosoya T, Onimaru M, Yoshida A, et al. Peroral endoscopic myotomy for esophageal achalasia: technique, indication, and outcomes. Thorac Surg Clin. 2011;21(4):519–25.
26. Moura ETH, Farias GFA, Coutinho LMA, Delgado AA, Nasi A, Queiroz NSF, et al. Ensaio clínico randomizado comparando a miotomia endoscópica peroral (POEM) versus miotomia laparoscópica no tratamento da acalasia. Gastroenterol Endosc Dig. 2018;38(Suppl 1):420.
27. Zaninotto G, Bennett C, Boeckxstaens G, Costantini M, Ferguson MK, Pandolfino JE, Patti MG, et al. The 2018 ISDE achalasia guidelines. DisEsophagus. 2018;31(9) https://doi.org/10.1093/dote/doy071.
28. Souza DH, Vaz MG, Fonseca CR, Luquetti A, Rezende Filho J, Oliveira EC. Current epidemiological profile of Chagasic megaesophagus in Central Brazil. Rev Soc Bras Med Trop. 2013;46(3):316–21. https://doi.org/10.1590/0037-8682-0065-2013.
29. Rezende JM, Luquetti AO. Chagasic megavisceras. In: Chagas' disease and the nervous system. Scientific publication no. 547. Washington, DC: Pan American Health Organization; 1994. p. 149–71.
30. Ximenes CA, Rezende JM, Moreira H, Glória M. Técnica simplificada para o diagnóstico radiológico do megacólon chagásico. Rev Soc Bras Med Trop. 1984;17:23.
31. Moreira H, Rezende JM, Sebba F, Azevedo IF, Leite ACA, Soares EP. Megacolo Chagasico. Ver Bras Colo-Proct. 1983;3(4):152–62.
32. Oliveira EC, Menezes JG, Cardoso VK, Luquetti AO, Gabriel Neto S, Garcia SB. The relationship between megacolon and constipation in Chagas' disease. Neurogastroenterol Motil. 2009;21(Suppl 1):5. https://doi.org/10.1111/j.1365-2982.2009.01343_1.x.

33. Bafutto M, Luquetti AO, Gabriel Neto S, Penhavel FAS, Oliveira EC. Constipation is related to small bowel disturbance rather than colonic enlargement in acquired chagasic megacolon. Gastroenterol Res. 2017;10(4):213–7. https://doi.org/10.14740/gr872w.
34. Nahas SC, Habr-Gama A, Nahas CS, Araujo SE, Marques CF, Sobrado CW, Bocchini SF, Kiss DR. Surgical treatment of Chagasic megacolon by abdominal rectosigmoidectomy with immediate posterior end-to-side stapling (Habr-Gama technique). Dis Colon Rectum. 2006;49(9):1371–8. https://doi.org/10.1007/s10350-006-0639-6.
35. Nahas SC, Pinto RA, Dias AR, Nahas CS, Araújo SE, Marques CF, Cecconello I. Long-term follow up of abdominal rectosigmoidectomy with posterior end-to-side stapled anastomosis for Chagas megacolon. Color Dis. 2011;13(3):317–22. https://doi.org/10.1111/j.1463-1318.2009.02128.x.
36. Gama RC, Costa JH, Azevedo IF. Tratamento cirúrgico do megacolo chagásico pela técnica de Duhamel-Haddad. Experiência no Hospital Geral de Goiânia. Análise de 204 casos. Rev Bras Colo-Proct. 1986;6:84–8.
37. Reis-Neto JA, Pedroso MA, Lupinacci RA, Reis Júnior JA, Ciquini SA, Lupinacci RM, et al. Megacolo adquirido: perspectivas fisiopatológicas para o tratamento laparoscópico. Rev Bras Colo-proctol. 2004;24(1):49–62.
38. Araujo SE, Bertoncini AB, Nahas SC, Cecconello I. Minimally invasive approach to chagasic megacolon: laparoscopic rectosigmoidectomy with posterior end-to-sidelow colorectal anastomosis. Surg Laparosc Endosc Percutan Tech. 2014;24(3):207–12. https://doi.org/10.1097/SLE.0000000000000002.
39. Garcia RL, Matos BM, Féres O, Rocha JJ. Surgical treatment of Chagas megacolon. Critical analysis of outcome in operative methods. Acta Cir Bras. 2008;23(Suppl 1):83–92.
40. Gama AH, Haddad J, Simonsen O, Warde P, Manzione A, Hyppölito da Silva J, Ioshimoto M, Cutait D, Raia A. Volvulus of the sigmoid colon in Brazil: a report of 230 cases. Dis Colon Rectum. 1976;19(4):314–20.
41. Moreira H. Tratamento cirúrgico do vólvulo da sigmóide no megacolo chagásico. Nova técnica cirúrgica. Rev Goiana Med. 1979;25:73–6.

Challenges in Response to Treatment Evaluation and Progression of the Disease

11

M. Carmen Thomas, Adriana Egui, Elena Pérez-Antón, Inmaculada Gómez, and Manuel C. López

11.1 Parasite–Host Interactions: What We Know Up to Now

The protozoan parasite *Trypanosoma cruzi* is the etiological agent of Chagas disease (CD), which is a chronic infection that affects seven million people worldwide. Migration has globally distributed this sickness, which is now endemic from the southern USA to South America. CD is mainly transmitted by the bite of infected triatomine vectors, with congenital transmission, blood transfusion, organ transplantation, and ingestion of contaminated food/juices as other frequent routes of transmission. Chagas disease pathology has two phases, namely, an acute and a chronic phase. The acute phase occurs after infection and is characterized by high parasitemia in blood. The acute phase is generally asymptomatic, leading to a strong immune response that partially controls the infection. Patients evolve to chronic phase that can last decades. Approximately 30% of the patients develop a chronic phase with clinical disease manifestations such as cardiac and gastrointestinal tissue damage as a consequence of parasitism persistence. The continuous exposure to pathogen antigens leads to a dysfunctional response of the T cells, which exhibit a diminished capacity to produce cytokines and cytotoxic molecules, together with a progressive increase in the expression and coexpression of inhibitory receptors by the antigen-specific T cells. There are several commercially available and useful tests for serological diagnosis of *T. cruzi* infection that detect *T. cruzi*

M. C. Thomas (✉) · M. C. López
Instituto de Parasitología y Biomedicina López Neyra, Consejo Superior de Investigaciones Científicas (IPBLN-CSIC), Granada, Spain

NHEPACHA network, Barcelona, Spain
e-mail: mcthomas@ipb.csic.es; mclopez@ipb.csic.es

A. Egui · E. Pérez-Antón · I. Gómez
Instituto de Parasitología y Biomedicina López Neyra, Consejo Superior de Investigaciones Científicas (IPBLN-CSIC), Granada, Spain

M.-J. Pinazo Delgado, J. Gascón (eds.), *Chagas Disease*,
https://doi.org/10.1007/978-3-030-44054-1_11

antigen-specific antibodies with a long life. Parasite detection by PCR during the acute phase is easy due to the high parasitemia; however, detection is difficult during the chronic phase, as there is a low level of parasites, thereby resulting in false negative cases.

Currently, two drugs are used to treat the sickness (i.e., benznidazole and nifurtimox), although both cause important and frequent side effects. The effectiveness of benznidazole during the chronic phase of the disease remains controversial. Several studies have reported the benefit of benznidazole for patients at the chronic phase, preventing the development of cardiomyopathy and improving the parasite-specific immune response. Recently, data showing the reduction of the exhaustion process in $CD8^+$ and $CD4^+CD8^+$ T cells in chronic Chagas disease patients have also been reported [1]. To date, there are no reliable tools to evaluate antiparasitic treatment efficacy, which has to be correlated with seroconversion and the absence of parasite DNA. The long lives of specific antibodies against *T. cruzi* and the high percentage of false results in PCR create an urgent need for biotools to detect treatment success or failure.

11.2 Immune Response Associated with *Trypanosoma cruzi* Chronic Infection as Disease/Pathology Progression and Therapeutic Efficacy Biomarkers

11.2.1 Biomarkers of Humoral Response Against Specific Parasite Antigens

Antiparasitic treatment efficacy in Chagas disease can only be measured through seroconversion of conventional serological tests; however, this seroconversion may take several years or decades to assess. A study of serologic follow-up of 430 chronic Chagas disease patients showed that a complete seronegative status was achieved in a median of 11.7 years [2]. This finding points to the need to identify biomarkers that allow evaluation of the treatment efficacy in a short period of time, providing information on the progression of the disease.

Several efforts have been made by researchers to search for early markers of treatment efficacy. Treatment success will be related to seronegative conversion and/or decrease in *T. cruzi*-specific antibody titers. Viotti et al. [3] evaluated the correlation of changes in antibody levels specific for *T. cruzi* measured by the three conventional serological assays: indirect hemagglutination assay (IHA), indirect fluorescent immunoassay (IFA), and enzyme-linked immuno sorbent assay (ELISA), and by a multiplex assay using a set of 16 *T. cruzi* proteins [4]. Analyses were carried out in 53 benznidazole-treated and 89 untreated chronic Chagas disease patients with a median follow-up of 36 months. The obtained results indicated that conversion to negative results or drop in reactivity on one or more conventional serological tests was significantly greater in treated patients than in untreated subjects, with benznidazole treatment the only variable related to a decreased reactivity. With negative seroconversion in two or three conventional tests and a drop in antibody titers on two or three conventional tests, a positive impact of treatment was detected in 45% of the patients. Approximately 75% of the patients showed similar reactivity trends against the combined three conventional serological tests and the multiplex assay, although the multiplex test showed a superior detection capacity, as it detected changes that were not observed by conventional ELISAs [3, 4].

Another multiplex assay, using a selection of a panel of 12 different *T. cruzi* antigens, was evaluated in 391 well-characterized samples (248 *T. cruzi* positive samples, 94 unscreened blood donors' samples from nonendemic area, and 49 sero-negative blood donors) and on 10 subjects with inconsistent results in at least one out the three different commercially available tests [5]. This multi-cruzi assay showed the correct classification of all of the positive and negative samples, leading to the serological confirmation of Chagas disease. In addition, the reactivity obtained was significantly higher among PCR positive than PCR negative samples, which could correlate with parasitemia level in certain patients. A longitudinal follow-up study is still required for validation of this test for diagnosis purposes, including as indicated by the authors, false-reactive samples such as those from leishmaniasis patients [5]. A larger study using samples derived from 455 patients with oral testimony obtained through a questionary of benznidazole treatment and 1199 without any reported treatment was subsequently carried out by these authors [6]. Of the 15 different antigens that compose this multiplex serology assay, it was selected the Ag3, as it was detected with high titers by sera from most *T. cruzi*-infected individuals and exhibited the highest titers in patients who had a positive PCR result performed for parasite detection [6].

The *T. cruzi* KMP11, HSP70, and PFR2 recombinant proteins were proposed as follow-up biomarkers. The reactivity against these immunodominant recombinant antigens was analyzed through a prospective analysis in sera from 46 chronic Chagas disease patients during different clinical phases of the disease (15 patients in the indeterminate stage, 16 in the cardiomyopathy stage, and 16 in the digestive stage), in addition to 22 control sera from healthy donors [7]. Regardless of the stage of the illness, all of these recombinant antigens were recognized by Chagas disease patient sera with statistical significance compared with the sera from healthy donors, patients with autoimmune diseases, or patients suffering from tuberculosis, leprosy, or malaria. The dynamics of sera reactivity from 35 treated patients was also tested (before and after benznidazole administration). Interestingly, a short-term decrease in reactivity against KMP11, HSP70, and PFR2 was observed post-treatment. A statically significant decrease in reactivity against KMP11 was observed at 6 months posttreatment in 74% of patients and at 9 months posttreatment against PFR2 and HSP70 molecules. The drop in reactivity remained constant or continued decreasing during the posttreatment follow-up period (24 months). It was also observed that following benznidazole treatment, the differential reactivity against these antigens was correlated with the clinical status of the patients [7]. Most likely, analysis of the reactivity against these recombinant antigens could be used to detect *T. cruzi* infection in any clinical form of Chagas disease and for monitoring the effectiveness of benznidazole treatment. The mentioned set of serological biomarkers was employed in a blind posttreatment follow-up of twin brothers congenitally infected with *T. cruzi* (infecting strain DTU-V). Remarkably, treatment success in Brother I and treatment interruption in Brother II were detected by the use of this set of serological biomarkers, which was corroborated by PCR assays [8].

An immunodominant repetitive epitope, named 3973 (FGQAAAGDKPSL), that was recognized by sera from adult Chagas disease patients having different clinical forms of the disease was also identified [9]. The epitope belongs to the TcCA-2 parasite membrane protein [10]. The 3973 peptide showed a high sensitivity (>90%) and specificity (>98%) in sera of Chagas patients, as it was not recognized by either individuals with autoimmune and inflammatory processes or by patients with a non-chagasic cardiomyopathy. Remarkably, the reactivity against the 3973 peptide, detected in the sera from asymptomatic chronic Chagas disease patients, was an average of 40% lower than that detected in the sera from patients in the symptomatic phase of the disease, who had manifest cardiac or digestive alterations. This molecule was proposed as an indicator of the illness pathological stage and to be used as a serologic biomarker [9]. The 3973 *T. cruzi* repetitive sequence was also studied as a marker of therapeutic drug efficacy in patients with chronic Chagas disease. Thus, modifications in reactivity against 3973 were measured in sera from asymptomatic and chronic Chagas patients with cardiac symptomatology at 9 and 24 months after benznidazole treatment. Remarkably, although no downward trend in the antibody levels against soluble *T. cruzi* total antigens (S*Tc*A) was observed in the sera from Chagas patients at any time during the posttreatment follow-up period, the level of antibodies against the 3973 peptide significantly decreased. Thus, at a short time (9 and 24 months) after benznidazole treatment, a continuous and important drop in reactivity against 3973 (at least 40% lower than those detected before treatment) was observed in 49% of asymptomatic patients and 34% of cardiac chronic Chagas patients relative to the pretreatment antibody levels ($\rho < 0.001$) [11]. Interestingly, the patients who exhibited a decrease in reactivity against the 3973 peptide after treatment also showed significantly decreased C-reactive protein (CRP) levels. It had been reported that an elevated CRP level was associated with increased disease severity [12]. In summary, all these results supported that the 3973 molecule could be an acceptable biomarker for assessing therapeutic efficacy, in addition to being a useful biomarker for the prognosis of Chagas disease [9, 11]. The reactivity of sera from 66 adults with chronic indeterminate Chagas disease (IND) against a set of four *T. cruzi* antigens (KMP11, PFR2, HSP70, and 3973) was recently analyzed before and after benznidazole treatment. The results showed that 42.4% and 68.75% of IND patients met the established criteria of therapeutic efficacy (STEC) at 24 months and 48 months posttreatment, respectively [13]. Meeting the STEC implied that there was a continuous decrease in the reactivity of the patient sera against the four antigens after treatment and that there was a substantial decrease in the reactivity for at least two of the antigens. Interestingly, patients who showed treatment efficacy based on this set of serological biomarkers had an improved multifunctional antigen-specific $CD8^+$ T-cell responsiveness [13].

In addition to tests that include a group of parasite proteins, many have been tested individually in different trials. The serological reactivity against the F105 antigen, the F-III and F-IV fractions from *T. cruzi* extracts, and the exo-antigens from trypomastigote-infected mice (EXO) were studied in 42 chronic Chagas disease patients for 2–20 years after treatment with nifurtimox and benznidazole, and in 42 untreated patients. A significant decrease in serologic reactivity against the F-IV fraction and EXO antigen was observed in 64% and 44% of treated versus 33% and 8% of untreated patients, respectively [14]. A 24-kDa recombinant protein from *T. cruzi* (rTc24) is another protein evaluated by ELISA and Western blot tests to identify treated Chagas disease patients considered cured on the basis of persistently negative tests of hemocultures and lytic antibodies. Some of these patients were termed dissociated because their sera, although negative by the complement-mediated lysis test (CoML), were positive by conventional serology. For the dissociated patients, 80% of the sera were seronegative to rTc24 in the ELISA or in Western blots, in agreement with the CoML test. For the group of uncured patients, all sera tested recognized rTc24 in both tests [15]. Antibody level against P2β protein (*T. cruzi* ribosomal protein) was evaluated as a possible cure biomarker in a retrospective study. Treated ($n = 30$) and untreated ($n = 37$) Chagas disease patients with the indeterminate form of disease and untreated patients with chronic Chagas disease cardiomyopathy ($n = 11$) were followed up over more than 20 years. Interestingly, the levels of antibodies against P2β decreased from their initial values in the first group compared with the other groups (untreated patients) [16]. Short-term monitoring using a recombinant *T. cruzi* flagellar calcium-binding protein (F29) in an ELISA was used in a double-blind, randomized study to test the efficacy of benznidazole in children in the indeterminate phase of Chagas disease. The results indicated that 35.2% and 62.1% of the 44 benznidazole-treated children were seronegative for F29 antigen at 6 months and 48 months posttreatment, respectively [17]. In a subsequent study, the ELISA-F29 test was evaluated as an early marker of therapeutic efficacy in adults with chronic Chagas disease. The retrospective study compared the time at which negative seroconversion was detected against the F29 antigen and by conventional serology on a cohort of 29 patients treated with nifurtimox or benznidazole and 37 untreated patients. ELISA-F29 significantly anticipated negative seroconversion compared to conventional serology, observing this seroconversion only in the treated patients [18]. Likewise, antibody levels against recombinant antigens 1, 2, 13, 30, and 36 and Shed Acute Phase Antigen (SAPA) were assessed in sera from 18 Chagas disease patients before and after 3 years of posttreatment follow-up. The specific reactivity against these antigens before treatment decreased in 50% (9/18) of the patients after treatment. Antigen 13 obtained the best results as a good marker of treatment efficacy using ELISA, due to the fact of negative seroconversion in 67% (six out of nine) of patients [19].

The glycosylphosphatidylinositol (GPI)-anchored mucins purified from mammalian cell culture-derived trypomastigotes (tGPI mucins) of the Y strain are specifically recognized by protective chagasic anti-α-galactosyl antibodies (anti-α-Gal Abs) and are able to diagnose active Chagas disease. A chemiluminescent enzyme-linked immunosorbent-based assay using purified trypomastigote-derived α-Gal-containing tGPI mucins or synthetic neoglycoproteins (NGPs) has been proposed as promising BMKs for diagnosis and early assessment of cure following Chagas disease chemotherapy. A study in patients with indeterminate Chagas disease found a small but significant reduction in the titers of trypanolytic anti-α-Gal antibodies among benznidazole-treated patients versus placebo patients at 12 months. In this study, 9% of the treated patients negatively seroconverted at the end of the follow-up compared with 4% of the placebo-treated group. However, conventional serology showed no significant differences between active treatment and placebo at any time point [20].

The 5F2 and 5A9B11 monoclonal antibodies against *T. cruzi* microsomal fraction (Mc) and its cross-reactivity with mammalian tissues were used to study differences in the immune response of chronic Chagas disease patients with different degrees of cardiac dysfunction against parasite epitopes by blocking the recognition of these monoclonal antibodies to the target antigen. The reactivity between 5F2 or 5A9B11 and their target antigens was significantly inhibited by sera from Chagas disease patients compared to sera from healthy donors. Moreover, 5F2 was able to distinguish sera from patients in the early phase of the symptomatic form of the disease from those patients with severe disease ($p < 0.05$). The results suggested the potential use of these anti-Mc antibodies as a prognosis marker in Chagas disease patients [21]. The presence of different isotypes of antibodies against specific recombinant antigens was also analyzed in association with the pathology evolution of Chagas disease. Thus, antibodies against FRA (flagellar recombinant repetitive antigen) of IgG2 isotype were more abundant, with statistically significant differences in patients with Chagas disease cardiomyopathy versus patients with the indeterminate stage [22]. On the other hand, the IgA antibody response against FRA and CRA (cytoplasmic recombinant antigen) was related with digestive forms of Chagas disease. Although these antibodies were detected in patients with different clinical forms of Chagas disease, they were significantly higher in patients with digestive and cardiodigestive forms of the disease [23].

Additional validation of the refereed targets and general methodologies will require analyses of a larger set of subjects coming from a wider range of *T. cruzi* endemic areas.

11.2.2 Host Antigen-Specific Cellular Response as Biomarkers

11.2.2.1 CD8⁺ T-cell Immune Response as Biomarker of Pathology and Treatment Impact in Chronic Chagas Disease

Control of *T. cruzi* infection requires the activation of both $CD4^+$ and $CD8^+$ T cells. The importance of the $CD8^+$ T cells in the progression of Chagas disease has been demonstrated in different ways, such as $CD8^+$ depletion and use of mice deficient in the CD8 molecule or β2-microglobulin [24–28]. The activation of the $CD8^+$ T cells occurs after the recognition of short, antigenic peptides (epitopes of 8–10 amino acids in length) derived from *T. cruzi* proteins that are processed and presented in the context of major histocompatibility complex class I (MHC I) [29–31]. A specific cytokine environment, such as IL-21, IL-2, or IL-12, is required to elicit efficient $CD8^+$ T-cell immunity [32, 33]. Recently, IL-17RA signaling was shown to also be required for the maintenance of specific $CD8^+$ T-cell responses and that the absence of this signaling during *T. cruzi* infection alters the transcriptional program of effector CD8+ T cells and consequently their effector function [34]. Thus, the establishment of an anti-*T. cruzi* CD8+ immune response focused on the parasite's immunodominant epitopes is crucial in the control of infection caused by *T. cruzi* [35]. However, the complexity of the molecular mechanisms that regulate the *T. cruzi* biological processes has made it difficult to identify individual targets of CD8+ T-cell responses in *T. cruzi*-infected subjects.

Little is known regarding CD8+ T-cell responses in subjects with chronic *T. cruzi* infection. Because the HLA-A*0201 allele is highly frequent (approximately 45%) in the population from Chagas disease endemic areas, the use of transgenic mice expressing human MHC-class I from HLA-A2-restricted epitopes constituted a useful tool for identification of CD8+ T-cell immunodominant epitopes. The reported HLA-A2-restricted epitopes are encoded by genes that are from different *T. cruzi* antigens; the most relevant are those contained in the trans-sialidase proteins (TS) [36], amastigote-stage surface protein-1 (ASP-1) [37], trypomastigote-form surface antigen-1 (TSA-1) [38], Kinetoplastid Membrane Protein-11 (KMP-11) [39, 40], calcium-binding protein [41], ribosomal P2b protein [42], cruzipain and FL-160 flagellar protein [43], heat shock protein-70 [44], paraflagellar rod proteins [45], or the *Tc*CA-2 membrane protein [46]. Immunodominant CD8+ T-cell epitopes (TSKB20 and TSKB18) encoded by trans-sialidase gene family members have been recognized by T cells from patients in acute and chronic phases of *T. cruzi* infection [36]. However, although the development of a high frequency of TS epitope-specific CD8+ T cells contributes to the optimal control of acute infection, it is not enough to develop immune resistance against *T. cruzi* infection [47].

Secretion of pro-inflammatory cytokines (IFN-γ and TNF-α) and the cytotoxic molecule granzyme B was detected in antigen-specific CD8+ T cells from asymptomatic and cardiac chronic Chagas patients after in vitro stimulation with short peptides contained in *T. cruzi* Hsp70, PFR or TcCA-2 proteins [44–46]. In these responder patients, a high level of IL-6, which has been reported as a pleiotropic cytokine related to the inhibition of expansion and functionality of Treg lymphocytes, was also detected. For PFRs and TcCA-2 derived peptides, a higher level of IFN-γ was observed in epitope-specific CD8+ T cells from asymptomatic Chagas patients versus those with cardiac symptomatology [45, 46]. Moreover, cytotoxic activity was only detected in the epitope-specific CD8+ T cells from the asymptomatic patients [45]. Interestingly, a differential phenotypic profile of the TcCA-2-specific CD8+ T cells of Chagas disease patients related to the progression of the disease was described. Thus, the TcCA-2-specific CD8+ T cells from patients with cardiac symptoms were mainly effector memory cells (T_{EM} and T_{EMRA}), while those present in patients at the asymptomatic phase were predominantly naive cells (T_{NAIVE}). Likewise, in patients with cardiac symptoms, the percentage of CD8+ T cells with senescence features (CD8+/CD44+/CD57+) was significantly higher than that detected in patients at the asymptomatic phase of the disease [46].

The K1/TcTLE HLA-A*0201-restricted peptide contained in *T. cruzi* KMP11 protein is an epitope that is efficiently processed, presented, and recognized by CD8+ T cells during the natural course of the disease, since it induced a high expression of INF-γ in $CD8^+$ T cells from asymptomatic and cardiac Chagas disease patients [39, 48]. The K1/TcTLE $CD8^+$ T epitope is highly recognized by Chagas disease patients in the context of more than one HLA-I subtype (HLA-A*0205, HLA-A*0222, HLA-A*0226, HLA-A*0259, and HLA-A*0287), which indicates that it is a promiscuous epitope presented by different HLA-A molecules [48, 49]. In the same way, several trans-sialidase derived promiscuous epitopes able to bind to HLA-A02, HLAA03, and HLA-A24 have also been described [50]. These results suggest that these HLA class I peptides can be used for evaluating the frequency and functional activity of antigen-specific $CD8^+$ T cells and, consequently, be used as biomarkers of pathology and progression of the disease. Since only a proportion of patients responded to particular peptides, due to the existence of high variability regarding antigen processing and presentation, the use of several epitopes belonging to different *T. cruzi* antigens to perform an accurate disease follow-up is likely necessary.

$CD8^+$ T cells control the *T. cruzi* infection by different methods, including the secretion of cytokines that induce microbicidal activity in host cells and the lytic activity through the perforin/granzyme pathway. Focusing on the immune response generated after activation of $CD8^+$ T cells, it was observed that protection against lethal *T. cruzi* infection in a murine experimental infection model was dependent on IFN-γ production [51]. In chronic Chagas disease patients, an inverse correlation has been described between the cardiac Chagas disease severity and IFN-γ secretion level by $CD8^+$ T cells stimulated with *T. cruzi*-infected autologous dendritic cells [52]. It has also been reported that the frequency of patients who respond to a parasite lysate producing IFN-γ, as measured by ELISPOT analysis, is very high among patients with mild clinical disease and very low among those with the most severe form of the disease [53]. The antigen-specific $CD8^+$ T cells from these patients with severe cardiac symptomatology forms are predominantly of the fully differentiated memory phenotype ($CD45RA^-CD27^-CD28^-$). In addition, there exists a significantly higher level of total effector/memory $CD8^+$ T cells ($CD45RA^-CCR7^-$) in cardiac symptomatic patients versus asymptomatic and healthy donors, and a decreased level of this subpopulation in patients with a more severe symptomatic cardiac stage of the disease. This observation could reflect the failure of the host immune response to control parasite replication at target tissues, concomitant with the onset of heart symptoms [52]. Other authors have described that the attenuated effector functions of $CD8^+$ T cells detected in symptomatic Chagas disease patients are related to a phenotype of terminal effector/memory $CD8^+$ T cells and late differentiation ($CD27^-CD127^-$) [48, 54]. It was also reported that the cytotoxic activity (measured by the expression of granzyme B and perforin) was higher in $CD8^+$ T cells in a late stage of differentiation, while the production of IFN-γ and IL-2 was lower in $CD8^+$ T cells from patients with more severe forms of cardiac disease [54]. In chronic Chagas disease patients with cardiac severe disease, a lower frequency of T_{SCM} cells than that observed in asymptomatic patients and healthy subjects along with an increased frequency of T_{TE} cells has been observed [55]. The T_{SCM} cells from human were described as an early differentiated and long-lived memory T-cell population with an enhanced capacity for self-renewal and a multipotent ability to generate other subsets of memory cells, even in the absence of antigen [56]. All these data suggest that $CD8^+$ T cells undergo a progressive dysfunction during *T. cruzi* infection, which may be because long-term parasite persistence can result in a specific T lymphocyte population with low capacity for self-renewal. Interestingly, it has also been reported that asymptomatic chronic Chagas patients have a higher frequency of multifunctional parasite-specific $CD8^+$ T cells that could lead to more effective parasite control, while the mono- or bifunctional cytotoxic parasite-specific $CD8^+$ T cells are predominant in symptomatic patients [54]. Concomitant with the loss of multifunctional T-cell response observed in parallel to evolution of severity of chronic Chagas disease, an increase in the frequency of $CD8^+$ T cells expressing inhibitory receptors (such as PD-1, CTLA-4, CD160, 2B4 or TIM-3) and especially coexpressing some of them, has also been observed [54]. Inhibitory receptor expression on $CD8^+$ T cells appears to contribute, in many chronic infections, to a poor pathogen control [57].

Interestingly, after benznidazole treatment, a decrease in the expression of inhibitory receptors and in the proportion of $CD8^+$ T cells coexpressing 2B4, CD160 and TIM-3 or PD-1, CTLA-4 and TIM-3 has been described in asymptomatic treated patients. An enhanced multifunctional capacity of these antigen-specific $CD8^+$ T cells (measured by the production of IFN-γ, IL-2, TNF-α, granzyme B, and perforin) was also found after treatment of these patients [1]. Likewise, a statistically significant increase in the frequency of $CD8^+$ T_{CM} and T_{TE} cells was also observed after treatment of these asymptomatic chronic Chagas disease patients [1]. These data are consistent with those described in a mouse model of chronic *T. cruzi* infection, which showed that after benznidazole treatment, a stable and protective $CD8^+$ T_{CM} cell population was detected [58]. In addition, another study carried out with cells from asymptomatic chronic patients showed that a sequential combined treatment of allopurinol and benznidazole restored the levels of total naive ($CD45RA^+CCR7^+CD62L^+$) $CD4^+$ and $CD8^+$ T cells. In addition, the frequency of *T. cruzi*-specific interferon-γ-producing T cells significantly increased after allopurinol treatment and decreased to background levels following benznidazole administration in a substantial proportion of subjects evaluated [59].

All together, these findings indicate that the functional and phenotypic cellular patterns of $CD8^+$ T cells are modified during the progression of chronic Chagas disease, as the therapeutic treatment capable of improving the functionality of these antigen-specific $CD8^+$ T cells by the partial reversion of the cell exhaustion process. Therefore, we propose that the functional and phenotypic patterns of the host $CD8^+$ T-cell immune response could serve both as biomarkers of disease progression and for monitoring the effectiveness of antiparasitic treatment in the early chronic stage of the disease.

11.2.2.2 Implication of $CD4^+$ T cells and Their Functional Response in CD Patients at Different Stages of Chronic Chagas Disease and Their Modulation After Antiparasitic Therapy

The lymphocyte subpopulation of $CD4^+$ T cells plays essential roles in mediating adaptive immune response to many pathogens. Naïve $CD4^+$ T cells may differentiate into T helper (Th) cells, with the ability to produce different cytokines [60]. The $CD4^+$ T cells are also involved in the primary and secondary activation of cytotoxic $CD8^+$ T cells (CTL), as well as in their persistence [61, 62]. $CD4^+$ T cells also have properties similar to those of the cytotoxic-$CD8^+$ T cells, such as secretion of perforin and granzyme [63]. A significant fraction of this population of $CD4^+$ T cells, known as Treg, presents specialized immunoregulatory functions that suppress the immune system when an exacerbated response occurs in order to maintain the immune homeostasis [64].

In the infection caused by *T. cruzi*, the $CD4^+$ T cells play an important role. In *T. cruzi* experimental infection models, depletion of $CD4^+$ T lymphocytes produced in these mice a high susceptibility to the parasite, exhibiting a higher parasitemia compared to that of wild-type mice [28]. In Chagas disease patients coinfected with HIV, a reduction in the number of $CD4^+$ T cells was correlated to a higher parasitemia [65]. Following *T. cruzi* experimental infection, a higher number of $CD4^+$ T-cell infiltrates was found in cardiac tissue from a mouse strain resistant to *T. cruzi* infection than in a mouse strain susceptible to infection. In the cardiac tissue of these resistant mice, higher levels of Th1-like cytokines, such as IFN-γ, TNF-α, and IL-1α, were detected than in the tissue of the mice susceptible to the infection [66]. The relevance of a Th1 response versus Th2 has also been reported from in vitro infection models of *T. cruzi*-infected macrophages. Thus, $CD4^+$ T cells that have a Th1-like response produce an effective activation of macrophages that are capable of killing intracellular parasites, while the Th2 $CD4^+$ T cells only cause a marginal reduction of the parasite load in these cells [67]. The addition of IFN-γ neutralizing antibodies significantly suppressed the detected antiparasitic activity in the macrophages, highlighting the importance of the production of Th1-type cytokines by $CD4^+$ T cells for *T. cruzi* control [67].

In Chagas disease patients coinfected with HIV, a reduction in the number of $CD4^+$ T cells was correlated with higher parasitemia [65]. On the other hand, dissimilar findings have been reported by different authors regarding the IFN-γ production by *T. cruzi* antigen-specific $CD4^+$ T cells and the severity of the chronic sickness. Some authors have observed an inverse correlation between the frequency of $CD4^+$ T cells expressing IFN-γ and the severity of cardiac pathology of chronic Chagas disease patients [53, 68]. In contrast, other authors observed that patients with severe cardiopathology have a higher frequency of $CD4^+$ T cells secreting IFN-γ in response to *T. cruzi* antigens than those who are asymptomatic [69, 70]. The $CD4^+$ T helper cell subsets are also essential for maintaining an effective cytotoxic memory of $CD8^+$ T-cell repertory, which is critical in the control of *T. cruzi* infection. Remarkably, it was reported that antigen-specific $CD4^+$ T cells were able to have cytotoxic functions, such as production of granzyme B and degranulation factor CD107a, against *T. cruzi* antigens ex vivo, with higher frequency of these cells in asymptomatic Chagas disease patients than in those with cardiac manifestations [71].

Several studies have analyzed the effect of the treatment on the Th1 response and the importance of the function of the $CD4^+$ T cells in the control of the *T. cruzi* infection. The drop of serological titers against the *T. cruzi* parasite, which occurs after treatment in some chronic Chagas disease patients, was associated with a rebound of IFN-γ production by antigen-specific $CD4^+$ T cells. Moreover, the expression of CD154 and its coexpression with IFN-γ ($CD154^+IFN\text{-}\gamma^+$) and TNF-α ($CD154^+IFN\text{-}\gamma^+TNF\text{-}\alpha^+$) was also increased in these treated patients [72]. An increase in the frequency of $CD4^+$ T cells producing IFN-γ and TNF-α was also detected in treated Chagas patients, while no modification in the production of IL-4 was detected [73]. These results suggest that trypanocidal treatment improved the functional and multifunctional response of $CD4^+$ T cells against *T. cruzi* infection, leading the immune response to a Th1 axis that is beneficial in controlling infection associated with the drop of parasite load. On the other hand, the $CD4^+$ T helper 17 cells, which are producers of IL-17, are correlated with a better cardiac function in patients with symptomatic chronic Chagas disease because a higher expression of IL-17 was observed in patients with the absence of cardiomyopathy [74, 75]. Treatment with antiparasitic drugs has a potential improvement effect on the $CD4^+$ T-cell function because a gradual increase in the frequency of the Th17 cells was observed in asymptomatic chronic patients after treatment [73].

Several processes can impair the functionality of T-cell populations, compromising the infection control by the host. The cellular exhaustion process can produce a dysfunctional response in $CD4^+$ T cells against pathogens [76, 77]. In the context of *T. cruzi* infection, little is reported about the exhaustion process of $CD4^+$ T cells. During chronic *T. cruzi* infection, the $CD4^+$ T-cell population shows signs of replicative senescence and apoptosis. These processes were measured by positive expression of CD57 or caspase 3, with the patients who presented a more advanced cardiac pathology having the highest frequency of $CD4^+$ T cells expressing these markers [68]. Likewise, a higher percentage of $CD4^+$ T cells expressing the LIR-1 and CTLA-4 inhibitory receptors was observed in symptomatic versus asymptomatic chronic Chagas disease patients [78]. Interestingly, the frequency of $CD4^+LIR\text{-}1^+$ in most of the chronic patients decreased in a short-medium period of time after treatment (at 2–6 months, and maintained at 24 months) [78]. In addition, at the cardiac tissue level, the expression markers of CTLA-4 and CD57 were located in the explant of chronic Chagas disease patients with severe cardiomyopathy, and the exhaustion and senescence processes were associated with a poor control of the infection at the tissue level that could lead to cardiopathology [68]. In heart tissues of patients with end-stage chronic Chagas disease, most infiltrating cells displayed markers of antigen-experienced T cells with a low grade of differentiation and a significant proliferative capacity [79]. Thus, cells with proliferative potential were observed in tissue samples from patients with severe myocarditis in areas with high inflammation. The authors concluded that the quality of T-cell responses and immunoregulatory mechanisms might determine the pattern of the cellular response and the severity of disease in *T. cruzi* infection [79]. Recent laboratory results showed that the population of $CD4^+$ T cells undergoes a cellular exhaustion process during the chronic infection by *T. cruzi*, which is enhanced in the symptomatic chronic phase of Chagas disease. Moreover, analyses of the impact of benznidazole treatment in the exhaustion process on this cellular population showed that the expression and coexpression of the inhibitory receptors, hallmarks of the exhaustion process, decrease after treatment, with statistical significance, in asymptomatic and cardiac chronic patients (Pérez-Antón et al., unpublished). This fact indicates that benznidazole treatment could partially reverse the exhaustion process of this T-cell population, which could improve the functional response of this important compartment of the adaptive cellular immune system against the parasite.

The differentiation stages of $CD4^+$ T-cell subsets in the course of chronic *T. cruzi* infection have been evaluated by different researchers. Superior frequencies of late differentiated memory ($CD45RA^-CD27^-CD28^-$) and terminally differentiated effector $CD4^+$ T cells ($CD45RA^+CD27^-CD28^-$) were found in patients with more severe disease versus asymptomatic patients [68]. This phenotype of later differentiation is inversely correlated with the production of IFN-γ by $CD4^+$ T cells and directly correlated with the senescence process [68]. Moreover, Chagas disease patients with cardiac symptoms showed a superior frequency of the $CD4^+CD62L^-$ T-cell subset and a lower lymphoproliferative capacity compared to that from asymptomatic patients [80]. Following antiparasitic treatment of asymptomatic patients, a modulation of the $CD4^+$ differentiation profiles showing an increase in the number of naïve $CD4^+$ T cells and a reduction in the frequency of terminally differentiated memory $CD4^+$ T cells is observed. Remarkably, these findings are associated with an improvement in the lymphoproliferative capacity of these cells [59].

11.2.2.3 Role of the Circulating $CD4^+CD8^+$ T-cell Compartment in Chronic Chagas Disease Patients

The population of $CD4^+CD8^+$ T cells in the peripheral blood from healthy subjects is not an abundant T-cell subset, reaching 1–3% of the total population of circulating lymphocyte cells [81]. Their frequency increases to 8–16% in immune disorders, such as those generated by infectious pathogens [82]. These cells are considered mature lymphocytes and present an activated memory phenotype, the ability to migrate to inflamed tissues and a very great ability to produce cytokines and lytic enzymes when compared to other T-cell subsets [83]. In chronic patients infected with *T. cruzi*, the frequency of $CD4^+CD8^+$ T cells expressing and coexpressing HLA-DR and CD38 activation markers as well as the integrin VLA-4, crucial for T-cell migration, is increased versus healthy donors, exhibiting an activated phenotype [84–87]. In addition, it has been demonstrated that $CD4^+CD8^+$ T cells underwent a cellular exhaustion process in chronic Chagas disease patients, since these cells upregulated the expression and coexpression of inhibitory receptors in their membranes. Furthermore, the exhaustion process was more severe in patients who presented cardiac symptoms than in those at the indeterminate phase [86]. This exhaustion process occurred as a consequence of a continuous activation of antigen-specific T cells against antigens of the parasite, denoting the active battle of these cells against *T. cruzi* antigens. The capacity of the $CD4^+CD8^+$ T cells to activate their functional response against total *T. cruzi* antigens was detected using in vitro assays. These findings demonstrated the ability of these cells to secrete cytokines and cytotoxic molecules against *T. cruzi*. The antigen-specific $CD4^+CD8^+$ T cells were able to secrete perforin and the degranulation factor CD170a/b, exhibiting specific cytotoxic activity against the parasite [84]. Analyses of the multifunctional capacity of this *T. cruzi* antigen-specific T-cell population showed its ability to coproduce the cytotoxic molecules, granzyme B, and perforin, together with IL-2, IFN-γ and TNF-α cytokines. The multifunctional response of $CD4^{+}CD8^{+}$ T cells was higher [86] than that observed by the $CD8^{+}$ T-cell population [1, 54].

Several different subsets of $CD4^{+}CD8^{+}$ T cells have been identified according to the high or low expression of the differentiation clusters (CD), CD4 or CD8 [88]. The $CD4^{+}CD8^{high}$ T cells are associated with a positive regulation of functional capacities and with a highly activated phenotype [89]. The $CD4^{+}CD8^{low}$ T cells represent a subset of $CD4^{+}CD8^{+}$ T cells derived from terminally differentiated $CD4^{+}$ T cells [82]. The ratio of $CD4^{+}CD8^{low}/CD4^{+}CD8^{high}$ was significantly higher in chronic Chagas disease patients than in healthy donors [84, 86]. Furthermore, the highest values of $CD4^{+}CD8^{low}$ and lowest proportion of $CD4^{+}CD8^{high}$ were detected in symptomatic patients versus asymptomatic patients [86] as an indication of the severity of the pathology associated with a deterioration of the immune response against *T. cruzi*. Interestingly, it was recently shown that following treatment with benznidazole, many patients modulate the referred imbalance of the $CD4^{+}CD8^{+}$ T-cell subsets towards an increase in the proportion of the activated subpopulation ($CD4^{+}CD8^{high}$) and a decrease in the number of $CD4^{+}CD8^{low}$, reaching similar proportions to those found in healthy donors [86]. Thus, all these data suggest that the population of $CD4^{+}CD8^{+}$ T cells could have an important role in the control of the *T. cruzi* infection, due to its active behavior, functional capacities, and increased frequency during parasite infection.

11.3 Final Remarks

- Many efforts are being made to identify serological biomarkers that provide information on the progression of the chronic Chagas disease. There are results that support that some molecules can be useful biomarkers for assessing therapeutic efficacy.
- During chronic Chagas disease, the $CD8^{+}$ T cells undergo gradual dysfunction characterized by an impaired multifuntional activity, late-stage cell differentiation, increased inhibitory receptor coexpression, and increased expression of cytolytic mediators.
- The frequency of $CD4^{+}CD8^{+}$ T cells is substantially increased in chronic Chagas disease patients versus healthy donors, exhibiting an activated phenotype. The ratio of $CD4^{+}CD8^{low}/CD4^{+}CD8^{high}$ is significantly higher in chronic Chagas disease patients than in healthy donors [84, 86]. Furthermore, the highest values of $CD4^{+}CD8^{low}$ (subset of $CD4^{+}CD8^{+}$ T cells derived from terminally differentiated $CD4^{+}$ T cells) and lowest proportion of $CD4^{+}CD8^{high}$ (T cells associated with a positive regulation of functional capacities and with a highly activated phenotype) is detected in symptomatic patients versus asymptomatic patients.
- After treatment there is a clear reversion of the exhaustion process of the T cells population based on the reduction of coexpression of the inhibitory receptors and

the increase in the antigen-specific multifunctional response (coproduction of cytokines and cytotoxic molecules).

- The phenotypic and functional patterns of the host T-cell immune response seem to be useful biomarkers for evaluating the therapeutic efficacy of known and new drugs.

References

1. Mateus J, Perez-Anton E, Lasso P, Egui A, Roa N, Carrilero B, Gonzalez JM, Thomas MC, Puerta CJ, Lopez MC, Cuellar A. Antiparasitic treatment induces an improved CD8+ T cell response in chronic chagasic patients. J Immunol. 2017;198:3170. https://doi.org/10.4049/jimmunol.1602095.
2. Viotti R, Vigliano C, Lococo B, Bertocchi G, Petti M, Alvarez MG, Postan M, Armenti A. Long-term cardiac outcomes of treating chronic Chagas disease with benznidazole versus no treatment: a nonrandomized trial. Ann Intern Med. 2006;144(10):724–34.
3. Viotti R, Vigliano C, Alvarez MG, Lococo B, Petti M, Bertocchi G, Armenti A, De Rissio AM, Cooley G, Tarleton R, Laucella S. Impact of aetiological treatment on conventional and multiplex serology in chronic Chagas disease. PLoS Negl Trop Dis. 2011;5(9):e1314. https://doi.org/10.1371/journal.pntd.0001314.
4. Cooley G, Etheridge RD, Boehlke C, Bundy B, Weatherly DB, Minning T, Haney M, Postan M, Laucella S, Tarleton RL. High throughput selection of effective serodiagnostics for Trypanosoma cruzi infection. PLoS Negl Trop Dis. 2008;2(10):e316. https://doi.org/10.1371/journal.pntd.0000316.
5. Granjon E, Dichtel-Danjoy ML, Saba E, Sabino E, Campos de Oliveira L, Zrein M. Development of a novel multiplex immunoassay multi-cruzi for the serological confirmation of Chagas disease. PLoS Negl Trop Dis. 2016;10(4):e0004596. https://doi.org/10.1371/journal.pntd.0004596.
6. Zrein M, Granjon E, Gueyffier L, Caillaudeau J, Liehl P, Pottel H, Cardoso CS, Oliveira CDL, de Oliveira LC, Lee TH, Ferreira AM, Ribeiro ALP, Busch MP, Sabino EC. A novel antibody surrogate biomarker to monitor parasite persistence in Trypanosoma cruzi-infected patients. PLoS Negl Trop Dis. 2018;12(2):e0006226. https://doi.org/10.1371/journal.pntd.0006226.
7. Fernandez-Villegas A, Pinazo MJ, Maranon C, Thomas MC, Posada E, Carrilero B, Segovia M, Gascon J, Lopez MC. Short-term follow-up of chagasic patients after benzonidazole treatment using multiple serological markers. BMC Infect Dis. 2011;11:206. https://doi.org/10.1186/1471-2334-11-206.
8. Fernandez-Villegas A, Thomas MC, Carrilero B, Tellez C, Maranon C, Murcia L, Moralo S, Alonso C, Segovia M, Lopez MC. The innate immune response status correlates with a divergent clinical course in congenital Chagas disease of twins born in a non-endemic country. Acta Trop. 2014;140:84–90. https://doi.org/10.1016/j.actatropica.2014.08.006.
9. Thomas MC, Fernandez-Villegas A, Carrilero B, Maranon C, Saura D, Noya O, Segovia M, Alarcon de Noya B, Alonso C, Lopez MC. Characterization of an immunodominant antigenic epitope from *Trypanosoma cruzi* as a biomarker of chronic Chagas' disease pathology. Clin Vaccine Immunol. 2012;19(2):167–73. https://doi.org/10.1128/CVI.05566-11.
10. Buschiazzo A, Campetella OE, Macina RA, Salceda S, Frasch AC, Sanchez DO. Sequence of the gene for a Trypanosoma cruzi protein antigenic during the chronic phase of human Chagas disease. Mol Biochem Parasitol. 1992;54(1):125–8.
11. Fernandez-Villegas A, Thomas MC, Carrilero B, Lasso P, Egui A, Murcia L, Segovia M, Alonso C, Lopez MC. A 12-mer repetitive antigenic epitope from *Trypanosoma cruzi* is a potential marker of therapeutic efficacy in chronic Chagas' disease. J Antimicrob Chemother. 2016;71(7):2005–9. https://doi.org/10.1093/jac/dkw090.

12. Lopez L, Arai K, Gimenez E, Jimenez M, Pascuzo C, Rodriguez-Bonfante C, Bonfante-Cabarcas R (2006) [C-reactive protein and interleukin-6 serum levels increase as Chagas disease progresses towards cardiac failure]. Rev Esp Cardiol 59 (1):50-56
13. Egui A, Thomas MC, Fernandez-Villegas A, Perez-Anton E, Gomez I, Carrilero B, Del Pozo A, Ceballos M, Andres-Leon E, Lopez-Ruz MA, Gainza E, Oquinena E, Segovia M, Lopez MC. A parasite biomarker set for evaluating benznidazole treatment efficacy in patients with chronic asymptomatic Trypanosoma cruzi infection. Antimicrob Agents Chemother. 2019;63(10) https://doi.org/10.1128/AAC.02436-18.
14. Moretti E, Cervetta L, Basso B, Castro I, Santamarina N. [Chronic Chagas' disease: effects of treatment nn the levels of antibodies to crude and partially purified Trypanosoma cruzi antigens]. Bol Chil Parasitol 1998;53(1–2):3–9
15. Krautz GM, Galvao LM, Cancado JR, Guevara-Espinoza A, Ouaissi A, Krettli AU. Use of a 24-kilodalton Trypanosoma cruzi recombinant protein to monitor cure of human Chagas' disease. J Clin Microbiol. 1995;33(8):2086–90.
16. Fabbro DL, Olivera V, Bizai ML, Denner S, Diez C, Marcipar I, Streiger M, Arias E, del Barco M, Mendicino D, Bottasso O. Humoral immune response against P2beta from Trypanosoma cruzi in persons with chronic Chagas disease: its relationship with treatment against parasites and myocardial damage. Am J Trop Med Hyg. 2011;84(4):575–80. https://doi.org/10.4269/ajtmh.2011.10-0261.
17. Sosa Estani S, Segura EL, Ruiz AM, Velazquez E, Porcel BM, Yampotis C. Efficacy of chemotherapy with benznidazole in children in the indeterminate phase of Chagas' disease. Am J Trop Med Hyg. 1998;59(4):526–9.
18. Fabbro D, Velazquez E, Bizai ML, Denner S, Olivera V, Arias E, Pravia C, Ruiz AM. Evaluation of the ELISA-F29 test as an early marker of therapeutic efficacy in adults with chronic Chagas disease. Rev Inst Med Trop Sao Paulo. 2013;55(3) https://doi.org/10.1590/S0036-46652013000300005.
19. Sanchez Negrette O, Sanchez Valdez FJ, Lacunza CD, Garcia Bustos MF, Mora MC, Uncos AD, Basombrio MA. Serological evaluation of specific-antibody levels in patients treated for chronic Chagas' disease. Clin Vaccine Immunol. 2008;15(2):297–302. https://doi.org/10.1128/CVI.00106-07.
20. Torrico F, Gascon J, Ortiz L, Alonso-Vega C, Pinazo MJ, Schijman A, Almeida IC, Alves F, Strub-Wourgaft N, Ribeiro I, Group ES. Treatment of adult chronic indeterminate Chagas disease with benznidazole and three E1224 dosing regimens: a proof-of-concept, randomised, placebo-controlled trial. Lancet Infect Dis. 2018;18(4):419–30. https://doi.org/10.1016/S1473-3099(17)30538-8.
21. Laucella SA, de Titto EH, Segura EL. Epitopes common to Trypanosoma cruzi and mammalian tissues are recognized by sera from Chagas' disease patients: prognosis value in Chagas disease. Acta Trop. 1996;62(3):151–62.
22. Vercosa AF, Lorena VM, Carvalho CL, Melo MF, Cavalcanti MG, Silva ED, Ferreira AG, Pereira VR, Souza WV, Gomes YM. Chagas' disease: IgG isotypes against cytoplasmic (CRA) and flagellar (FRA) recombinant repetitive antigens of Trypanosoma cruzi in chronic Chagasic patients. J Clin Lab Anal. 2007;21(5):271–6. https://doi.org/10.1002/jcla.20186.
23. Vasconcelos RH, Amaral FN, Cavalcanti MG, Silva ED, Ferreira AG, Morais CN, Gomes YM. Increased levels of IgA antibodies against CRA and FRA recombinant antigens of Trypanosoma cruzi differentiate digestive forms of Chagas disease. Hum Immunol. 2010;71(10):964–7. https://doi.org/10.1016/j.humimm.2010.07.004.
24. Junqueira C, Caetano B, Bartholomeu DC, Melo MB, Ropert C, Rodrigues MM, Gazzinelli RT. The endless race between *Trypanosoma cruzi* and host immunity: lessons for and beyond Chagas disease. Expert Rev Mol Med. 2010;12:e29. https://doi.org/10.1017/S1462399410001560.
25. Tarleton RL. Depletion of $CD8^+$ T cells increases susceptibility and reverses vaccine-induced immunity in mice infected with *Trypanosoma cruzi*. J Immunol. 1990;144(2):717–24.

26. Tarleton RL, Grusby MJ, Postan M, Glimcher LH. Trypanosoma cruzi infection in MHC-deficient mice: further evidence for the role of both class I- and class II-restricted T cells in immune resistance and disease. Int Immunol. 1996;8(1):13–22.
27. Tarleton RL, Koller BH, Latour A, Postan M. Susceptibility of beta 2-microglobulin-deficient mice to *Trypanosoma cruzi* infection. Nature. 1992;356(6367):338–40. https://doi.org/10.1038/356338a0.
28. Tarleton RL, Sun J, Zhang L, Postan M. Depletion of T-cell subpopulations results in exacerbation of myocarditis and parasitism in experimental Chagas' disease. Infect Immun. 1994;62(5):1820–9.
29. Gromme M, Neefjes J. Antigen degradation or presentation by MHC class I molecules via classical and non-classical pathways. Mol Immunol. 2002;39(3-4):181–202.
30. Martin D, Tarleton R. Generation, specificity, and function of CD8+ T cells in *Trypanosoma cruzi* infection. Immunol Rev. 2004;201:304–17. https://doi.org/10.1111/j.0105-2896.2004.00183.x.
31. Yewdell JW, Reits E, Neefjes J. Making sense of mass destruction: quantitating MHC class I antigen presentation. Nat Rev Immunol. 2003;3(12):952–61. https://doi.org/10.1038/nri1250.
32. Cox MA, Harrington LE, Zajac AJ. Cytokines and the inception of CD8 T cell responses. Trends Immunol. 2011;32(4):180–6. https://doi.org/10.1016/j.it.2011.01.004.
33. Kaech SM, Cui W. Transcriptional control of effector and memory CD8+ T cell differentiation. Nat Rev Immunol. 2012;12(11):749–61. https://doi.org/10.1038/nri3307.
34. Tosello Boari J, Araujo Furlan CL, Fiocca Vernengo F, Rodriguez C, Ramello MC, Amezcua Vesely MC, Gorosito Serran M, Nunez NG, Richer W, Piaggio E, Montes CL, Gruppi A, Acosta Rodriguez EV. IL-17RA-signaling modulates CD8+ T cell survival and exhaustion during Trypanosoma cruzi infection. Front Immunol. 2018;9:2347. https://doi.org/10.3389/fimmu.2018.02347.
35. Higuchi Mde L, Benvenuti LA, Martins Reis M, Metzger M. Pathophysiology of the heart in Chagas' disease: current status and new developments. Cardiovasc Res. 2003;60(1):96–107.
36. Martin DL, Weatherly DB, Laucella SA, Cabinian MA, Crim MT, Sullivan S, Heiges M, Craven SH, Rosenberg CS, Collins MH, Sette A, Postan M, Tarleton RL. CD8+ T-Cell responses to *Trypanosoma cruzi* are highly focused on strain-variant trans-sialidase epitopes. PLoS Pathog. 2006;2(8):e77. https://doi.org/10.1371/journal.ppat.0020077.
37. Santos MA, Garg N, Tarleton RL. The identification and molecular characterization of Trypanosoma cruzi amastigote surface protein-1, a member of the trans-sialidase gene superfamily. Mol Biochem Parasitol. 1997;86(1):1–11.
38. Wizel B, Nunes M, Tarleton RL. Identification of *Trypanosoma cruzi* trans-sialidase family members as targets of protective CD8+ TC1 responses. J Immunol. 1997;159(12):6120–30.
39. Diez H, Lopez MC, Del Carmen Thomas M, Guzman F, Rosas F, Velazco V, Gonzalez JM, Puerta C. Evaluation of IFN-gamma production by CD8 T lymphocytes in response to the K1 peptide from KMP-11 protein in patients infected with *Trypanosoma cruzi*. Parasite Immunol. 2006;28(3):101–5. https://doi.org/10.1111/j.1365-3024.2005.00815.x.
40. Maranon C, Thomas MC, Planelles L, Lopez MC. The immunization of A2/K(b) transgenic mice with the KMP11-HSP70 fusion protein induces CTL response against human cells expressing the *T. cruzi* KMP11 antigen: identification of A2-restricted epitopes. Mol Immunol. 2001;38(4):279–87.
41. Engman DM, Krause KH, Blumin JH, Kim KS, Kirchhoff LV, Donelson JE. A novel flagellar Ca2+-binding protein in trypanosomes. J Biol Chem. 1989;264(31):18627–31.
42. Garcia F, Sepulveda P, Liegeard P, Gregoire J, Hermann E, Lemonnier F, Langlade-Demoyen P, Hontebeyrie M, Lone YC. Identification of HLA-A*0201-restricted cytotoxic T-cell epitopes of Trypanosoma cruzi TcP2beta protein in HLA-transgenic mice and patients. Microbes Infect. 2003;5(5):351–9.
43. Fonseca SG, Moins-Teisserenc H, Clave E, Ianni B, Nunes VL, Mady C, Iwai LK, Sette A, Sidney J, Marin ML, Goldberg AC, Guilherme L, Charron D, Toubert A, Kalil J, Cunha-Neto E. Identification of multiple HLA-A*0201-restricted cruzipain and FL-160 CD8+ epitopes recognized by T cells from chronically *Trypanosoma cruzi*-infected patients. Microbes Infect. 2005;7(4):688–97. https://doi.org/10.1016/j.micinf.2005.01.001.

44. Maranon C, Egui A, Carrilero B, Thomas MC, Pinazo MJ, Gascon J, Segovia M, Lopez MC. Identification of HLA-A *02:01-restricted CTL epitopes in *Trypanosoma cruzi* heat shock protein-70 recognized by Chagas disease patients. Microbes Infect. 2011;13(12-13):1025–32. https://doi.org/10.1016/j.micinf.2011.05.010.
45. Egui A, Thomas MC, Morell M, Maranon C, Carrilero B, Segovia M, Puerta CJ, Pinazo MJ, Rosas F, Gascon J, Lopez MC. *Trypanosoma cruzi* paraflagellar rod proteins 2 and 3 contain immunodominant CD8(+) T-cell epitopes that are recognized by cytotoxic T cells from Chagas disease patients. Mol Immunol. 2012;52(3-4):289–98. https://doi.org/10.1016/j.molimm.2012.05.021.
46. Egui A, Thomas MC, Carrilero B, Segovia M, Alonso C, Maranon C, Lopez MC. Differential phenotypic and functional profiles of TcCA-2 -specific cytotoxic CD8+ T cells in the asymptomatic versus cardiac phase in Chagasic patients. PLoS One. 2015;10(3):e0122115. https://doi.org/10.1371/journal.pone.0122115.
47. Rosenberg CS, Martin DL, Tarleton RL. CD8+ T cells specific for immunodominant trans-sialidase epitopes contribute to control of *Trypanosoma cruzi* infection but are not required for resistance. J Immunol. 2010;185(1):560–8. https://doi.org/10.4049/jimmunol.1000432.
48. Lasso P, Mesa D, Cuellar A, Guzman F, Bolanos N, Rosas F, Velasco V, Thomas MC, Lopez MC, Gonzalez JM, Puerta CJ. Frequency of specific $CD8^+$ T cells for a promiscuous epitope derived from *Trypanosoma cruzi* KMP-11 protein in chagasic patients. Parasite Immunol. 2010;32(7):494–502. https://doi.org/10.1111/j.1365-3024.2010.01206.x.
49. Lasso P, Beltran L, Guzman F, Rosas F, Thomas MC, Lopez MC, Gonzalez JM, Cuellar A, Puerta CJ. Promiscuous recognition of a Trypanosoma cruzi CD8+ T cell epitope among HLA-A2, HLA-A24 and HLA-A1 supertypes in chagasic patients. PLoS One. 2016;11(3):e0150996. https://doi.org/10.1371/journal.pone.0150996.
50. Alvarez MG, Postan M, Weatherly DB, Albareda MC, Sidney J, Sette A, Olivera C, Armenti AH, Tarleton RL, Laucella SA. HLA Class I-T cell epitopes from trans-sialidase proteins reveal functionally distinct subsets of CD8+ T cells in chronic Chagas disease. PLoS Negl Trop Dis. 2008;2(9):e288. https://doi.org/10.1371/journal.pntd.0000288.
51. Kumar S, Tarleton RL. Antigen-specific Th1 but not Th2 cells provide protection from lethal *Trypanosoma cruzi* infection in mice. J Immunol. 2001;166(7):4596–603.
52. Albareda MC, Laucella SA, Alvarez MG, Armenti AH, Bertochi G, Tarleton RL, Postan M. *Trypanosoma cruzi* modulates the profile of memory CD8+ T cells in chronic Chagas' disease patients. Int Immunol. 2006;18(3):465–71. https://doi.org/10.1093/intimm/dxh387.
53. Laucella SA, Postan M, Martin D, Hubby Fralish B, Albareda MC, Alvarez MG, Lococo B, Barbieri G, Viotti RJ, Tarleton RL. Frequency of interferon- gamma -producing T cells specific for Trypanosoma cruzi inversely correlates with disease severity in chronic human Chagas disease. J Infect Dis. 2004;189(5):909–18. https://doi.org/10.1086/381682.
54. Lasso P, Mateus J, Pavia P, Rosas F, Roa N, Thomas MC, Lopez MC, Gonzalez JM, Puerta CJ, Cuellar A. Inhibitory receptor expression on CD8+ T cells is linked to functional responses against Trypanosoma cruzi antigens in chronic chagasic patients. J Immunol. 2015;195(8):3748–58. https://doi.org/10.4049/jimmunol.1500459.
55. Mateus J, Lasso P, Pavia P, Rosas F, Roa N, Valencia-Hernandez CA, Gonzalez JM, Puerta CJ, Cuellar A. Low frequency of circulating CD8+ T stem cell memory cells in chronic chagasic patients with severe forms of the disease. PLoS Negl Trop Dis. 2015;9(1):e3432. https://doi.org/10.1371/journal.pntd.0003432.
56. Gattinoni L, Lugli E, Ji Y, Pos Z, Paulos CM, Quigley MF, Almeida JR, Gostick E, Yu Z, Carpenito C, Wang E, Douek DC, Price DA, June CH, Marincola FM, Roederer M, Restifo NP. A human memory T cell subset with stem cell-like properties. Nat Med. 2011;17(10):1290–7. https://doi.org/10.1038/nm.2446.
57. Wherry EJ. T cell exhaustion. Nat Immunol. 2011;12(6):492–9.
58. Bustamante JM, Bixby LM, Tarleton RL. Drug-induced cure drives conversion to a stable and protective CD8+ T central memory response in chronic Chagas disease. Nat Med. 2008;14(5):542–50. https://doi.org/10.1038/nm1744.

59. Perez-Mazliah DE, Alvarez MG, Cooley G, Lococo BE, Bertocchi G, Petti M, Albareda MC, Armenti AH, Tarleton RL, Laucella SA, Viotti R. Sequential combined treatment with allopurinol and benznidazole in the chronic phase of Trypanosoma cruzi infection: a pilot study. J Antimicrob Chemother. 2013;68(2):424–37. https://doi.org/10.1093/jac/dks390.
60. Zhu J, Yamane H, Paul WE. Differentiation of effector CD4 T cell populations (*). Annu Rev Immunol. 2010;28:445–89. https://doi.org/10.1146/annurev-immunol-030409-101212.
61. Janssen EM, Lemmens EE, Wolfe T, Christen U, von Herrath MG, Schoenberger SP. CD4+ T cells are required for secondary expansion and memory in CD8+ T lymphocytes. Nature. 2003;421(6925):852–6. https://doi.org/10.1038/nature01441.
62. Kalams SA, Walker BD. The critical need for CD4 help in maintaining effective cytotoxic T lymphocyte responses. J Exp Med. 1998;188(12):2199–204.
63. Takeuchi A, Saito T. CD4 CTL, a cytotoxic subset of CD4(+) T cells, their differentiation and function. Front Immunol. 2017;8:194. https://doi.org/10.3389/fimmu.2017.00194.
64. Belkaid Y, Tarbell K. Regulatory T cells in the control of host-microorganism interactions (*). Annu Rev Immunol. 2009;27:551–89. https://doi.org/10.1146/annurev.immunol.021908.132723.
65. de Freitas VL, da Silva SC, Sartori AM, Bezerra RC, Westphalen EV, Molina TD, Teixeira AR, Ibrahim KY, Shikanai-Yasuda MA. Real-time PCR in HIV/Trypanosoma cruzi coinfection with and without Chagas disease reactivation: association with HIV viral load and CD4 level. PLoS Negl Trop Dis. 2011;5(8):e1277. https://doi.org/10.1371/journal.pntd.0001277.
66. Sanoja C, Carbajosa S, Fresno M, Girones N. Analysis of the dynamics of infiltrating CD4(+) T cell subsets in the heart during experimental Trypanosoma cruzi infection. PLoS One. 2013;8(6):e65820. https://doi.org/10.1371/journal.pone.0065820.
67. Rodrigues MM, Ribeirao M, Boscardin SB. CD4 Th1 but not Th2 clones efficiently activate macrophages to eliminate Trypanosoma cruzi through a nitric oxide dependent mechanism. Immunol Lett. 2000;73(1):43–50.
68. Albareda MC, Olivera GC, Laucella SA, Alvarez MG, Fernandez ER, Lococo B, Viotti R, Tarleton RL, Postan M. Chronic human infection with Trypanosoma cruzi drives CD4+ T cells to immune senescence. J Immunol. 2009;183(6):4103–8. https://doi.org/10.4049/jimmunol.0900852.
69. Cuellar A, Rojas F, Bolanos N, Diez H, Del Carmen Thomas M, Rosas F, Velasco V, Lopez MC, Gonzalez JM, Puerta C. Natural CD4(+) T-cell responses against Trypanosoma cruzi KMP-11 protein in chronic chagasic patients. Immunol Cell Biol. 2009;87(2):149–53. https://doi.org/10.1038/icb.2008.76.
70. Gomes JA, Bahia-Oliveira LM, Rocha MO, Martins-Filho OA, Gazzinelli G, Correa-Oliveira R. Evidence that development of severe cardiomyopathy in human Chagas' disease is due to a Th1-specific immune response. Infect Immun. 2003;71(3):1185–93.
71. Keesen TS, Gomes JA, Fares RC, de Araujo FF, Ferreira KS, Chaves AT, Rocha MO, Correa-Oliveira R. Characterization of CD4(+) cytotoxic lymphocytes and apoptosis markers induced by Trypanossoma cruzi infection. Scand J Immunol. 2012;76(3):311–9. https://doi.org/10.1111/j.1365-3083.2012.02730.x.
72. Alvarez MG, Bertocchi GL, Cooley G, Albareda MC, Viotti R, Perez-Mazliah DE, Lococo B, Castro Eiro M, Laucella SA, Tarleton RL. Treatment success in *Trypanosoma cruzi* infection is predicted by early changes in serially monitored parasite-specific T and B cell responses. PLoS Negl Trop Dis. 2016;10(4):e0004657. https://doi.org/10.1371/journal.pntd.0004657.
73. Vallejo A, Monge-Maillo B, Gutierrez C, Norman FF, Lopez-Velez R, Perez-Molina JA. Changes in the immune response after treatment with benznidazole versus no treatment in patients with chronic indeterminate Chagas disease. Acta Trop. 2016;164:117–24. https://doi.org/10.1016/j.actatropica.2016.09.010.
74. Guedes PM, Gutierrez FR, Silva GK, Dellalibera-Joviliano R, Rodrigues GJ, Bendhack LM, Rassi A Jr, Rassi A, Schmidt A, Maciel BC, Marin Neto JA, Silva JS. Deficient regulatory T cell activity and low frequency of IL-17-producing T cells correlate with the extent of cardiomyopathy in human Chagas' disease. PLoS Negl Trop Dis. 2012;6(4):e1630. https://doi.org/10.1371/journal.pntd.0001630.

75. Magalhaes LM, Villani FN, Nunes Mdo C, Gollob KJ, Rocha MO, Dutra WO. High interleukin 17 expression is correlated with better cardiac function in human Chagas disease. J Infect Dis. 2013;207(4):661–5. https://doi.org/10.1093/infdis/jis724.
76. Crawford A, Wherry EJ. The diversity of costimulatory and inhibitory receptor pathways and the regulation of antiviral T cell responses. Curr Opin Immunol. 2009;21(2):179–86. https://doi.org/10.1016/j.coi.2009.01.010.
77. Morou A, Palmer BE, Kaufmann DE. Distinctive features of CD4+ T cell dysfunction in chronic viral infections. Curr Opin HIV AIDS. 2014;9(5):446–51. https://doi.org/10.1097/COH.0000000000000094.
78. Arguello RJ, Albareda MC, Alvarez MG, Bertocchi G, Armenti AH, Vigliano C, Meckert PC, Tarleton RL, Laucella SA. Inhibitory receptors are expressed by Trypanosoma cruzi-specific effector T cells and in hearts of subjects with chronic Chagas disease. PLoS One. 2012;7(5):e35966. https://doi.org/10.1371/journal.pone.0035966.
79. Arguello RJ, Vigliano C, Cabeza-Meckert P, Viotti R, Garelli F, Favaloro LE, Favaloro RR, Laguens R, Laucella SA. Presence of antigen-experienced T cells with low grade of differentiation and proliferative potential in chronic Chagas disease myocarditis. PLoS Negl Trop Dis. 2014;8(8):e2989. https://doi.org/10.1371/journal.pntd.0002989.
80. Chaves AT, de Assis Silva Gomes Estanislau J, Fiuza JA, Carvalho AT, Ferreira KS, Fares RC, Guimaraes PH, de Souza Fagundes EM, Morato MJ, Fujiwara RT, da Costa Rocha MO, Correa-Oliveira R. Immunoregulatory mechanisms in Chagas disease: modulation of apoptosis in T-cell mediated immune responses. BMC Infect Dis. 2016;16:191. https://doi.org/10.1186/s12879-016-1523-1.
81. Overgaard NH, Jung JW, Steptoe RJ, Wells JW. CD4+/CD8+ double-positive T cells: more than just a developmental stage? J Leukoc Biol. 2015;97(1):31–8. https://doi.org/10.1189/jlb.1RU0814-382.
82. Parel Y, Chizzolini C. CD4+ CD8+ double positive (DP) T cells in health and disease. Autoimmun Rev. 2004;3(3):215–20. https://doi.org/10.1016/j.autrev.2003.09.001.
83. Clenet ML, Gagnon F, Moratalla AC, Viel EC, Arbour N. Peripheral human CD4+CD8+ T lymphocytes exhibit a memory phenotype and enhanced responses to IL-2, IL-7 and IL-15. Sci Rep. 2017;7(1):11612. https://doi.org/10.1038/s41598-017-11926-2.
84. Giraldo NA, Bolanos NI, Cuellar A, Guzman F, Uribe AM, Bedoya A, Olaya N, Cucunuba ZM, Roa N, Rosas F, Velasco V, Puerta CJ, Gonzalez JM. Increased CD4+/CD8+ double-positive T cells in chronic Chagasic patients. PLoS Negl Trop Dis. 2011;5(8):e1294. https://doi.org/10.1371/journal.pntd.0001294.
85. Morrot A, Terra-Granado E, Perez AR, Silva-Barbosa SD, Milicevic NM, Farias-de-Oliveira DA, Berbert LR, De Meis J, Takiya CM, Beloscar J, Wang X, Kont V, Peterson P, Bottasso O, Savino W. Chagasic thymic atrophy does not affect negative selection but results in the export of activated CD4+CD8+ T cells in severe forms of human disease. PLoS Negl Trop Dis. 2011;5(8):e1268. https://doi.org/10.1371/journal.pntd.0001268.
86. Perez-Anton E, Egui A, Thomas MC, Puerta CJ, Gonzalez JM, Cuellar A, Segovia M, Lopez MC. Impact of benznidazole treatment on the functional response of Trypanosoma cruzi antigen-specific CD4+CD8+ T cells in chronic Chagas disease patients. PLoS Negl Trop Dis. 2018;12(5):e0006480. https://doi.org/10.1371/journal.pntd.0006480.
87. Perez AR, Morrot A, Berbert LR, Terra-Granado E, Savino W. Extrathymic CD4+CD8+ lymphocytes in Chagas disease: possible relationship with an immunoendocrine imbalance. Ann N Y Acad Sci. 2012;1262:27–36. https://doi.org/10.1111/j.1749-6632.2012.06627.x.
88. Zloza A, Al-Harthi L. Multiple populations of T lymphocytes are distinguished by the level of CD4 and CD8 coexpression and require individual consideration. J Leukoc Biol. 2006;79(1):4–6. https://doi.org/10.1189/jlb.0805455.
89. Sullivan YB, Landay AL, Zack JA, Kitchen SG, Al-Harthi L. Upregulation of CD4 on CD8+ T cells: CD4dimCD8bright T cells constitute an activated phenotype of CD8+ T cells. Immunology. 2001;103(3):270–80.

12 Access to Comprehensive Chagas Disease Care: A Global Effort

Francisco Javier Sancho, Oscar Bernal, and Jeffrey V. Lazarus

12.1 Introduction

When considering the difficulties of accessing health care, one can picture a long, winding road. In many endemic and non-endemic settings, in both low- and high-income countries, that road is not just a simple metaphor but represents a real barrier to accessing services. In fact, some studies in maternal and newborn health, for example, have found that improved transportation is the single most important factor for better health outcomes [1]. But in the end, roads and transportation are just one of a myriad of obstacles that people face in their encounter with the health system. In this chapter we will focus on global efforts to eliminate Chagas disease, a neglected, highly stigmatized tropical disease, and highlight how health systems can reorient to be more efficient, effective and, ultimately, people-centered.

One morning, at the end of April 2018, in the Colombian region of Arauca, the health staff involved in the Chagas vector control program coordinated by the Ministry of Health set out at 2 AM towards a community at risk of acquiring Chagas. They drove through muddy dirt roads in order to meet with the members of a rural community who do not have access to any health services other than those offered in the main town 7 h away.

F. J. Sancho (✉)
Global Chagas Disease Coalition, Barcelona, Spain
e-mail: javier.sancho@chagascoalition.org

O. Bernal
University of Andes, Bogotá, Colombia
e-mail: obernal@uniandes.edu.co

J. V. Lazarus
Barcelona Institute for Global Health (ISGlobal), Hospital Clinic, University of Barcelona, Barcelona, Spain
e-mail: jeffrey.lazarus@isglobal.org

M.-J. Pinazo Delgado, J. Gascón (eds.), *Chagas Disease*,
https://doi.org/10.1007/978-3-030-44054-1_12

This is the first visit by the health staff of the Chagas Program after 2 years. It is part of the follow-up plan after the training provided by the vector control staff to this community. When the staff arrive, a group of 15 farmers are already waiting inside the house of one of the community leaders. Only one of the elder leaders remember some of the main topics learned during their training about how to identify and prevent the presence of "los pitos" (the local name for "kissing bugs"). Acquiring the resources needed to reach and run neglected tropical disease (NTD) programs in these settings is extremely difficult for any health system, but essential to elimination efforts.

The same story, the same landscape, and the same lack of resources can be found in endemic countries like Argentina, Bolivia, Mexico, and Paraguay. It is also an issue in some areas within countries, like along the West Coast of the United States, where distance is a barrier for the people in need of diagnosis and treatment of Chagas disease. In California, for instance [2], which has a large population at risk (mainly Central American migrants), public transportation to reach services is not always available or accessible.

In European countries, like Spain, some patients living with Chagas disease face the real risk of losing their jobs [3] if they need to attend at regular medical services, which is often the case during treatment. Many of them are women and are usually working in uncertain job conditions, which do not allow flexibility to attend medical consultations during working hours.

Some of the people affected by Chagas disease in the United States are recognized as active community members, playing a role in helping to identify people at risk for Chagas disease. Some communities are more close-knit than others and this leads to a much higher proportion of patients from these specific countries or regions of origin represented when examining the burden of disease, which may lead to a distorted understanding of the burden of the disease among other communities.

The main step to overcome this difficulty starts with the affected people. It is crucial that they are aware of the importance of being treated before the infection progresses and causes other complications. They are also the key to reaching and engaging their communities in order to scale up access to care. As more affected populations become fully aware of this issue, the probability of controlling the disease will increase.

In 2009, on the 100th anniversary of the discovery of Chagas disease, several campaigns denounced the scant progress made [4] to date in regard to the care received by the affected populations. Since then, some action has been taken in different spheres (medical, scientific, and political) to counter the stagnated efforts. In endemic countries, positive vector control results have now been achieved and several areas have received certification of interruption of vectoral transmission [5]. In affected countries, almost 100% of the blood banks are now being screened for Chagas disease. These types of measures have reduced the number of new cases and must be sustained. The next decisive step is to achieve the same support for the diagnosis and treatment of patients with the disease.

12.1.1 A Global Challenge

We need realistic timeframes and frameworks to eliminate Chagas disease. The ambitious Sustainable Development Goals (SDGs) set out a road map over the next decade, with concrete targets, to advance towards the control of Chagas and ultimately its elimination as a public health challenge. Chagas disease is included implicitly in some targets of SDG Goal 3, which is related to health issues. More specifically, Chagas disease is directly addressed in the target 3.3, on the end of epidemics of Neglected Tropical Diseases by 2030 and indirectly in other targets, such as target 3.4 (reduce mortality from noncommunicable diseases …), target 3.8 (achieve universal health coverage …), because of congenital transmission, target 3.7 (ensure universal access to sexual and reproductive health care services), and target 11.1 (adequate, safe, and affordable housing).

In 2010, the World Health Organization (WHO) adopted the resolution WHA63.20 [6] on control and elimination of Chagas disease. It was an important recognition of the need to tackle the disease through a comprehensive strategy which considered the transmission control in endemic and non-endemic countries, the access to diagnosis and treatment starting at the primary health level, and the need to increase the resources and networking among national, regional, and international public and private partners.

In 2012, the road map for implementation "Accelerating work to overcome the global impact of neglected tropical diseases" was approved by the WHO [7]. It prioritized the objectives to interrupt transmission via intra-domiciliary vectors in endemic countries as well as the transmission via blood transfusion in endemic and non-endemic countries by 2015; and to eliminate peri-domiciliary infestation in Latin America by 2020.

Also, in 2013, WHO launched a global "tricycle" strategy, with four components [8]:

1. Interruption of transmission (acquired through vector, oral, transfusions, organ transplants, or congenital modes of transmission)
2. Providing health care to affected populations
3. Implementing a global information and surveillance system
4. Providing information, education, and communication to control Chagas disease for key actors and affected populations

The imperative strategy that still remains unmet is accessibility and availability of treatment for all those who need it. Diagnosis and treatment remain unacceptably low worldwide: less than 1% of the population at risk currently receives treatment for Chagas disease [9]. To date, many efforts have been dedicated to increasing access to comprehensive care, but have largely failed. For example, the Bolivian health system has not been able to integrate comprehensive care across the country despite several initiatives and around US $45 million invested from the Inter-American Development Bank (IDB) in the last decade. Patients and their families are faced with a bewildering array of medical, institutional, and social obstacles that make it impossible for them to find an adequate solution.

We have structural, psychosocial, clinical, and systemic barriers, which make access a challenging issue for those affected by the disease. Compared to other neglected diseases that particularly target the world's poorest communities, the silence—both political and in the media—surrounding Chagas disease is striking and has delayed the implementation of available solutions.

Various advances have made the control of Chagas disease more feasible today than ever before. Firstly, two companies are now manufacturing the first-line treatment benznidazole, and a pediatric formula of the drug is available [9]. Secondly, new scientific evidence supporting the importance and usefulness of treatment has emerged and proposals have been made to simplify diagnosis in rural areas [10]. Although more research into new diagnostic tools and drugs is needed, Chagas disease can be successfully treated today.

In August 2017, the U.S. Food and Drug Administration (FDA) approved benznidazole for use in children ages 2–12 years old but not for adults. It is the first treatment approved in the United States for the treatment of Chagas disease. The FDA granted the company making the medicine with a priority review voucher (PRV) [11], which is a mechanism that aims to promote access to diagnosis and treatment of neglected diseases. This incentive gives pharmaceutical companies access to an expedited FDA review process. However, companies are not required to use a PRV for the drug for which the voucher was originally issued. The voucher can instead be used for another product or even be sold. Part of the funds from the PRV will be dedicated to an access plan led by Mundo Sano and DNDi in collaboration with other allies such as the Chagas Global Coalition. Nifurtimox (the other drug option for Chagas disease) has been donated by Bayer Pharmaceutical to national control programs until the year 2020, as set out in a commitment by the company.

Like many other infectious diseases, Chagas disease must be controlled through comprehensive programs that offer a broad range of interventions in addition to case management (diagnosis and treatment); these include Chagas disease information and education and training of health care personnel.

The urgent need to increase the number of people diagnosed and treated for Chagas disease must be a priority in endemic countries. Success stories have clearly demonstrated that good results can be achieved. Some programs run by private institutions and NGOs in coordination with health systems have increased the number of people receiving treatment [12]. Lastly, in a program piloted by the Drugs for Neglected Diseases Initiative (DNDi) and the Colombian Ministry of Health, we have witnessed a huge increase of people properly diagnosed when access is facilitated in some regions of Colombia—overcoming the barriers [13] linked to decades of conflict. Additionally, in Spain, the leadership of the Spanish Aid Agency and the support from entities like the Barcelona Institute for Global Health (ISGlobal) have led to an increase in the number of patients treated both in Spain and in Bolivia through an innovative platform of Chagas control and care led by ISGlobal, the Bolivian foundation CEADES, and the Ministry of Health of Bolivia. Other experiences in Argentina where the Mundo Sano Foundation has scaled-up access to health for those affected by Chagas, and in the United States where the first center of excellence for Chagas disease was created in Los Angeles, demonstrate that the scaling up of access is not just a pipedream. The programs implemented in these three countries combine strong leadership, economic resources, and creativity, to offer service for vulnerable populations [12].

Three key components are required to ensure that a health system is capable of properly identifying and treating people infected with *T. cruzi*, the parasite that causes Chagas disease:

- Integration of the diagnosis and treatment of Chagas disease into the health system
- Screening of pregnant and at-risk women and follow-up care of newborn infants with *T. cruzi* infection
- Referral for treatment of all positive cases detected at blood banks

The main challenge facing health systems in low-income countries (LICs) is that most people with Chagas disease live in remote rural areas with limited access to resources far from any laboratory capable of analyzing blood samples and at distances that make post-treatment follow-up very difficult. To overcome this problem, the Pan American Health Organization (PAHO) recommends the incorporation of diagnosis and treatment of Chagas disease into the primary health care system to ensure access to patients. Integrating NTD programms are also a pillar of the draft of the new NTDs road map (2021–2030) that has to be approved and implemented by the member countries of WHO.

In this chapter, we analyze how to improve access from different perspectives:

- People at risk
- Policy makers and health systems
- Researchers and Academia
- Innovation: a new player

All of these players are a crucial part of the global and comprehensive response to the global health problem of Chagas disease. These are the main actors that drive through that "long, winding" road to arrive closer to the affected people.

12.2 How to Improve Access: Starting from the People at Risk

It does not explode like a bomb, nor sound like a shot. Like hunger, it kills silently. Like hunger it kills the silent, those who live condemned to be silent and die condemned to oblivion. (Eduardo Galeano)

Some six to seven million people worldwide are estimated to be infected with *T. cruzi*, and there are another 65 million people at risk in endemic and non-endemic countries.

The definition of risk (in the context of Chagas disease) means: *to be under the circumstances of being affected by the different modes of transmission: by a vector (through the feces of an insect), through vertical or congenital transmission (from mother to her child during pregnancy), by way of blood transfusions or organ transplants, by ingesting contaminated food or drink* [14].

In addition to the concept of risk, we must consider the concept of affected population. This is a wider and different approach for understanding the real dimensions of the consequences of Chagas disease. The affected populations are not only those suffering the infection and the disease but also those under risk and those who live or support their closest relatives and neighbors living with the infection.

In 2010, more than 20 associations [15] from all over the world gathered to create the International Federation of Associations of People Affected by Chagas Disease (FINDECHAGAS). We must also keep in mind how language choice matters. The member associations that are a part of FINDECHAGAS chose to be identified as "affected" people and not "infected." In deciding to be represented through non-stigmatizing word choice, the members of FINDECHAGAS are promoting a transformation in the way we traditionally look at those who suffer these kinds of diseases. We can look at the lessons learned during the early years of the HIV/AIDS pandemic, for example, to value the importance of changing the vocabulary we associate with certain diseases as a tool to reduce stigma and discrimination. The term "people living with HIV" (PLHIV) was agreed upon in consensus after it was decided that the use of those "infected" by HIV perpetuated stigma and negatively affected the health outcomes of PLHIV [16].

The communities fighting against Chagas disease face not only the different ways of transmission, but also other traditional stigmas related to this disease: silence, neglect, and poverty.

In the last Global Assembly of FINDECHAGAS (October 2018, Xalapa, Veracruz, Mexico), the members of the associations discussed the connotations of some common words used when speaking about Chagas disease, especially those commonly used among health staff to address the affected populations, or those used on social media. Traditionally associated to poor rural areas, Chagas disease used to be a synonym for two very different and, in some ways, contradictory concepts: on the one hand, it may have meant "death" and on the other hand, "invisibility" because it is an infection that can be imperceptible for the people who live with the disease all their lives. This must drive to strengthen the efforts to promote awareness and training for health staff, and in particular, those who care directly for patients or affected populations.

The need of increasing awareness about Chagas disease is a common gap for both patients and health staff. To fill the gap, both sectors must work together. On 28 April 2015, the FDA held a public meeting [17] to hear the perspectives of people with Chagas disease about their condition, its impact on their daily life, and their perspectives on approaches to treating the disease. The meeting heard from various stakeholders in the drug development process, patients, patient advocacy organizations, health care providers, academic experts, and the industry. From the patient's perspective, the most important input provided was the significant lack of awareness and understanding of Chagas disease in the health care community and provided examples of the challenges they face due to the overall lack of Chagas disease awareness, including the lengthy and often confusing process of establishing a diagnosis. The patient participants identified struggling with fear of future symptoms, social isolation, difficulty in finding others to discuss their experiences with, and the frustration of living with a condition that was not well understood.

When the associations of affected people from endemic and non-endemic countries joined FINDECHAGAS in 2010, they agreed to promote the principles of the Declaration of Uberaba, the Brazilian town where they previously met during the 100-year anniversary of the discovery of the disease by Dr. Carlos Chagas in 1909. This declaration reflects the main demands by the people affected by Chagas disease to overcome the barriers associated with inadequate comprehensive care. FINDECHAGAS aims to strengthen collective action by the people at risk of Chagas disease. The members are families, friends, and people affected by the disease working in collaboration with multidisciplinary teams. These are the main principals proposed in the Uberaba Declaration [15] (see Box 12.1).

Box 12.1 Main Principals of the Uberaba Declaration

- We (the affected people associations) recognize that information, education, and communication are essential tools to increase awareness of Chagas disease, to fight stigmatization and indifference, and to guarantee human rights, including prevention and treatment of Chagas disease.
- We demand continuity and strengthening of prevention initiatives.
- We advocate for universal access to clinical and laboratory diagnosis.
- We urge universal access to full treatment and follow-up during the acute and chronic phases of the disease.
- We support research and development of optimal technologies for the prevention, treatment, and control of Chagas disease.
- We maintain that Chagas disease is not a death sentence. It is necessary to know, face, and spread the challenges of people living with the disease. Initiatives to break the cycle that links poverty and disease should be implemented.

Since then, the different associations have requested a World Chagas Disease Day as a symbolic call to action. Some countries together with the Chagas Disease Coalition members (composed of both private and public organizations) supported associations in expanding their petition. Finally, in May 2019, the World Health Assembly unanimously approved the new World Chagas Disease Day, which will be observed on 14 April each year, because it was on that date in 1909 when the first human being, a Brazilian girl named Berenice Soares, was diagnosed with Chagas, by Doctor Carlos Chagas.

Finally, the actions to comply with the target 3.3 agreed in the SDGs ("By 2030, end the epidemics of AIDS, tuberculosis, malaria and neglected tropical diseases") must consider the leading role of the affected population to guide researchers and health community to address their real needs. Other experiences have shown that there is no other way to reach the goal of eliminating this disease as a global health problem. The aim is to reverse the description of the disease, as so eloquently written by Uruguayan author Eduardo Galeano in a book edited by Doctors Without Borders (MSF) in 2005 [18]:

> *It does not explode like a bomb, nor sound like a shot. Like hunger, it kills silently. Like hunger it kills the silent, those who live condemned to be silent and die condemned to oblivion. Tragedy that rings no bells, patients who do not pay, a disease that does not sell. Mal de Chagas is not a business that attracts the pharmaceutical industry, nor is it a subject that interests politicians or journalists. It chooses its victims from the throngs of the poor, bites them and lingeringly, bit by bit, finishes them off. Its victims have no right — nor money — to buy the rights they do not have. They do not even have the right to know what they die of.*

12.3 Challenges and Key Recommendations for Comprehensive Chagas Disease Care from a Health Systems Perspective

Chagas disease caused by *Trypanosoma cruzi* parasite is estimated to affect eight million people worldwide [19] resulting an annual financial burden of US $7.2 billion [20] taken on by health systems and affected individuals and their families. If left untreated, *T. cruzi* infection becomes chronic and can be life-threatening. Thirty percent of infected people will develop cardiac complications as a result of becoming infected by the parasite [21].

Chagas disease is no longer restricted to the region of Latin America. Due to migratory flows and globalization, it is now present in some non-endemic countries. While epidemiological data for Spain is varying, relative estimates calculated that 5.5% of adult migrants in the country were living with Chagas disease [22]. In comparison, estimates of Chagas disease in the USA are also unclear and range between 300,000 to over one million, with some estimates reporting 250,000 cases in the state of Texas alone [23]. Other estimates have even calculated the effects the parasite has on homeland security dogs [24].

Today, less than 1% of those infected with *T. cruzi* around the world receive adequate treatment even though the treatment exists. In Latin America, where the prevalence is highest, most people lack access to diagnosis and treatment, whereas in non-endemic countries like the USA, limited knowledge of and experience with the disease hinders access and an appropriate response to this neglected growing issue [25].

Currently, in Colombia, screening coverage is estimated at 1.2% of the population at risk and etiological treatment only covers 0.3–0.4% of the infected population [26]. In Bolivia, specifically, the Ministry of Health estimated that 150,000 people needed treatment in 2015 [27] but only 5515 received treatment in 2017 [28].

Despite the many efforts dedicated to increasing access to comprehensive Chagas disease care, the Bolivian health system has repeatedly failed to successfully integrate comprehensive care across the country even after US $45 million invested [29].

After many failed initiatives to scale up comprehensive Chagas disease care in the country, a comprehensive analysis of the health system is needed to facilitate overcoming the barriers associated with the unsuccessful management of Chagas disease in endemic countries like Bolivia.

A health system is defined by WHO as all organizations, institutions, resources, and people whose primary purpose is to improve health. As mentioned before, there are three key components required for health systems to be able to properly identify and treat people infected with Chagas disease:

- Integration of the diagnosis and treatment of Chagas disease into the formal health system
- Screening of pregnant and at-risk women and follow-up care of newborn infants with *T. cruzi* infection
- Referral for treatment of all positive cases detected at blood banks

The WHO health systems framework has six core components: *service delivery, health workforce, health information systems, access to essential medicines, financing, and leadership/governance* (see Fig. 12.1).

For the system to be effective, and Chagas disease case management successful, it must address equity, access, and stakeholder engagement.

Given the often-siloed nature of health system components, comprehensive Chagas disease care should be systematically addressed across all elements of the systems. Moreover, the strategies used to strengthen health systems in order to ensure effective Chagas disease case management must incorporate the patient's perspective as well as that of populations at risk [30].

12.3.1 Service Delivery

Service delivery is the provision of health care to people. All inputs to the health system, for example, health workforce, medical procurement and health information systems are intended to enhance service delivery. WHO categorizes good

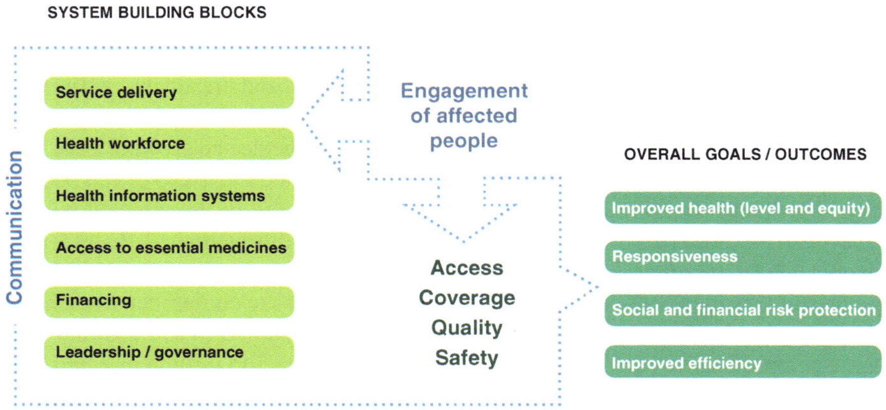

Fig. 12.1 Proposed modified WHO health systems framework. Source: *Lazarus JV, France T.* A new era for the WHO health systems building blocks? Health Systems Global 2014

service delivery as possessing the following key characteristics: comprehensiveness, accessibility, coverage, continuity, quality, person-centeredness, coordination, and accountability and efficiency.

Currently, the Chagas disease care pathway requires intervention from different specialties from all levels of the health system to ensure a continuum of care. However, PAHO/WHO recommends that medical care for Chagas patients be provided at the primary health care (PHC) level [31].

The main barriers to treating Chagas disease are [32]:

1. Delays and loss to follow-up during the diagnostic process
2. Insufficient disease diagnosis and treatment at the PHC level
3. Lack of expertise and training in Chagas disease among health care workers
4. Delays involved in diagnosis confirmation and treatment authorization
5. Provision of care limited to specialists instead of PHC physicians who are more accessible to patients

12.3.2 Health Workforce

The health workforce is defined as "all people engaged in actions whose primary intent is to enhance health." WHO identifies human resources as clinical staff, as well as management and support staff, that is, those who do not deliver services directly but are essential to the performance of health systems.

Programs promoting the scaling up of health care interventions to control Chagas disease and other emerging infectious diseases often encounter structural weaknesses in the health system in which they aim to be integrated [33].

12.3.3 Health Information Systems

Health information systems are the foundation of decision-making across the health system. They enable decision makers to identify problems and needs, make evidence-based decisions on health policy, and allocate resources optimally.

Despite epidemiological estimates relating to Chagas disease prevalence and burden of disease among people of low socioeconomic status, there are still gaps in research and monitoring data. Addressing evidence gaps and improving methods for data collection is a priority for scaling up appropriate and effective Chagas disease case management.

12.3.4 Medical Procurement

According to WHO, a well-functioning health system ensures equitable access to essential medical products and technologies of assured quality, safety, efficacy, and cost-effectiveness [34].

The drugs used to treat infection with *T. cruzi* are nifurtimox and benznidazole. Benznidazole is approved in Bolivia and has experienced significant supply chain disruptions and access problems along with reported adverse drug reactions [31, 32]. In August 2017, the U.S. Food and Drug Administration (FDA) approved benznidazole for use in children ages 2–12 years old but not for adults.

12.3.5 Financing

Health financing is fundamental to the functionality of the health system. It involves both revenue generation/collection and purchasing/provision of services. Optimal health care financing allows access to needed services through efficient resource utilization.

12.3.6 Leadership/Governance

Effective health system leadership and governance enables strategic policy frameworks, effective service delivery oversight, coalition-building, regulation, attention to system design and accountability. As a cross-cutting component of the health systems framework, leadership and governance is an integral part of improving health outcomes.

Bolivia has the highest prevalence of Chagas disease in the world with more than 600,000 people (6.1% of the population) estimated to be infected with an additional 586,000 people at risk of infection and 60% the country's territory declared endemic [20]. Chagas disease was declared a national priority over a decade ago yet in 2018, the country still does not have a protocol for the proper diagnosis, treatment, and management of Chagas disease in adults.

12.4 Access to a Comprehensive Care in Chagas Disease: A Global Effort—Innovation: The New Player

We need to innovate with what we have today and research for new solutions. Today, the following priorities are validated: the use of RDTs for screening; to simplify Chagas disease confirmation guidelines; the implementation of people-centered Chagas disease treatment; to test different doses and duration of benznidazole (BNZ); to integrate Chagas disease management at Primary Health Care Level; and to recover the vector control achieved in the 1990s.

For new solutions, the research priorities defined by the WHO's Reference Group included a new diagnostic for case detection, tests of cure, new drugs, new vector control technologies, including markers of successful vector control, integrated disease and vector control and research to assess the importance of asymptomatic infection [35].

Benznidazole (BNZ) solubility and tolerability improvements are possible using techniques that include microemulsion, nanoparticles, solid dispersions, to increase the water solubility of BNZ by 4–25-fold on dissolution and 85% release with efficacy in only a few minutes [36].

Since the introduction of BNZ and nifurtimox (NFX), allopurinol and the azoles itraconazole, fluconazole, ketoconazole, posaconazole, and ravuconazole have been studied in clinical trials [37]. Recent clinical trials such as posaconazole, benznidazole/E1224 combination, fexinidazole, and BENEFIT have contributed new data and highlighting some key issues in the field [38].

Chagas care requires intervention from different specialties and from all levels (primary to tertiary) to ensure a continuum of care. However, medical care for Chagas patients without complications should be mainly provided at the primary health care (PHC) level, as recommended by PAHO/WHO [31]. The natural evolution of the disease is a key criterion for establishing medical care at primary level. Previously this has occurred with diseases such as diabetes mellitus, hypertension, and more recently, human immunodeficiency virus infection [32].

The integration of Chagas disease management in the PHC needs a good planning, political commitment from the health authorities at national and local level, training of all health personnel, laboratory equipment, Chagas disease drugs and drugs for adverse events, good referral and contra referral system, adequate IEC, community participation, vector control and the adaptation of the health information system.

There is not one model that fits all Chagas disease treatment, because the contexts, different health system, local regulations, and access to health care are different in each country, state, autonomous community, or province. However, we can exchange experience about pilot projects and find the way to transform local success in national guidelines.

We will summarize experiences treating Chagas disease patients in endemic (Bolivia and Colombia) and non-endemic countries (Spain and the USA). These experiences involved among others, the Barcelona Institute for Global Health

(ISGlobal), Doctors Without Borders, Drugs for Neglected Diseases initiative (DNDi), Fundación Mundo Sano (Spain), CEADES (Bolivia), and the Center of Excellence for Chagas Disease at Olive View-UCLA.

12.4.1 The Colombian Experience

In Colombia, after a long negotiation among stakeholders from government, academia, nongovernmental organizations, and patient associations a pilot project in four endemic communities is beginning in 2016. The Chagas disease integration model included a patient road map that simplifies diagnostic and treatment processes, shifting them from specialists to primary care facilities. The patient road map was implemented with the goal of testing and refining the model, so it can be implemented nationally. The Chagas disease Colombian model is summarized in Fig. 12.2.

The Ministry of Health (MoH) and the Chagas program provide leadership and coordination at the national level and are entrusted with importing benznidazole and other critical medications. The Ministry of Health provides technical assistance to departmental and municipal entities and monitors key indicators. Departmental health programs coordinate actors on a departmental level, ensure medication is efficiently distributed to municipal health centers, assure quality, and monitor results. Both departmental and municipal health entities conduct screening of expectant mothers and minors [32]. The core ingredients to increase access are summarized in Fig. 12.3.

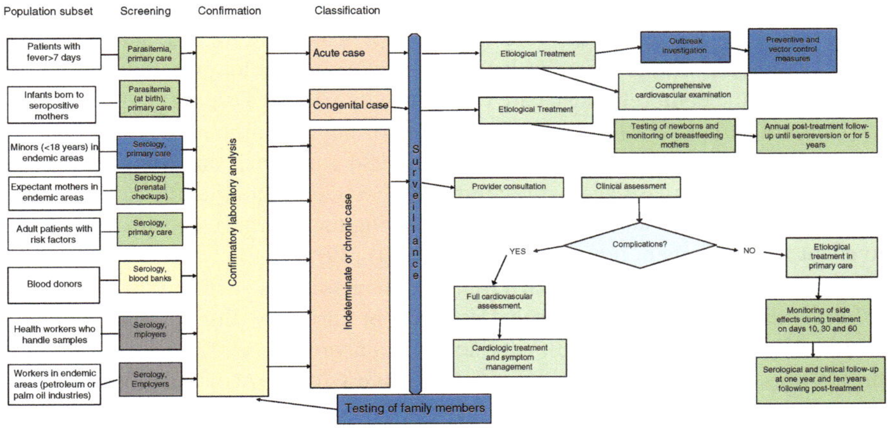

Fig. 12.2 Colombia's patient road map for Chagas disease. (Source: Marchiol A, Forsyth C, Bernal O, Valencia Hernández C, Cucunubá Z, Pachón Abril E, et al. Increasing access to comprehensive care for Chagas disease: development of a patient-centered model in Colombia)

Fig. 12.3 Core ingredients for increasing treatment access for patients with Chagas disease. (Source: Marchiol A, Forsyth C, Bernal O, Valencia Hernández C, Cucunubá Z, Pachón Abril E, et al. Increasing access to comprehensive care for Chagas disease: development of a patient-centered model in Colombia)

12.4.2 The Bolivian Experience

In 2008, the Bolivian NGO CEADES and ISGlobal initiated a program to tackle the challenge of treating adults with *T. cruzi* infection and scale up care by integrating it into the primary health care system. Between 2008 and 2015, more than 26,000 people received consultations, more than 8500 patients were treated, and more than 1600 health care workers were trained at the six centers belonging to the Bolivian platform for the comprehensive care of adults with Chagas Disease [39].

The project helped to increase the number of adults with Chagas disease diagnosed and treated, produce evidence-based clinical guidelines, and bring changes in policy that will increase access to comprehensive care among adults with Chagas disease. The Chagas National Program is studying the platform's health care model to adapt and implement it nationwide (Fig. 12.4) [39].

Doctors Without Borders (MSF) treated patients in Bolivia from 2000 to 2016. MSF started with a parallel, vertical program in Tarija for children under 5, then increased age groups up to adults in Sucre, Cochabamba, and Chuquisaca. The last project was in Monteagudo, with a joint program with the MoH and integration at the PHC level. MSF proved that treatment of Chagas disease was possible, the use of RDTs to screen was valid and the treatment was safe. Among more than 2000 patients, permanent benznidazole treatment suspension occurred in 211 patients (10.2%). Benznidazole treatment was safe and a large proportion of patients were able to complete a full course of benznidazole treatment under close treatment surveillance [40].

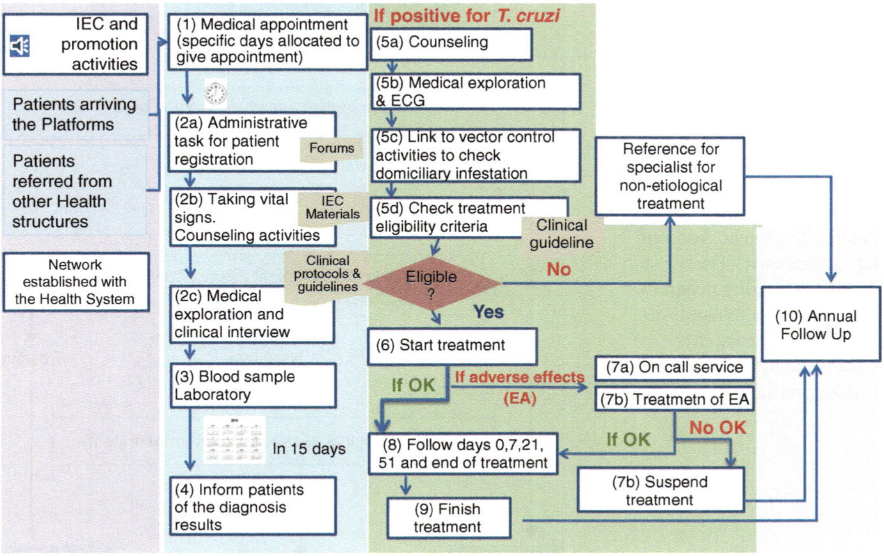

Fig. 12.4 Bolivia Chagas Disease treatment model. (Source: Pinazo M-J, Pinto J, Ortiz L, Sánchez J, García W, Saravia R, et al. A strategy for scaling up access to comprehensive care in adults with Chagas disease in endemic countries: The Bolivian Chagas Platform)

The Bolivian MoH and MSF developed a Chagas disease treatment program in Monteagudo, Bolivia. In this program, it was possible to promote articulation between the vector control, IEC program and treatment, reducing the infestation rate of the vector from 10 to 4 communities at high risk. There has also been a need to reinforce reference and counter-referral systems between different levels of care (first, second, and third).

The prenatal control program has been adequately articulated to the Chagas connatal component achieving excellent results. The Chagas program competes for the same resources with other diseases transmitted by vectors such as Dengue or Zika. Greater empowerment of the counterpart of the Municipality of Monteagudo and initiatives to introduce Chagas activities in other Municipalities of the Department was achieved.

12.4.3 The Spanish Experience

Mundo Sano, aware of the difficulties that migrants face coming from areas where Chagas disease is endemic in accessing health care services, has been boosting screening for Chagas disease in asymptomatic Latin American adults living in Spain since 2011.

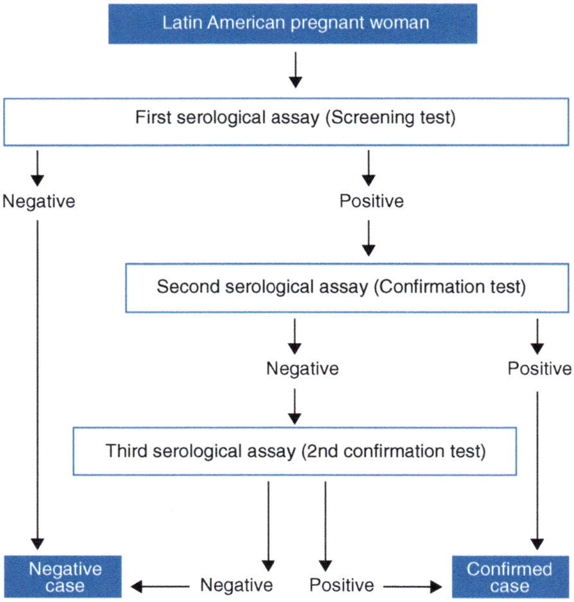

Fig. 12.5 Serological screening of pregnant Latin American women for Chagas disease, Catalonia. (Source: Basile L, Oliveira I, Ciruela P, Plasencia A, working group for developing the Catalonian screening program for congenital transmission of Chagas disease. The current screening program for congenital transmission of Chagas disease in Catalonia, Spain. Eurosurveillance)

More than 5200 Chagas disease patients have been treated in 160 Spanish health care centers following the normal procedures of the national and regional health care system in the past 4 years (unpublished, Navarro M), representing around 10% of all estimated Chagas disease cases in Spain [22].

Catalonia estimated that between 10,000 and 20,000 *T. cruzi*-infected migrants were living in this region in 2010 and the pregnancies in women from endemic countries during that year was of 6795. For this reason the Catalonian Health Department has recently implemented a screening program for preventing congenital transmission, targeting Latin American pregnant women who attend antenatal consultations (Fig. 12.5) [41].

Based on the incremental cost-effectiveness ratio of €6841 per QALY gained, Requena-Méndez and colleagues conclude that the testing for Chagas is cost-effective and should be supported in Europe [42].

12.4.4 The USA Experience

The Center of Excellence for Chagas Disease (CEChagas disease) at Olive View-UCLA Medical Center in Los Angeles is one of the few US providers offering treatment for Chagas disease. Patients are evaluated for chronic Chagas cardiomyopathy and undergo requisite labs (renal and hepatic function and complete blood count) to determine eligibility for etiological treatment [43].

The CEChagas disease has screened 4755 Latin American immigrants in Los Angeles and found that 1.31% were positive for Chagas disease. This effort, which

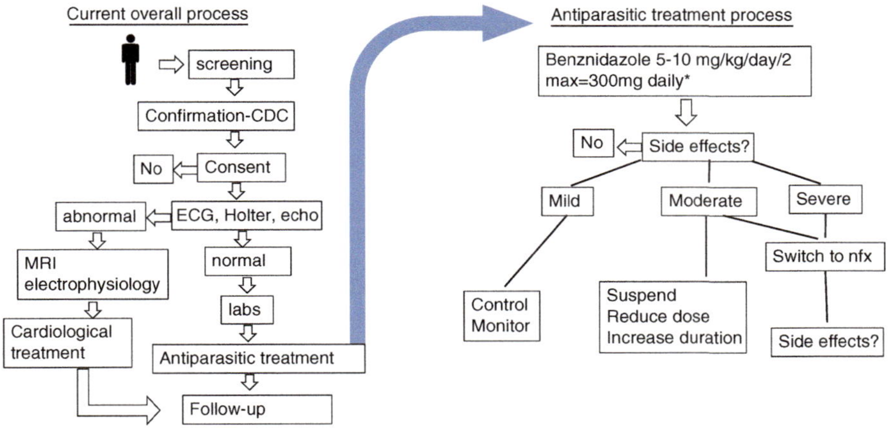

Fig. 12.6 Chagas disease treatment model. Center of Excellence for Chagas Disease, Los Angeles, USA. (Source: Meymandi S, Hernandez S, Park S, Sanchez DR, Forsyth C. Treatment of Chagas Disease in the United States. Curr Treat Options Infect Dis)

involved 89 health fairs conducted with volunteers over a period of 6 years, has only reached a minute proportion of the nearly 5.8 million Latinos in Los Angeles, 42% of whom are foreign born (Fig. 12.6) [44].

12.5 Conclusions

Innovative experiences in low- and high-endemic countries have shown that it is feasible to improve access to comprehensive Chagas disease care. However, we need to scale up and transform pilot experiences. It is essential to involve all relevant stakeholders to integrate Chagas disease management at all stages of the health system with a people-centered approach. This will lead to a greater response at the global level, ultimately ending the disease as a public health problem in 2030.

References

1. Pathmanathan I, Liljestrand J, Martins JM, Rajapaksa LC, Lissner C, de Silva A, Selvaraju S, Joginder Singh P. Investing in maternal health: learning from Malaysia and Sri Lanka (English). Health, nutrition, and population series. Washington, DC: World Bank; 2003. Available from: http://documents.worldbank.org/curated/en/367761468760748311/Investing-in-maternal-health-learning-from-Malaysia-and-Sri-Lanka.
2. Manne-Goehler J, Umeh CA, Montgomery SP, Wirtz VJ. Estimating the burden of Chagas disease in the United States. PLoS Negl Trop Dis. 2016;10(11):e0005033. https://doi.org/10.1371/journal.pntd.0005033.
3. Rovira B. "El mal de los pobres". A vivir que son dos días. Cadena SER Radio. Uploaded 25/02/2017. Available from: https://cadenaser.com/programa/2017/02/24/a_vivir_que_son_dos_dias/1487954102_748040.html.

4. Medecins Sans Frontieres, MSF. México: "Chagas: es hora de romper el silencio". MSF México. 2009. Available from: https://www.msf.mx/article/chagas-es-hora-de-romper-el-silencio.
5. WHO. Chagas' disease in Latin America: an epidemiological update based on 2010 estimates. Wkly Epidemiol Rec. 2015;90:33–40. Available from: http://www.who.int/wer/2015/wer9006.pdf?ua=1.
6. WHO. WHA63.20 resolution. "Chagas disease: control and elimination". 63 World Health Assembly. Geneva: WHO; 2010. Available from: https://www.who.int/neglected_diseases/mediacentre/WHA_63.20_Eng.pdf?ua=1.
7. WHO. Accelerating work to overcome the global impact of neglected tropical diseases. A roadmap for implementation. Geneva: WHO Press; 2012. Available from: https://www.who.int/neglected_diseases/NTD_RoadMap_2012_Fullversion.pdf.
8. WHO. Chagas disease. In: WHO. Integrating neglected tropical diseases into global health and development: fourth WHO report on neglected tropical diseases. 5.2. Geneva: WHO; 2017. p. 155–62. Available from: https://www.who.int/neglected_diseases/resources/9789241565448/en/.
9. Costa Chaves G, Abi-Saab Arrieche M, Rode J, Mechali D, Ouverney Reis P, Vieira Alves R, et al. Estimación de la demanda de medicamentos antichagásicos: una contribución para el acceso en América Latina. Rev Panam Salud Publica. 2017;41:e45.
10. Roddy P, Goiri J, Flevaud L, Palma P, Morote S, Lima N, Villa L, Torrico F, Albajar-Viñas P. Field evaluation of a rapid immunochromatographic assay for detection of Trypanosoma cruzi infection by use of whole blood. J Clin Microbiol. 2008;46:2022–7. Available from: https://www.ncbi.nlm.nih.gov/pmc/articles/PMC2446863/.
11. FDA. FDA approves first U.S. treatment for Chagas disease. 2017. Available from: https://www.fda.gov/newsevents/newsroom/pressannouncements/ucm573942.htm. Updated 27 Mar 2018.
12. Moriana S, Ortiz G, Fanjul G. Breaking the silence: an opportunity for patients with Chagas. Chagas disease global coalition by creative commons. English online edition. 2016. Available from: http://www.coalicionchagas.org/documents/5415804/5524305/breaking+the+silence_report/65091404-85cf-4796-bebb-64c120a26216.
13. DNDi. Ending the Neglect of Chagas. DNDi. 2018. Available from: https://www.dndi.org/2018/media-centre/news-views-stories/stories/visualstory/chagas-colombia/. Accessed 21 Jan 2019.
14. Infochagas. Chagas coalition. How is it transmitted? Available from: http://www.infochagas.org/en/como-se-transmite. Accessed 21 Jan 2019.
15. Federación Internacional de Asociaciones de Personas Afectadas por la Enfermedad Chagas. FINDECHAGAS. 2018. Available from: http://findechagas.org/. Accessed 22 Jan 2019.
16. UNAIDS. Terminology guidelines. 2015. Available from: http://www.unaids.org/sites/default/files/media_asset/2015_terminology_guidelines_en.pdf. Accessed 22 Jan 2019.
17. FDA (The U.S. Foods and Drug Administrations). The voice of the patient. Center for Drug Evaluation and Research (Chagas diseaseER) U.S. Food and Drug Administration (FDA). Report uploaded in November 2015. Available from: https://www.fda.gov/downloads/Drugs/NewsEvents/UCM474051.pdf. Accessed 23 Jan 2019.
18. Doctors Without Borders (MSF). Chagas, a silent tragedy. Buenos Aires: Editorial Losada; 2005.
19. World Health Organization. Weekly epidemiological record - Chagas diseases in Latin America: an epidemiological update based on 2010 estimates. Geneva: WHO; 2015. p. 5–13.
20. Lee BY, Bacon KM, Bottazzi ME, Hotez PJ. Global economic burden of Chagas disease: a computational simulation model. Lancet Infect Dis. 2013;13:342–8.
21. Villalba R, Fornes G, Alvarez MA, Roman I, Rubio V, Fernandez M, Garcia JM, Vinal M, Torres A. Acute Chagas' disease in a recipient of a bone marrow transplant in Spain: case report. Clin Infect Dis. 1992;14:594–5.
22. Navarro M, Navaza B, Guionnet A, López-Vélez R. Chagas disease in Spain: need for further public health measures. PLoS Negl Trop Dis. 2012;6(12):e1962. https://doi.org/10.1371/journal.pntd.0001962.

23. Hotez PJ, Dumonteil E, Betancourt Cravioto M, Bottazzi ME, Tapia-Conyer R, Meymandi S, et al. An unfolding tragedy of Chagas disease in North America. PLoS Negl Trop Dis. 2013;7(10). Available from: https://www.ncbi.nlm.nih.gov/pmc/articles/PMC3814410/. Accessed 9 May 2018.
24. Chagas Disease affects 300,000 people (and homeland security dogs) in U.S. Available from: https://www.idse.net/From-Zoo-to-You/Article/11-18/Chagas-Causing-Heart-Problems-in-Homeland-Security-Dogs/53321?sub=&enl=true.
25. Alvarez MG, Bertocchi GL, Cooley G, Albareda MC, Viotti R, Perez-Mazliah DE, et al. Treatment success in Trypanosoma cruzi infection is predicted by early changes in serially monitored parasite- specific T and B cell responses. PLoS Negl Trop Dis. 2016;10(4). Available from: https://www.ncbi.nlm.nih.gov/pmc/articles/PMC4851297/. Accessed 1 Apr 2018.
26. Cucunubá ZM, et al. How universal is coverage and access to diagnosis and treatment for Chagas disease in Colombia? A health systems analysis. Soc Sci Med. 2017;175:187. https://doi.org/10.1016/j.socscimed.2017.01.002.
27. Ministerio de Salud Bolivia. Plan sectorial de desarrollo 2011-2015. 2009. Available from: https://www.minsalud.gob.bo/images/Documentacion/dgp/Plan_Sectorial_de_Desarrollo_2011- 2015_metas2020.pdf. Accessed 19 Apr 2018.
28. Ministerio de Salud de Bolivia "La presencia del vector transmisor del Chagas en viviendas se redujo al 1,8% en Bolivia." Available from: https://www.minsalud.gob.bo/3205-la-presencia-de-chagas-en-viviendas-se-redujo-al-1-8-en-bolivia. Accessed 18 Feb 2019.
29. IDB. Inter-American Development Bank approves $45 million to improve health in Bolivia. Available from: https://www.iadb.org/en/news/news-releases/1999-02-10/idb-approves-45-million-to-improve-health-in-bolivia%2C1401.html. Accessed 25 May 2018.
30. Day E, Hellard M, Treloar C, et al. On behalf of the International Network on Hepatitis in Substance Users (INHSU). Hepatitis C elimination among people who inject drugs: challenges and recommendations for action within a health systems framework. Liver Int. 2019;39:20–30.
31. PAHO WHO. Chagas disease. Available from: http://www.paho.org/hq/index.php?option=com_topics&view=article&id=10&Itemid=40743&lang=en. Accessed 17 Apr 2018.
32. Marchiol A, Forsyth C, Bernal O, Valencia Hernández C, Cucunubá Z, Pachón Abril E, et al. Increasing access to comprehensive care for Chagas disease: development of a patient-centered model in Colombia. Rev Panam Salud Publica. 2017;41:e153. Available from: http://iris.paho.org/xmlui/handle/123456789/34506. Accessed 1 Apr 2018.
33. Gyapong JO, et al. Integration of control of neglected tropical diseases into health-care systems: challenges and opportunities. Lancet. 2010;375:160–5.
34. WHO. The world health report 2006 – working together for health. Geneva: WHO; 2006.
35. TDR Disease Reference Group on Chagas Disease, Human African Trypanosomiasis and Leishmaniasis, Weltgesundheitsorganisation. Research priorities for Chagas disease, human African trypanosomiasis and leishmaniasis: technical report of the TDR Disease Reference Group on Chagas Disease, Human African Trypanosomiasis and Leishmaniasis. Geneva: World Health Organization; 2012. 100 p. (WHO technical report series).
36. Sales Junior PA, Molina I, Fonseca Murta SM, Sánchez-Montalvá A, Salvador F, Corrêa-Oliveira R, et al. Experimental and clinical treatment of Chagas disease: a review. Am J Trop Med Hyg. 2017;97(5):1289–303. Available from: https://www.ncbi.nlm.nih.gov/pmc/articles/PMC5817734/. Accessed 8 Nov 2017.
37. Brener Z, Cançado JR, Galvão LM, Da Luz ZM, Filardi LS, Pereira ME, Santos LM, Cançado CB. An experimental and clinical assay with ketoconazole in the treatment of Chagas disease. Mem Inst Oswaldo Cruz. 1993;88:149–53.
38. Chatelain E. Chagas disease research and development: is there light at the end of the tunnel? Comput Struct Biotechnol J. 2016;15:98–103. Available from: https://www.ncbi.nlm.nih.gov/pmc/articles/PMC5196238/. Accessed 18 Nov 2018.
39. Pinazo M-J, Pinto J, Ortiz L, Sánchez J, García W, Saravia R, et al. A strategy for scaling up access to comprehensive care in adults with Chagas disease in endemic countries: the Bolivian Chagas platform. PLoS Negl Trop Dis. 2017;11(8):e0005770.

40. Sperandio da Silva GM, Mediano MFF, Hasslocher-Moreno AM, Holanda MT d, Silvestre de Sousa A, Sangenis LHC, et al. Benznidazole treatment safety: the Médecins Sans Frontières experience in a large cohort of Bolivian patients with Chagas' disease. J Antimicrob Chemother. 2017;72(9):2596–601.
41. Basile L, Oliveira I, Ciruela P, Plasencia A, Working group for developing the Ca collective. The current screening programme for congenital transmission of Chagas disease in Catalonia, Spain. Eurosurveillance. 2011;16(38). Available from: http://www.eurosurveillance.org/content/10.2807/ese.16.38.19972-en. Accessed 18 Nov 2018.
42. Castillo-Riquelme M. Chagas disease in non-endemic countries. Lancet Glob Health. 2017;5(4):e379–80. Available from: https://www.thelancet.com/journals/langlo/article/PIIS2214-109X(17)30090-6/abstract. Accessed 18 Nov 2018.
43. Meymandi S, Hernandez S, Park S, Sanchez DR, Forsyth C. Treatment of Chagas disease in the United States. Curr Treat Options Infect Dis. 2018;10(3):373–88. Available from: https://www.ncbi.nlm.nih.gov/pmc/articles/PMC6132494/. Accessed 18 Nov 2018.
44. Center of Excellence for Chagas Disease at Olive View UCLA Medical Center. 100&Change Solutions Bank. 100 and Change. Available from: http://100andchange.foundationcenter.org/profiles/811. Accessed 15 Nov 2018.

Chagas Disease: A Parasitic Infection in an Immunosuppressed Host

13

Maria Aparecida Shikanai-Yasuda,
Eros Antonio de Almeida, Manuel Carlos López,
and María-Jesús Pinazo Delgado

13.1 Introduction: Immunological Response of Human Host Against *T. cruzi* Infection and Biological Implication of Immunosuppression

Trypanosoma cruzi is the etiologic agent of Chagas disease (CD), also known as American trypanosomiasis. From a clinical point of view, CD has a variable clinical presentation and progression. In most individuals, the acute phase of infection is asymptomatic or presents a nonspecific symptomatology. The individuals exhibit patent parasitemia, which induces a strong activation of the immune system, including high levels of plasmatic cytokines, intense activation of B and T lymphocytes, and inflammatory reactions in the infected tissues [1–3]. Although the activation of

---

Maria Aparecida Shikanai-Yasuda and Eros Antonio de Almeida contributed equally to this work.

M. A. Shikanai-Yasuda
Department of Infectious and Parasitic Diseases, São Paulo, Brazil

E. A. de Almeida
State University of Campinas-UNICAMP, Campinas, São Paulo, Brazil
e-mail: masyasuda@usp.br

M. C. López
Instituto de Parasitología y Biomedicina López-Neyra, Consejo Superior de Investigaciones Científicas, Granada, Spain

NHEPACHA network, Barcelona, Spain
e-mail: mclopez@ipb.csic.es

M.-J. Pinazo Delgado (✉)
NHEPACHA network, Barcelona, Spain

Barcelona Institute of Global Health Research (ISGlobal), Hospital Clínic, Universidad de Barcelona, Barcelona, Spain
e-mail: mariajesus.pinazo@isglobal.org

M.-J. Pinazo Delgado, J. Gascon (eds.), *Chagas Disease*,
https://doi.org/10.1007/978-3-030-44054-1_13

host immune response results in a dramatic reduction in parasite load, this response fails to complete *T. cruzi* clearance, and the parasite manages to hide in tissues that are less accessible to the immune response, resulting in infection chronicity [4]. In the chronic phase, there exists a fragile balance between the host immune response and parasite replication that keeps the patients in a clinically silent asymptomatic stage over years or even decades (roughly 60% of patients). The loss of this balance is crucial for the progression of the sickness; approximately 30% of chronically infected individuals will develop cardiac symptoms (associated with severe myocarditis), and up to 10% of patients will present a neurological dysfunction and/or damage in the digestive system over time [3].

The infection by *T. cruzi* parasite triggers multiple immune mechanisms in their host to combat the pathogen. These are given at the innate and adaptive level, as well as on a humoral and cellular scale. Because these are intracellular parasites that replicate within cells, the cell-mediated response of the host adaptive immunity plays a critical role. This function is mainly orchestrated by T lymphocytes, which recognize parasite antigens and promote specific functions to control the infection [5]. The lymphocyte subpopulation of $CD4^+$ T cells, together with $CD8^+$ T cells, is one of the main participants of the adaptive immune response. $CD8^+$ T lymphocytes with cytotoxic abilities, known as CTL, are essential for controlling the intracellular infection, but require the help provided by the cross-priming of $CD4^+$ T cells to reach the memory phenotype and be autonomous in a secondary expansion following re-encounter with the antigen [6]. This subset of $CD8^+$ T cells is critical in the control of the *T. cruzi* intracellular infection [7–9]. It has also been reported that the antigen-specific T cells that produce Th1-like cytokine networks, such as IFN-γ, TNF-α, and IL-2, and not Th2-type cytokines, are beneficial for antiparasitic action. These Th1 cytokines activate macrophages for eliminating the intracellular parasites and thus disrupting the progression of the infection [10]. $CD4^+$ T cells are also able to develop cytotoxic qualities, producing molecules such as perforin and granzyme B, similar to cytotoxic-$CD8^+$ T cells [11]. Recent reports have shown that the $CD4^+$ T helper 17 cells exhibit a protective role against the parasite and help to mitigate the outcome of the Chagas pathology [12, 13], as higher expression of IL-17 was found in those patients without cardiomyopathy. A subset of $CD4^+$ T cells known as regulatory T cells (Treg) are also critical for maintaining a homeostatic environment, avoiding an exacerbation of the immune response that causes tissue damage. Thus, a deficient suppressor function of Treg was correlated with the extent of cardiomyopathy [14]. However, there are some controversial results in the context of the *T. cruzi* infection that demonstrate the need for more knowledge [15]. Several groups have shown that $CD4^+CD8^+$ T cells comprise mature T cells that are capable of being activated, to respond to specific antigen-producing cytokines and cytotoxic molecules and to migrate to inflamed tissues [16, 17]. An increased frequency of this circulating subset of cells expressing greater amounts of activation markers and cytotoxic molecules are present in chronic CD patients versus healthy subjects, and there is also a higher frequency in asymptomatic versus cardiac CD patients [18]. These cells were also found in the inflammatory infiltrates of cardiac tissue from a patient who underwent cardiac transplant [19]. Thus, it was suggested that the balance between a regulatory and effector immune responses is crucial to avoid the development of the symptomatic chronic phase.

The cellular immune response of the host has been demonstrated to be compromised after a long period of chronic infection by *T. cruzi*. The immune cells undergo several processes that affect their functional capacities, with an impaired ability to produce cytokines and cytotoxic molecules against the infectious agent that lead to an uncontrolled infection and the evolution of the CD [18, 20, 21]. The persistence of pathogen causes a gradual loss of the response capacity of antigen-specific T cells due to an increase in the expression and coexpression of inhibitory receptors in the membrane of the host immune cells in a manner known as exhaustion process. A higher level of expression and coexpression of inhibitory receptors was detected in the different populations of T cells from chronic patients, which was related to the progression and severity of the chronic disease [3]. Thus, symptomatic patients had a significantly higher frequency of T cells coexpressing inhibitory receptors than the asymptomatic patients, which were markedly superior in those patients with severe cardiac alterations [18, 20, 21]. Furthermore, an inverse correlation between the degree of T-cell exhaustion process and the multifunctional response of *T. cruzi*-specific T cells was described [3]. The expression of some inhibitory receptors, such as PD-1 and CTLA-4, was also detected in the infiltrated cells of myocardial explants of patients with severe cardiomyopathy [22, 23]. These results suggest that the cellular exhaustion process and, consequently, the failure in the T-cell response could lead to the loss of parasite control and evolution of the disease towards an advanced severe disease stage [3]. The quality of the host immune responses against *T. cruzi* correlated with the disease severity in chronic CD patients.

Other alterations of the immune system related to the differentiation level and the phenotypic profile of the T-cell subsets have been associated with the progression of the disease or related to a more advanced severe disease. Thus, reported data showed that chronic CD patients presented a lower frequency of central memory cells (T_{CM}), lower numbers of T naïve ($T_{NAÏVE}$) cells, and a higher percentage of effector memory (T_{EM}) $CD8^+$ T cells than healthy donors. The chronic patients were also enriched with cells in a late stage of differentiation compared to healthy donors [21, 24]. Furthermore, the highest percentage of T_{EM} cells and lowest of T_{CM} are present in those patients with a severe more advanced pathology [21]. In addition, the symptomatic patients have a statistically superior percentage of late differentiation cells than that observed in asymptomatic patients [21]. In the $CD4^+$ T-cell compartment, the frequencies of $T_{NAÏVE}$ and T_{CM} cells were significantly decreased in patients with cardiac disorders compared to those patients without apparent symptoms [24].

Remarkably, it was recently shown that treatment induces an improvement in the quality of the antigen-specific response in all of the different subsets of $CD8^+$ T cells in these patients, increasing the frequency of antigen-specific $CD8^+$ T_{CM} and T_{TE} cells, decreasing the coexpression of inhibitory receptors, and improving the multifunctional activity of these cells [25]. Likewise, treatment increases the subpopulation of $CD4^+CD8^{high}$ T cells (described as a more activated population of this subset of T cells) while simultaneously reducing the $CD4^+CD8^{low}$ T-cell subpopulation that derives from the terminal senescence of the $CD4^+$ T-cell population [18]. At the heart level, a lower percentage of T cells with a low degree of differentiation was found in heart explants of cardiac chronic Chagas disease patients than in noninfected patients with similar myocardial alterations [22]. These cells have a good proliferative potential and express the T-bet transcription factor associated to Th1 profile, and do not express senescence markers [22]. All these results suggest that the trypanocide treatment improves the quality of the functional response of different T-cell subsets against the *T. cruzi* infection [3].

Even though the treatment goal for infectious diseases is pathogen elimination, some researchers suggested that control and reduction of the *T. cruzi* burden would be an important therapeutic outcome to be considered as it will convert CD into a controlled chronic disease [26]. However, reactivation of the disease could occur along all the chronic phase of the disease [27]. Infection reactivation takes place when the immune system of a *T. cruzi* chronically infected host is compromised, resulting in a reduced ability of the host to control the infection [28]. In CD, the most commonly detected effect after reactivation of the infection is a significant increase of parasitemia [29], sometimes only detectable after symptoms and signs, such as rapid weight loss, general malaise, myocarditis, and cutaneous and neurological manifestations [30–33]. Reactivation of chronic Chagas disease is commonly linked to immunosuppressive treatments which have to be administered previously to organ transplant in those patients that require it as a consequence of the CD [30]. In addition, HIV infected individuals, patients suffering a neoplasia, and those with an autoimmune disease who have been treated with immunosuppressive chemotherapy also have the potential to reactivate CD [34]. Reactivation of CD led to a high rate of morbidity and mortality, and complicates the clinical management of CD [27]. Thus, it is recommended that patients with positive *T. cruzi* serology must be monitored during immunosuppressive treatment period [27, 29, 34].

Recently, it has been reported that the interactions between the parasite and the immune system of pregnant mice play an important role in *T. cruzi* congenital transmission [35]. Using an experimental murine model of *T. cruzi* chronic infection, it was observed that the mothers that transmitted the parasite had a higher frequency of T cells that expressed and co-expressed inhibitory receptors as well as a lower frequency of multifunctional parasite-specific $CD4^+$ and $CD8^+$ T cells than those that did not transmit the infection, even though the parasitemia was similar in both groups [35]. Thus, although congenital *T. cruzi* infection involves multiple factors, the quality of the T-cell response can be considered as a determinant factor in the congenital transmission of CD [3].

13.2 *Trypanosoma cruzi* Infection in the Immunosuppressed Host

13.2.1 Reactivation of Chagas Disease: Definition

The acute symptoms of *T. cruzi* infection and/or the increase in parasitemia in patients with chronic infection is called **reactivation** of the disease. Reactivation should appear in immunosuppression situations. Inclusion criteria could be parasitological and/or clinical findings.

Parasitological reactivation usually precedes clinical reactivation, and inclusion criteria are positive direct microscopy on peripheral blood or any biological material and/or presence of abundant amastigote nests in the middle of acute inflammatory infiltrate.

Clinical reactivation should be considered with an unexplained febrile illness, skin involvement (including panniculitis and erythema nodosum) [36–38], myocarditis, or meningoencephalitis [38–40], but symptoms usually appear after a parasitological burden increase in the site of reactivation, sometimes non-expressed in the peripheral blood. Involvement of digestive tract, uterus, cervix, peritoneum, and ocular muscle was seldom described [38, 39]. Vertical transmission of severe disease in the newborn was reported [41], including an asymptomatic reactivation during pregnancy [38].

Meningoencephalitis in Chagas disease reactivation is usually presented as febrile syndrome, intracranial hypertension syndrome (headache), and cerebral involvement syndrome (seizures, focal signs, sensory and consciousness disturbances, progressing to coma). Signs of encephalitis are predominant with rare signs of meningitis. Computed Tomography (CT) and Magnetic Resonance (MNR) show usually pseudotumor lesions (>90%), with mass effect and ring uptake of contrast in the white matter of the cerebral hemispheres, thalamus, base nuclei, and cerebellum. MNR is more sensitive in the brainstem and cerebellum and CT rarely could be normal. In the cerebrospinal fluid (CSF), there is a mild increase of lymphomononuclear cells (less than 100 cells/mm^3), mild or moderate increase of proteins, and normal or mild decrease of glycorrachia. As differential diagnosis toxoplasmosis, tuberculosis, cryptococcosis, primary CNS lymphoma, and progressive multifocal leukoencephalopathy should be considered.

Myocarditis in Chagas disease reactivation might be similar to the natural evolution of chronic cardiopathy, making this diagnosis difficult, usually not suspected when meningoencephalitis is present. It is characterized by severe arrhythmia (ventricular extra systoles, supraventricular arrhythmia, flutter and atrial fibrillation), and signs or symptoms of different degrees of heart failure (tachycardia, generalized edema, hepatomegaly, dyspnea). Pericarditis or increased cardiac volume could be shown on the echocardiogram. Clinically the most important differential diagnosis is congestive heart failure or severe arrhythmia due to chronic chagasic cardiopathy.

13.2.1.1 Diagnosis of Chagas Disease Reactivation

Chagas disease reactivation *is confirmed* by the presence of the parasite on direct microscopy (Gold Standard) of the peripheral blood (CSF or other biological secretions) fresh preparations or by more sensitive concentration methods (microhematocrit, Strout methods, or buffy coat, see Chap. 5). CSF was positive in 78.1% of suspected cases and 31.6% in the peripheral blood.

Indirect parasitological methods and qualitative PCR (even fewer sensitive primers) or quantitative PCR are not recommended for diagnosis of reactivation since they are positive in chronic Chagas disease cases without reactivation. Very high number of parasite copies by quantitative PCR could suggest the diagnosis of reactivation, but intermediate high levels similar in coinfected patients without reactivation occurred in non-severe cases of Chagas disease reactivation [42].

On histopathology, parasites are also seen in the central nervous system (CNS) in the macrophages and in microglia cells associated with necrohemorrhagic lesions in the subcortical regions (white matter). The parasites are also observed in the cardiac myocytes and in the interstitial tissue associated.

Immunohistochemistry could help to confirm the diagnosis since several opportunistic infections can coexist. Lesions are diffuse and the inflammatory exudate is mononuclear. Acute inflammation of the epicardium, endocardium, and excito-conductive system may also occur and more rarely, trypomastigotes were found in the pericardial fluid.

13.2.1.2 Etiological Treatment of Chagas Disease Reactivation

Benznidazole (BNZ) is the drug of choice, employed in 87% of treated patients with Chagas disease reactivation [38]. Early diagnosis and treatment are crucial since the prognosis is poor in cases of meningoencephalitis and/or severe myocarditis even when early introduced. Nifurtimox (NFX) used as a second-line drug. Dose for BNZ was 5 mg/kg/day and for NFX 8 mg/kg, 2–3×/day for at least 60 days. Involution of parasitemia by direct methods in the first 2 weeks and by indirect parasitological and molecular methods later on is associated with clinical improvement with involution of signs and symptoms.

Adverse effects may occur after the first week and were represented by exanthema, pancytopenia, particularly granulocytopenia and later on, peripheral neuropathy. These events represent about 18% of the patients treated in a prospective study [38] but other unspecific signs and symptoms are more common like gastrointestinal discomfort nauseas, vomiting, and intestinal pain. Agranulocytosis and Stevens–Johnson syndrome are rarely described.

13.2.2 Coinfection *Trypanosoma cruzi*/HIV

Around 7.3 million of individuals living with HIV infection in 2017 were registered worldwide; about 1.8 in Latin America (2017) [43, 44]. Migration of HIV infected groups from big cities to sub-urban and rural areas, in parallel with the urbanization of CD in endemic and non-endemic areas contribute to the coexistence of coinfection *T. cruzi*/HIV. This coinfection usually occurs in people who have acquired Chagas disease previously to HIV infection, mainly via vector transmission and rarely via transfusion of blood and blood products [38]. Chagas disease affects about six million people in Latin America. The prevalence of *T. cruzi*/HIV is underestimated and it is based on unicenter studies and case reports. In Brazil, it ranges from 1.3% to 5.0% [39, 45] in infected HIV individuals and in Argentina 2.6–8.9%, the latter in endovenous drug users [46]. A map pointing the reported cases of coinfected persons around the world globally is accessible on WHO site [47].

Almeida et al. [39] reported 291 cases of coinfection HIV/*T. cruzi*, most of them from Brazil up to March 2010. Transmitted in the majority of the cases by the vector, predominant chronic form of Chagas disease in these patients is the indeterminate form (50.8%), followed by cardiac form (37.3%,), digestive (5.1%) and cardiac plus digestive forms (6.8%). In Spain, 15 additional cases of coinfection were reported [28, 34] and in Argentina [48] 80 cases.

As Chagas disease and HIV infection are underdiagnosed, a National Network for Attention and Studies on *T. cruzi*/HIV coinfection was created in Brazil in 2006, and later become an International Network, open for all countries [49]. This network aims to contribute for structuring a comprehensive care network and provide a continuing education network for health care professionals and to identify priorities for research on this subject.

13.2.2.1 Diagnosis of Chronic Chagas Disease and HIV Infection: Reactivation Criteria

Both *T. cruzi* and HIV infection diagnosis are usually based on positivity of serological methods. For *T. cruzi* infection, at least two serological reactions are necessary to confirm the diagnosis. Qualitative PCR is not recommended for diagnosis since it is much less sensitive than serology [42].

T. cruzi/HIV interaction, and increase of IL4/IFN-γ ratio in patients with CD and HIV infection in comparison to patients with CD indicate an imbalance of TH2 response in coinfection [50]. In SCID mice *T. cruzi* infected, increased number of parasite and decrease of IFN-γ were associated with lower tissue inflammation and early mortality [51]. Additionally, bidirectional influences in this interaction shows high levels of parasites in tissues and peripheral blood in human and mice as well as decreased HIV replication in *T. cruzi* infected human and in placenta [52] and macrophage cultures.

Reactivation risk in patient underwent HIV infection is associated with CD4 <200 cells/mm^3, explaining the inability of macrophages to kill and control parasite multiplication.

The reactivation of Chagas disease was recognized as the defining condition of AIDS in Brazil since 2004 and since 2005, by Pan American Health Organization and World Health Organization [49], due to the following characteristics: (a) severe morbidity and high mortality; (b) worse response to antiparasitic treatment; (c) severe immunosuppression expressed by CD4$^+$ T cells <200 cells/mm^3 and increased parasitemia [39, 40]; (d) increased rate of maternofetal transmission of Chagas disease.

Reactivation of Chagas disease occurs in about 10–20% of patients with *T. cruzi*/HIV coinfection [38, 40]. In a prospective study, the reactivation rate observed was 21.7% [38], but this rate changed to 10.25% when only patients followed up by at least 1 month were considered (therefore excluding reactivation at the first assessment). Reactivation of Chagas disease in an AIDS patient was first described [53] in 1990 by Del Castillo et al. and in 1922, other case was registered in the USA, and it was communicated in a Congress before 1990, according to Gluckstein et al. [54]. However, in 1988, Spina França et al. reported both parasites and anti-*T. cruzi* antibodies in the CSF of a patient with HIV and Chagas disease reactivation [55]. Most of the cases were further registered in Brazil [38, 39, 56] and Argentina [41, 48, 57].

Clinical Aspects of Chagas Disease Reactivation in HIV Infected Patients
In 2011, Almeida et al. revised 120 cases of reactivation of Chagas diseases in HIV infected patients, 2/3 of them in the central nervous system, and about 17% in the myocardium [39], even if myocarditis cases seemed to be underestimated as indicated by necropsy series. Low CD4 levels (mean 98 cells/mm^3) and 73.3% of the deaths described were associated with reactivation of Chagas disease; 88.9% of them related to meningoencephalitis and 12% in those with myocarditis. Since then, few cases of reactivation of Chagas disease were recorded in Brazil, Colombia, Chile, and the USA, including congenital disease due to maternal-fetal transmission [33, 57–61]. Recently, nine cases of reactivation (11.25%) among 80 coinfected patients were described in Argentina [48].

Other Laboratorial Tests
CD4$^+$ T cells <200/mm^3 in the peripheral blood is found in more than 80% of the cases of reactivation. A few cases were described in patients with >350 cells/mm^3. High viral load associated with Chagas disease reactivation possibly expresses non-adherence to HAART; but its importance as cofactor or reactivation is a matter of discussion.

13.2.2.2 Mother to Child Transmission

Maternofetal transmission in pregnant women HIV infected with reactivation of Chagas disease [38, 41] is reported in more than 50% of the cases, in comparison with patients without HIV infection (1–13.8%). Mortality is over 70% and severe morbidity was associated with abortus, low weight for the gestational age, and even pre-term neonate with meningoencephalitis or sepsis. It is possible that the higher parasitemia described in coinfection in comparison with Chagas disease without HIV infection [42] might be responsible by the increased rate of transmission of Chagas disease to the newborn. The influence of HAART on the maternofetal transmission of Chagas disease is not well known.

13.2.2.3 Etiological Treatment and Prognosis in Patients with Chagas Disease-HIV Infections

In case of reactivation, early diagnosis and treatment is crucial since the prognosis is poor in cases of meningoencephalitis and/or severe myocarditis even when early introduced. For patients treated for at least 30 days, survival rate is approximately 80% vs. 20% when this period time was less than 20 days [40]. Since this is a life-threatening disease for pregnant women with Chagas disease reactivation and her baby, these patients should receive benznidazole as antiparasitic treatment. As alternative when benznidazole could not be prescribed, nifurtimox was reported for treatment for 14% of the patients [38]. Although both fluconazole and itraconazole were employed for the treatment of less than 3% of the patients, sometimes after benznidazole, there are not enough evidence to recommend them as first-line therapy.

To avoid immune reconstitution syndrome, the introduction of HAART is recommended after 2 weeks of the start of antiparasitic treatment. Maintenance therapy with benznidazole is not a consensus but 5 mg/kg/day 3× per week may be prescribed [40].

There are no reliable data about results of treatment of coinfected patients with low parasitemia. But when parasitemia is high or persistent, evolution shows negativization of parasitemia in more than 60% [40, 62]. Similar to Chagas disease patients, antiparasitic treatment may be offered informing the possible results for the patients.

Prognosis

Mortality in cases of reactivation is 73.3% on dependence of severity of the involvement, high in meningoencephalitis and no death in oligosymptomatic cases [38]. Considering coinfection without Chagas disease reactivation, the average of survival time-period was 34 months in relationship to 11 months in patients with reactivation [39]. The role of HAART in reducing the incidence of reactivation of Chagas disease in the last decade seems to result from the recovery of cellular immunity and macrophagic activity, expressed by the increased number of CD^+ T4 cells, and consequent viral and parasite control.

Predictive Factors

The only known factor predictive of reactivation is the increase of parasitemia as reported in a prospective study in a coinfected patient who presented simultaneously a mild decrease of CD4 T cells/mm^3 and febrile illness with monolike syndrome [38]. Patients with higher parasitemia by xenodiagnoses (higher than 20% of nymphs positive) will present reactivation in the following 5 years [38]. Prospective monitoring with quantitative PCR is recommended to detect increased levels of parasitemia.

13.2.3 Neoplasia and *Trypanosoma cruzi* Infection

Chagas disease can be associated to other diseases. While this may happen on both stages of the disease, it is at its chronic stage that the comorbidities happen more frequently, due to the long evolution and aging of the carriers in regions where the main forms of transmission are controlled. Changes in the epidemiology of Chagas disease, previously rural and nowadays also urban, as well as better conditions for access to better health services, allowing for the diagnosis and treatment, contribute for the concomitance of diseases connected to these places and of adults and elderly people. It has been observed that comorbidities with Chagas disease can change the natural history of the parasitosis, as well as being influenced by it on its natural course [63, 64]. Potentially, Chagas disease, as it happens with any other infectious disease, can present a modification of its natural history in case of comorbidities that cause immunodepression, which became very evident after the surfacing of AIDS [39]. The comorbidity of Chagas disease with malignant neoplasia, however, happened before the coming of the infection by HIV and has been evaluated through time, although infrequently.

The association between Chagas disease and neoplasia must be focused under some specific aspects. The first of them corresponds to the association of the two diseases and the corresponding influence of one on the other. Old studies on necropsy material and on patients [65, 66] seeking to verify the frequency of Chagas disease and malignant neoplasias did not verify a higher frequency of either. It was noted that necroscopic results were those habitually found in Chagas disease, chronic stage, as well as those of neoplasias. The same thing happened with respect to the clinical observation, in which the frequency of Chagas disease, on individuals with neoplasia or not, did not present significant differences. Likewise, the frequency of malignant neoplasia in carriers of Chagas disease was not different from that of non-carriers. Therefore, both diseases followed their natural courses. More recently, this data was confirmed at a better-structured clinical study [66], in which carriers of Chagas disease with malignant neoplasia were accompanied for a period, compared to Chagas carriers without neoplasia. The clinical and parasitological situation presented no differences between the two groups. Likewise, during the necroscopic study of cases of patients with malignant neoplasia without a chemotherapy treatment, there was an in-depth search for changes to the natural history of the Chagas disease, which were not found [67].

Therefore, the behavior of Chagas disease in carriers of malignant neoplasia does not undergo modifications, even if it is observed with respect to neoplasia. While there is no controversy about this specific aspect involving the association of Chagas disease to neoplasia, a higher incidence of uterine leiomyoma in carriers of Chagas disease has been described [68]. However, another study did not verify a higher frequency of gynecological neoplasia under necropsy, concluding that there are no risk factors between them. Both in older studies and in more recent ones evaluating the association of the two diseases, the types of neoplasia analyzed involve all lineages of cancer, the chronic stage and the many clinical forms of Chagas disease [64, 65, 67, 69].

Another situation involving the association of both diseases refers to the immunosuppressor treatment for malignant neoplasia, which corresponds to a new, complicating factor. Infectious diseases generally worsen in the presence of immunosuppression, reactivating those that are latent and flaring up the chronic ones, which has been demonstrated in carriers of the coinfection between *T. cruzi* and AIDS [39]. Thus, there have been accounts of isolated cases of leukemia or lymphoma associated to Chagas disease and treated with immunosuppressor drugs that developed serious encephalitis or myocarditis, resulting in death [70, 71]. The necropsy reveals the marks of acute Chagas disease, with the presence of the parasite in the lesions, generally in places different from the usual chronic disease. These clinical and anatomopathological presentations were, indeed, reactivation of the chronic Chagas disease. Again, in the *T. cruzi* and AIDS coinfection, this became more evident, but it already occurred in the association between Chagas disease, malignant neoplasia, and immunosuppressor treatment, having been described in the nervous system, heart, and other regions of the body, such as larynx, esophagus, and stomach [72]. The frequency with which this reactivation occurs associated to Chagas disease, malignant neoplasia, and chemotherapy treatment is unknown. In literature, there are descriptions of cases and, when the samples are reported, the reactivation is not always found [73]. Some factors are pointed as responsible for this situation, such as the type and dosage of the chemotherapy medicine, the stage of Chagas disease, the cycle and genetics of the parasite, the specific immunological condition of the patient, and the type of neoplasia [74]. The methodologies used in the reports of the literature do not systematically analyze all of these possibilities and, thus, the differences between the studies can be understood. However, some data must be highlighted as relevant. There are no reports of reactivation events on patients associating Chagas disease to malignant neoplasia without an immunosuppressor treatment. Thus, immunological alterations inherent to neoplasia would not suffice to modify the one necessary for controlling the parasitemia. The reactivation of Chagas disease on carriers of malignant neoplasia has only been described in literature in association with leukemias or lymphomas and chemotherapy. When the association to solid tumors of all origins is reported, even in the immunosuppressor treatment of different types, the flare-up has not been observed, with the natural evolution of Chagas disease maintained. Thus, the immunodepression that happens in hematological and lymphatic neoplasia, associated to the therapeutic immunosuppression would be needed to compromise the immunological control of the *T. cruzi*, allowing for the reactivation. Such specific immunodepression for hematological neoplasia has been reportedly related to the modifications of the immunological functions of the B and T cells, observed in Hodgkin's lymphoma [75].

Another situation involving the association between the two diseases is that in which Chagas evolves with megaesophagus and megacolon. The digestive mega would potentially favor the surfacing of malignant neoplasia, since the stasis of the bolus in the esophagus and of the feces in the colon is considered to be a risk factor for the neoplasia becoming malignant [76]. In the esophagus, the stasis would be responsible for inflammation, ulcers, and pseudo polyps on the lining, causing accelerated cellular proliferation, with the possibility of a malignant transformation. The presence of gastroesophageal reflux would also contribute to causing chronic esophagitis. In the colon, the fecal stasis would allow for a higher duration of the contact between the carcinogenic substances present in the feces with the lining, with predisposition to neoplasia. There have been descriptions on the literature regarding the relation between malignant neoplasia and megaesophagus, but in very varied frequencies, and as the casuistry increases, the neoplasia frequency decreases; thus, it is impossible to discard the hypothesis of these being random factors [77, 78]. It is also noticeable, likewise, that this relation occurs regardless of whether or not the cause of the megaesophagus is Chagas-related, as it is reported in regions that are not endemic for Chagas disease, in idiopathic achalasia, with the same frequency. Thus, the association would happen more with the dilation of the organ than with the etiology. Cases of this association have been observed in the service of care to carriers of Chagas disease by the author, but in a frequency similar to that of other malignant neoplasia; it does not seem to be a cause-and-effect relation, as it happens with other literature reports. The same cannot be considered for the relation between megacolon and colon cancer, which does not seem to occur. Extensive reviews available in the literature for chagasic megacolon evaluating the relation with colon cancer have demonstrated its nonexistence [79, 80]. People have attributed it to factors that block the action of fecal carcinogenic, such as a diet that is rich on vegetable fibers and anaerobic microbiotics, which could be exacerbated in the cases of megacolon. It can be interpreted that the negative relation between megacolon and colon cancer would be with the dilation, not the etiology. The verification that, in Wistar rat with megacolon produced by benzalkonium and treated with a chemical inducer of colon cancer, tumour ocurred less frequently in animals with a megacolon, contributing to the sense that the enteric denervation would have a fundamental role in chronic carcinogenesis, aside from those mentioned and not with the Chagas disease etiology [81]. However, the same does not hold true for chagasic mega-esophagus, although the denervation exists with the same intensity. Thus, there are carcinogenic factors linked to chagasic megacolon and anti-carcinogenic factors well documented in literature [82], but there is no definition yet of which of them would be fundamental; the authors conclude that more investigations should be performed.

Experimental studies on chemically induced colon tumors demonstrate a lesser frequency of those in Wistar rats infected with *T. cruzi*, without a megacolon, suggesting the participation of the parasite in the protection against colon cancer [83]. A factor that protects against cancers in general and that depends on *T. cruzi* has been proposed since the last century, when Russian researchers published reports of cure and reduction of tumors in experiment animals treated with extracts of the parasite [84]. However, other authors at the same occasion did not manage to repeat such experiments, as well as not verifying a higher frequency of malignant tumors in clinic studies with chagasic patients [64, 85]. Considering the possibility that infections caused by different microorganisms present antitumoral activities, including protozoa, the study of the carcinolytic action of *T. cruzi* kept on being explored and the evidence of this situation has been verified more recently [86]. The mechanisms proposed as responsible for this tumoricidal effect would involve the immunological response the *T. cruzi* triggers in the host, which would also act on the tumor [87] and properties intrinsic to the parasite [88]. A protein with low molecular weight named calreticulin, which has properties of interaction with components of the complement and the capacity to inhibit the angiogenesis, which have an inhibitory relation with malignant neoplasia, has been identified in the *T. cruzi* [89]. Such protein is mainly in the endoplasmic reticulum and has several effects on the organism, being present in animals and plants, with an identity, greater or lesser, among the species. The activity of calreticulin in the *T. cruzi* has also been verified, in animals, to delay the tumoral growth of melanoma [90]. However, there is no proof yet that these tumoricidal actions of *T. cruzi* represent advances in preventing, curing, or delaying the growth of malignant neoplasia in carriers of Chagas disease.

13.2.4 Solid Organ Transplantation and Chagas Disease

Organ transplantation in patients with chronic CD and the use of organs from infected donors has been a matter of debate for many years in endemic countries, and in the last decade also in non-endemic countries. Despite the situation in both scenarios is different, there is a common concern: we assist to a disparity between the number of organs potentially available to be transplanted and the number of persons in the waiting list. Due to this situation, there is a need to consider the use of organs from donors with some specific infections whose transmission could be easily monitored as well as their potential risks, through protocols that facilitate effective prophylaxis or, if not possible, at least preemptive strategies that can minimize risk of disease in the recipient. Several guidelines on the indication and acceptability of solid organs coming from *T. cruzi* infected donors have been published [91, 92], as well as the management of *T. cruzi* infected receptors that needs to be immunosuppressed before transplantation [91, 92]. In this regard, the wider experience was based on heart transplantation [93, 94].

Transmission of *T. cruzi* infection is not universal after solid organ transplantation from *T. cruzi* infected donors, but acute forms of the disease may appear in immunosuppressed hosts that previously did not have the infection [95, 96]. Transmission occurs in 15–30% of recipients of an organ (other than the heart) from an infected donor. Patients at risk of having CD should be screened for disease before immunosuppression for two main reasons [91]: (1) to avoid and/or minimize the possibility of reactivation when immunosuppressed before organ donation; (2) to close monitor and follow up organ receipts, in order to early diagnose *T. cruzi* transmission.

Regarding recipients with CD, it is important to highlight that due to immunosuppression, they can also develop reactivation of CD, already described in Sect. 13.2.1 of this chapter, that has high mortality associated if untreated. The risk of reactivation in people suffering from CD at post-transplant period has been reported in different Latin American series, mainly in solid organ transplantation [97], and it shows differences by organ type. In case of kidney transplantation, reactivation risk was 9–16% [98]; 50–100% when heart was the organ transplanted [93, 99, 100]; and 17–40% in case of bone marrow transplantation [101].

In recipients with a negative PCR before transplant, a positive PCR after treatment confirms diagnosis. For recipients with a positive PCR before transplant, a quantitative PCR should be used to assess the increase in parasitic load after transplant.

13.2.4.1 Immunosuppression for Organ Transplantation in a Person with Chagas Disease: How to Deal with Reactivation Risk

When the receipt is a person with CD, it is recommended to establish a protocolized and close follow-up in order to early detect reactivation. Besides periodic clinical evaluation, PCR is the recommended test in immunosuppressed individuals in order to assess reactivation [102, 103]. Sequential PCR [104] (either quantitative or qualitative, but always better quantitative in order to assess an increase in parasitemia burden) weekly the first month after transplantation, every 2 weeks the next 2 months and then, monthly up to 6–12 months is the most common course used. After that, monitoring intervals can be lengthened, and any unexplained illness (e.g., fever of unknown origin) or suspected rejection event would be reason for more frequent monitoring [105]. After that, all infected transplanted recipients should be periodically (at least once a year) assessed for cardiomyopathy and gastrointestinal Chagas disease, together with a PCR test.

Immunosuppressive drugs in transplantation context, and its relation with reactivation episodes, are not well known. Among the reports on this regard, there is evidence of high incidence of parasite infection reactivation in Chagas disease patients on mycophenolate mofetil-based immunosuppressive regimen [106], compared to azathioprine.

In the context of CD in transplant recipients, the management of reactivation could be addressed by two main strategies:

1. Prevent reactivation. Prophylaxis recommendation post-transplant with BZD or NFX is not universal. There are reports in favor of BZD prophylaxis post-transplant, and there are reports in which there were no evidence of *T. cruzi* transmission in a negative recipient after a prophylaxis regime [107]. There is no prospective randomized evidence to support that trypanocide treatment before transplantation is useful to inhibit or to avoid post-transplant reactivation. However, infected candidates with proven parasitemia may benefit from trypanocide treatment before transplantation [108]. In case of infected living donors should receive trypanocide treatment for at least 30 days prior to donation to allow clearance of parasitemia. Donation should take place as soon as possible after completion of treatment [108].
2. Treat confirmed reactivations. A second strategy in order to manage reactivations is to ensure a close follow-up and to start a treatment with BNZ or NFX as soon as reactivation is diagnosed. CD reactivation etiologic treatment is indicated in people receiving solid organ transplantations from a *T. cruzi* infected donor with primary infection, confirmed clinical and/or parasitological reactivation and/or seroconversion (AII recommendation) [109]. There are several publications in which clinical and parasitological response to BNZ treatment was assessed, mainly after heart transplantation, but not exclusively [108]; and no differences in survival were found after 5 years of follow-up between recipients with and without Chagas disease after heart transplantation [110].

Regarding follow-up period after transplantation, there is no consensus, but several authors recommend at least 1 year, as the period in which the risk of reactivation is higher. Nevertheless, there are reports of late reactivation up to 6 years after transplantation (in a patient still with tacrolimus and mycophenolate) [111].

13.2.5 Systemic Autoimmune Diseases and Chagas Disease

The coexistence of CD and systemic autoimmune diseases, as systemic lupus erythematosus (SLE), psoriasis, and rheumatoid arthritis, has been rarely reported in scientific literature [48].

With the emergence of CD in non-endemic countries, the number of cases reported and the evidence on this topic has slightly increased [29, 48, 112]. To control systemic autoimmune diseases, chronic immunosuppressant treatment is necessary, and the effects of immunomodulatory drugs on *T. cruzi* parasitemia dynamics and CD course in humans are not well known, with only experimental evidence available from animal models with contradictory results [113–115]. There is no consensus on the follow-up strategy recommended. There is not enough evidence of an association between immunosuppressive doses of corticosteroids alone and higher rates of *T. cruzi* reactivation, and given the lack of evidence, no preemptive therapy is recommended [116].

Nevertheless, the early diagnosis of *T. cruzi* infection in patients underlying those diseases and the early indication of treatment improve prognosis. In immunosuppressed patients, trypanocidal treatment with BNZ or NFX is indicated (BII recommendation) [109]. No cases of reactivation have been reported up to now in patients treated timely [48, 117], and in a report of dos Santos-Neto et al., an accurate treatment shortly after a reactivation diagnosis was also successful [112]. Some authors suggest adding *T. cruzi* serology to initial screening studies in patients at risk of having CD before starting biologic therapies in CD and universal in CD endemic areas. There is no consensus on the follow-up strategy recommended.

13.3 Conclusions

Immunosuppression of any etiology is a health condition that could be intercurrent in people with *T. cruzi* infection, and is more and more frequent in endemic and non-endemic countries due to new epidemiological scenarios. In regions where the main forms of transmission of Chagas disease are controlled, the patients are at the chronic stage and more advanced ages, which allows for the presence of comorbidities. The evaluation of the parasitemia by *Trypanosoma cruzi* is considered necessary, as well as the presence of this parasite in locations where the reactivation of the disease is suspected. Early diagnosis of reactivation and accurate treatment and follow-up are key to overcome this severe and even fatal clinical situation, in immunosuppressed patients.

Emphasis on global surveillance is necessary to overcome the diagnostic limitation in HIV infected or *T. cruzi* infected patient and to establish control strategies. Sustainability in the training of human resources and the improvement of guidelines for the management of coinfected patients and for the early diagnosis, treatment, prophylaxis, and control of reactivation are highlighted, with emphasis on maternal-fetal transmission, in severe and in atypical reactivation.

Malignant neoplasia and systemic autoimmune diseases are among the comorbidities of concern, as they run their course with immunodepression and immunosuppression due to the chemotherapy treatment. Literature on the subject is limited, but enough to demonstrate that the natural evolution of Chagas disease is only modified in the presence of chemotherapy treatment, since there may be a reactivation. The occurrence is not frequent, but it is serious when the reactivation happens on the nervous system, as a pseudotumoral meningoencephalitis or in the heart, as myocarditis. In the association with the neoplasia of several origins, without chemotherapy, the natural evolution happens for both diseases.

Organs from donors with *T. cruzi* infection may save lives. Comprehensive guidelines in order to ensure a correct management in recipients with and without *T. cruzi* infection, and in donors with *T. cruzi* infection will minimize reactivation in these especially vulnerable groups.

In general, new tools to better manage people at immunosuppressive state, like molecular diagnostic methodologies and new treatment options will improve the comprehensive management of such potentially fatal health situations.

References

1. Rassi A Jr, Rassi A, Marin-Neto JA. Chagas disease. Lancet. 2010;375(9723):1388–402.
2. WHO. Chagas disease in Latin America: an epidemiological update based on 2010 estimates. Releve epidemiologique hebdomadaire. 2015;90(6):33–43.
3. Egui A, Lasso P, Perez-Antón E, Thomas MC, Lopez MC. Dynamics of T cells repertoire during *Trypanosoma cruzi* infection and its post-treatment modulation. Curr Med Chem. 2019;26:6519. https://doi.org/10.2174/0929867325666181101111819.
4. Cardoso MS, Reis-Cunha JL, Bartholomeu DC. Evasion of the immune response by *Trypanosoma cruzi* during acute infection. Front Immunol. 2015;6:659.
5. de Morais CG, Castro Lima AK, Terra R, dos Santos RF, Da-Silva SA, Dutra PM. The dialogue of the host-parasite relationship: *Leishmania* spp. and *Trypanosoma cruzi* infection. Biomed Res Int. 2015;2015:324915. https://doi.org/10.1155/2015/324915.
6. Novy P, Quigley M, Huang X, Yang Y. CD4 T cells are required for CD8 T cell survival during both primary and memory recall responses. J Immunol. 2007;179(12):8243–51.
7. Planelles L, Thomas MC, Alonso C, López MC. DNA immunization with the *Trypanosoma cruzi* HSP70 fused to KMP11 protein elicits a cytotoxic and humoral immune response against the antigen and leads to protection. Infect Immun. 2001;69(10):6558–63.
8. Martin D, Tarleton R. Generation, specificity, and function of $CD8^+$ T cells in *Trypanosoma cruzi* infection. Immunol Rev. 2004;201:304–17.
9. Morell M, Thomas MC, Caballero T, Alonso C, López MC. The genetic immunization with paraflagellar rod protein-2 fused to the HSP70 confers protection against late *Trypanosoma cruzi* infection. Vaccine. 2006;24:7046–55.

10. Rodrigues MM, Ribeirao M, Boscardin SB. CD4+ Th1 but not Th2 clones efficiently activate macrophages to eliminate *Trypanosoma cruzi* through a nitric oxide dependent mechanism. Immunol Lett. 2000;73(1):43–50.
11. Brown DM, Lampe AT, Workman AM. The differentiation and protective function of cytolytic CD4 T cells in influenza infection. Front Immunol. 2016;7:93.
12. Magalhaes LM, Villani FN, Nunes Mdo C, Gollob KJ, Rocha MO, Dutra WO. High interleukin 17 expression is correlated with better cardiac function in human Chagas disease. J Infect Dis. 2013;207(4):661–5.
13. Sousa GR, Gomes JA, Damasio MP, Nunes MC, Costa HS, Medeiros NI, Fares RC, Chaves AT, Corrêa-Oliveira R, Rocha MO. The role of interleukin 17-mediated immune response in Chagas disease: high level is correlated with better left ventricular function. PLoS One. 2017;12(3):e0172833.
14. Guedes PM, Gutierrez FR, Silva GK, Dellalibera-Joviliano R, Rodrigues GJ, Bendhack LM, Rassi A Jr, Rassi A, Schmidt A, Maciel BC, Marin Neto JA, Silva JS. Deficient regulatory T cell activity and low frequency of IL-17-producing T cells correlate with the extent of cardiomyopathy in human Chagas' disease. PLoS Negl Trop Dis. 2012;6(4):e1630.
15. de Araujo FF, Vitelli-Avelar DM, Teixeira-Carvalho A, Antas PR, Assis Silva Gomes J, Sathler-Avelar R, Otávio Costa Rocha M, Elói-Santos SM, Pinho RT, Correa-Oliveira R, Martins-Filho OA. Regulatory T cells phenotype in different clinical forms of Chagas' disease. PLoS Negl Trop Dis. 2011;5(5):e992.
16. Clenet ML, Gagnon F, Moratalla AC, Viel EC, Arbour N. Peripheral human $CD4^+CD8^+$ T lymphocytes exhibit a memory phenotype and enhanced responses to IL-2, IL-7 and IL-15. Sci Rep. 2017;7(1):11612.
17. Xie D, Hai B, Xie X, Liu L, Ayello J, Ma X, Zhang J. Peripheral $CD4^+CD8^+$ cells are the activated T cells expressed granzyme B (GrB), Foxp3, interleukin 17 (IL-17), at higher levels in Th1/Th2 cytokines. Cell Immunol. 2009;259(2):157–64.
18. Perez-Anton E, Egui A, Thomas MC, Puerta CJ, Gonzalez JM, Cuellar A, Segovia M, López MC. Impact of benznidazole treatment on the functional response of *Trypanosoma cruzi* antigen-specific $CD4^+CD8^+$ T cells in chronic Chagas disease patients. PLoS Negl Trop Dis. 2018;12(5):e0006480.
19. Giraldo NA, Bolanos NI, Cuellar A, Guzman F, Uribe AM, Bedoya A, Olaya N, Cucunubá ZM, Roa N, Rosas F, Velasco V, Puerta CJ, González JM. Increased $CD4^+/CD8^+$ double-positive T cells in chronic Chagasic patients. PLoS Negl Trop Dis. 2011;5(8):e1294.
20. Arguello RJ, Albareda MC, Alvarez MG, Bertocchi G, Armenti AH, Vigliano C, Meckert PC, Tarleton RL, Laucella SA. Inhibitory receptors are expressed by *Trypanosoma cruzi*-specific effector T cells and in hearts of subjects with chronic Chagas disease. PLoS One. 2012;7(5):e35966. https://doi.org/10.1371/journal.pone.0035966.
21. Lasso P, Mateus J, Pavia P, Rosas F, Roa N, Thomas MC, López MC, González JM, Puerta CJ, Cuéllar A. Inhibitory receptor expression on $CD8^+$ T cells is linked to functional responses against *Trypanosoma cruzi* antigens in chronic chagasic patients. J Immunol. 2015;195(8):3748–58.
22. Arguello RJ, Vigliano C, Cabeza-Meckert P, Viotti R, Garelli F, Favaloro LE, Laguens R, Laucella SA. Presence of antigen-experienced T cells with low grade of differentiation and proliferative potential in chronic Chagas disease myocarditis. PLoS Negl Trop Dis. 2014;8(8):e2989.
23. Albareda MC, Olivera GC, De Rissio AM, Postan MA. Assessment of CD8(+) T cell differentiation in *Trypanosoma cruzi*-infected children. Am J Trop Med Hyg. 2010;82(5):861–4.
24. Fiuza JA, Fujiwara RT, Gomes JA, Rocha MO, Chaves AT, de Araujo FF, Fares RC, Teixeira-Carvalho A, Martins-Filho OA, Cançado GG, Correa-Oliveira R. Profile of central and effector memory T cells in the progression of chronic human Chagas disease. PLoS Negl Trop Dis. 2009;3(9):e512.

25. Mateus J, Perez-Anton E, Lasso P, Egui A, Roa N, Carrilero B, González JM, Thomas MC, Puerta CJ, López MC, Cuéllar A. Antiparasitic treatment induces an improved CD8+ T cell response in chronic chagasic patients. J Immunol. 2017;198(8):3170–80.
26. Viotti R, Alarcón de Noya B, Araujo-Jorge T, Grijalva M, Guhl F, López MC, Ramsey J, Ribeiro I, Schijman A, Sosa-Estani S, Torrico F, Gascon J. Towards a paradigm shift in the treatment of chronic Chagas disease. Antimicrob Agents Chemother. 2014;58(2):635–339.
27. Perez CJ, Lymbery AJ, Thompson RCA. Reactivation of Chagas disease: implications for global health. Trends Parasitol. 2015;31(11):595–603.
28. da Costa SC. Immunocompromised host: from the early events until the impact of acquired immunodeficiency syndrome. Mem Inst Oswaldo Cruz. 2000;95(Suppl 1):141–4.
29. Pinazo MJ, Espinosa G, Cortes-Lletget C, Posada Ede J, Aldasoro E, Oliveira I, Muñoz J, Gállego M, Gascon J. Immunosuppression and Chagas disease: a management challenge. PLoS Negl Trop Dis. 2013;7(1):e1965.
30. Campos SV, Strabelli TM, Amato Neto V, Silva CP, Bacal F, Bocchi EA, Stolf NA. Risk factors for Chagas' disease reactivation after heart transplantation. J Heart Lung Transplant. 2008;27:597–602.
31. Camargos S, Vieira Moreira MD, Meneses Cury Portela DM, Imperes Lira JP, Santos Modesto FV, Marques Miranda Menezes G, Ribeiro Moreira D. CNS chagoma: reactivation in an immunosuppressed patient. Neurology. 2017;88(6):605–6.
32. Ferraresso MG, Torre AC, Piva MMM, Barcan L. Chagas disease reactivation: cutaneous manifestations in a transplanted patient. An Bras Dermatol. 2018;93(6):890–2.
33. Gomez CA, Banaei N. *Trypanosoma cruzi* reactivation in the brain. N Engl J Med. 2018;378(19):1824.
34. Salvador F, Sánchez-Montalvá A, Valerio L, Serre N, Roure S, Treviño B, Pou D, Sulleiro E, Bocanegra C, Molina I. Immunosuppression and Chagas disease; experience from a non-endemic country. Clin Microbiol Infect. 2015;21(9):854–60.
35. Egui A, Lasso P, Thomas MC, Carrilero B, González JM, Cuellar A, Segovia M, Puerta CJ, López MC. Expression of inhibitory receptors and polyfunctional responses of T cells are linked to the risk of congenital transmission of *Trypanoosma cruzi*. PLoS Negl Trop Dis. 2017;11(6):e0005627.
36. de Oliveira Carneiro da Motta J, de Oliveira KF, Caldas NG, de Lima Nogueira Guimarães A, Carvalho Costa IM, de Paula CD, Soares Takano GH. Cutaneous presentation of Chagas' disease reactivation in a heart-transplant patient. J Eur Acad Dermatol Venereol. 2017;31(2):e120–1.
37. Campos FP, Pansard HM, Arantes LC, Rodrigues AT, Daubermann MF, Azambuja MF, Argenta LC, Silva LA. A case of Chagas' disease panniculitis after kidney transplantation. J Bras Nefrol. 2016;38(1):127–31.
38. Sartori AM, Ibrahim KY, Nunes Westphalen EV, Braz LM, Oliveira OC Jr, Gakiya E, et al. Manifestations of Chagas disease (American trypanosomiasis) in patients with HIV/AIDS. Ann Trop Med Parasitol. 2007;101(1):31–50.
39. Almeida EA, Ramos AN Jr, Correia D, Shikanai-Yasuda MA. Co-infection *Trypanosoma cruzi*/HIV: systematic review (1980 - 2010) Coinfecção *Trypanosoma cruzi*/HIV: revisão sistemática (1980 - 2010). Rev Soc Bras Med Trop. 2011;44(6):762–70.
40. Rocha A, Ramos AN Jr, Sartori AMC, Correa D, Gontijo ED, Tatto E, et al. Recommendations for diagnosis, treatment and management of *Trypanosoma cruzi*: human immunodeficiency virus (HIV) coinfection. Rev Soc Bras Med Trop. 2006;39(4):392–415.
41. Freilij H, Altcheh J, Muchinik G. Perinatal human immunodeficiency virus infection and congenital Chagas' disease. Pediatr Infect Dis J. 1995;14(2):161–2.
42. de Freitas VL, da Silva SC, Sartori AM, Bezerra RC, Westphalen EV, Molina TD, et al. Real-time PCR in HIV/*Trypanosoma cruzi* coinfection with and without Chagas disease reactivation: association with HIV viral load and CD4 level. PLoS Negl Trop Dis. 2011;5(8):e1277.
43. Available from: https://unaids.org.br/wp-content/uploads/2018/07/2018_07_17_Fact-Sheet_miles-to-go.pdf2. Accessed 28 Sep 2018.
44. Epidemia de VIH nos países de língua oficial portuguesa -ONUSIDA Comunidade dos Países de Língua Portuguesa – CPLP. Org. Chequer P. UNAIDS, 4th ed. 2018. Available from: http://

www.aids.gov.br/pt-br/pub/2018/epidemia-de-vih-nos-paises-de-lingua-oficial-portuguesa. Accessed 28 Sep 2018.
45. Stauffert D, Silveira MF, Mesenburg MA, Manta AB, Dutra AS, Bicca LC, et al. Brief communication. Prevalence of *Trypanosoma cruzi*/HIVcoinfection in southernBrazil. Braz J Infect Dis. 2017;21(2):180–4.
46. Dolcini G, Ambrosioni J, Andreani G, Pando MA, Martínez Peralta L, Benetucci J. Prevalence of human immunodeficiency virus (HIV)-*Trypanosoma cruzi* co-infection and injectable-drugs abuse in a Buenos Aires health center. Rev Argent Microbiol. 2008;40(3):164–6.
47. WHO map of global distribution of *T. cruzi*/HIV infection. Available from: http://www.who.int/chagas/Global_distribution_*T_cruzi*_HIV_coinfection_2006_2010.pdf.
48. Benchetrit AG, Fernández M, Bava AJ, Corti M, Porteiro N, Martínez Peralta L. Clinical and epidemiological features of chronic *Trypanosoma cruzi* infection in patients with HIV/AIDS in Buenos Aires, Argentina. Int J Infect Dis. 2018;67:118–21.
49. Ramos AN Jr, Correia D, Almeida EA, Shikanai Yasuda MA. History, current issues and future of the Brazilian Network for attending and studying *Trypanosoma cruzi*/HIV coinfection. J Infect Dev Ctries. 2010;4(11):682–8.
50. Rodrigues DBR, Correia D, Marra MD, Giraldo LER, Silva EL, Silva-Vergara ML, et al. Cytokine serum levels in patients infected by human immunodeficiency virus with and without *Trypanosoma cruzi* coinfection. Rev Soc Bras Med Trop. 2005;38(6):483–7.
51. Silva JS, Barral-Netto M, Reed SG. Aggravation of both *Trypanosoma cruzi* and murine leukemia virus by concomitant infections. Am J Trop Med Hyg. 1993;49(5):589–97.
52. Dolcini GL, Solana ME, Andreani G, Celentano AM, Parodi LM, Donato AM, et al. *Trypanosoma cruzi* (Chagas' disease agent) reduces HIV-1 replication in human placenta. Retrovirology. 2008;5:53.
53. Del Castilho M, Mendoza G, Oviedo J, Branco RP, Anselmo AE, Silva M. AIDS and Chagas' disease with central nervous system tumor-like lesion. Am J Med. 1990;1990(88):693–4.
54. Gluckstein D, Ciferri F, Ruskin J. Chagas' disease: another cause of cerebral mass in the Acquired Immunodeficiency Syndrome. Am J Med. 1992;92:429–32.
55. Spina-França JA, Machado LR, Yassuda N. Anticorpos a *Trypanosoma cruzi* no líquido céfaloraqueano. Arq Neuro-Psiquiat. 1988;46:374–88.
56. Ferreira MS, Nishioka SA, Silvestre MT, Borges AS, Nunes-Araujo FR, Rocha A. Reactivation of Chagas' disease in patients with AIDS: report of three new cases and review of the literature. Clin Infect Dis. 1997;25:1397–400.
57. Cordova E, Boschi A, Ambrosioni J, Cudos C, Corti M. Reactivation of Chagas disease with central nervous system involvement in HIV-infected patients in Argentina, 1992–2007. Int J Infect Dis. 2008;12:587–92.
58. Nisida IV, Amato Neto V, Braz LM, Duarte MI, Umezawa ES. A survey of congenital Chagas' disease, carried out at three health institutions in São Paulo City, Brazil. Rev Inst Med Trop Sao Paulo. 1999;41(5):305–11.
59. Bisio M, Altcheh J, Lattner J, Moscatelli G, Fink V, Burgos JM, et al. Benznidazole treatment of chagasic encephalitis in pregnant woman with AIDS. Emerg Infect Dis. 2013;19(9):1490. Available from: https://wwwnc.cdc.gov/eid.
60. Agosti MR, Ercoli P, Dolcini G, Andreani G, Peralta LM, Ayala SG. Two cases of mother-to-child transmission of HIV and *Trypanosoma cruzi* in Argentina. Braz J Infect Dis. 2012;16:398–9.
61. Hernández C, Cucunubá Z, Parra E, Toro G, Zambrano P, Ramírez JD. Chagas disease (*Trypanosoma cruzi*) and HIV co-infection in Colombia. Int J Infect Dis. 2014;26:146–8.
62. Shikanai-Yasuda MA, Carvalho NB, Gallafrio CN, Medeji C, Silva SC, Bezerra R, et al. *Trypanosoma cruzi* parasitemia in immunosupressed patients with HIV infection or organ transplant recipientes. 63rd ASTHM Annual Meeting. New Orleans, LA, USA, November 2–6, 2014.
63. Guariento ME, Allegro FC, Almeida EA. Chagas' disease associated with chronic infirmities in outpatients fllowed in a university hospital. Rev Bras Clin Med. 2009;7:84–8.

64. Chapadeiro E, Lopes ER, Mesquita PM, Pereira FEL. Ocorrência de neoplasias malignas associadas à doença de Chagas. O Hospital. 1964;66:111–4.
65. de Lustig ES, Puricelli L, Lansetti JC. Association of Chagas disease and cancer. Medicina (Buenos Aires). 1980;40:43–5.
66. Almeida EA, Figueiredo DM, Guariento ME, Souza ML, Wanderley JS. Clinic profile and evolution in chronic chagasic patients with malign neoplasm. Rev Soc Bra Clin Med. 2008;6(1):1–7.
67. Almeida EA, Martins CBF, Pinto JE, Carvalhal SS. Doença de Chagas em portadores de neoplasias malignas. Estudo do comportamento das lesões chagásicas em sete (7) casos necropsiados. Anais da V Reunião de Pesquisa Aplicada em doença de Chagas. 116. 1988.
68. Murta EFC, Oliveira GP, Prado FO, Souza MAH, Murta BMT, Adad SJ. Association of uterine leiomyoma and Chagas' disease. Am J Trop Med Hyg. 2002;66(3):321–4.
69. Dominical VM, Cavellani CL, Rocha LP, Corrêa RRM, Pereira GA, Teixeira VPA. Chagas' disease and gynecologic neoplasias. Ann Diagn Pathol. 2010;14:337–41.
70. Almeida EA, Metze K, Teixeira MAB, Lopes ER. Encefalopatia chagásica em imunossuprimido por tratamento quimioterópico em portado de leucose crônica. Apresentação de caso com necropsia. Rev Soc Bras Med Trop. 1991;24(Suppl 1):33.
71. Metze K, Metze IL, Almeida EA, Moraes SL. Reactivation of Chagas' disease myocarditis during therapy of Hodgkin's disease. Trop Geogr Med. 1991;43(1&2):228–30.
72. Rezende REF, Lescano MA, Ramalho LNZ, Figueiredo JFC, Dantas RO, Meneghelli UG, et al. Reactivation of Chagas' disease in a patient with non-Hodgkin's lymphoma: gastric, oesophageal and laryngeal involvement. Trans Royal Trop Med Hyg. 2006;100:74–8.
73. Barrouse AF, Costa JA, Eposto M, Laplume H, Segura EL. Enfermedad de Chagas e inmunossupresion. Medicina (Buenos Aires). 1980;40(Suppl I):17–26.
74. Brener Z, Chiari E. The effects of some immunosuppressive agents in experimental chronic Chagas' disease. Trans R Soc Trop Med Hyg. 1971;65(5):629–36.
75. Bergman L, Mitrou PS, Demmer M, Ruhmann FT, Weidmann E. Impaired T and B cell functions in patients with Hodgkin's disease. Cancer Immunol Immunother. 1987;25:59–64.
76. Ford MB, Mitchell MF. Epidemiologia do câncer. In: Boyer KL, Ford MB, Judkins AF, Levin B, editors. Oncologia na Clínica Geral. Rio de Janeiro: Guanabara-Koogan; 2000. p. 2–16.
77. Huggins D. Carcinoma do esôfago associado ao megeesôfago chagásico (relato de um caso). An Instituto de Higine e Med Trop. 1976;4:1–4.
78. Rezende JM, Rosa H, MGM V, Andrade-Sá N, Porto JD, Neves Neto J, et al. Endoscopia no megaesôfago. Estudo prospectivo de 600 casos. Arq Gastroenterol. 1985;22:53–62.
79. Meneses ACO, Lopes MAB, Rocha A, Fatureto MC, Lopes GP, Lopes ER, et al. Megas e câncer. Câncer do intestino grosso em chagásicos com megacolon. Arq Gastroenterol São Paulo. 1989;26(12):13–6.
80. Garcia SB, Aranha AL, Garcia FRB, Basile FV, Pinto APM, Oliveira EC, et al. A retrospective study of histopathological findings in 894 cases of megacolon. What is the relationship between megacolon and colonic câncer? Rev Inst Med Trop Sao Paulo. 2003;45(2):91–3.
81. Kannen V, Oliveira EC, Motta BZ, Chaguri AJ, Brunalde MO, Garcia SB. Trypanosomiasis-induced megacolon ilustrates how myenteric neurons modulate the risk for colon câncer in rats and humans. PLoS Negl Trop Dis. 2015;17(4):1–11.
82. Manoel-Caetano FS, Borim AA, Caetano A, Cury PM, Silva AE. Cytogenetic alterations in chagasic achalasia compared to esophageal carcinoma. Cancer Genet Cytogenet. 2004;149:17–22.
83. Escalante ED, Oliveira EC, Cunha FQ, Vespúcio MVO, Ribeiro-Silva A, Aprilli F at al. *Trypanosoma cruzi* infection and/or administration of the nonsteroidal anti-inflammatory nimesulide increase the number of colonic crypts overexpressing metallothioneins in rat colon carcinogenesis. Braz J Med Biol Res. 2006;39:895–9.
84. Klyueva NG, Roskin G. Cancerolytic substance of Schuzitrypanum cruzi. Ann Rev Soviet Med. 1946;4:127–9.
85. Hauschka TS, Goodwin MB. *Trypanosoma cruzi* endotoxin (KR) in the treatment of malignant mouse tumors. Science. 1948;107:600–2.

86. Cabral HRA. The tumoricidal effect of *Trypanosom cruzi*: its intracellular cycle and the immune response of the host. Med Hypotheses. 2000;54(1):1–6.
87. Ubillas L, Freire T, Berriel E, Chiribao ML, Chiale C, Festari MF, et al. *Trypanosoma cruzi* extracts elicit protective immune response against chemically induce colon and mammary câncer. Int J Cancer. 2016;138:1719–31.
88. Ramirez-Toloza G, Abello P, Ferreira A. 2016. Is the antitumor property of *Trypanosoma cruzi* infection mediated by its calreticulin? Front Immunol. 2016;7:1–8.
89. López NC, Valck C, Ramirez G, Rodriguez M, Ribeiro C, Orellana J, et al. Antiangiogenic and antitumor effects of *Trypanosoma cruzi* calreticulin. PLOS Negl Trop Dis. 2010;4(7):e730.
90. Aguillar-Gúzman L, Lobos-González L, Rosas C, Vallejos G, Falcón C, Sosoniuk E, et al. Human survivin and *Trypanosoma cruzi* calreticulin act in synergy against a murine melanoma *in vivo*. PLoS One. 2014;9(4):e95457.
91. Pinazo MJ, Miranda B, Rodríguez-Villar C, Altclas J, Brunet Serra M, García-Otero EC, et al. Recommendations for management of Chagas disease in organ and hematopoietic tissue transplantation programs in nonendemic areas. Transplant Rev (Orlando). 2011;25(3):91–101.
92. Chin-Hong PV, Schwartz BS, Bern C, Montgomery SP, Kontak S, Kubak B, et al. Screening and treatment of Chagas disease in organ transplant recipients in the United States: recommendations from the Chagas in transplant working group. Am J Transplant. 2011;11(4):672–80.
93. Diez M, Favaloro L, Bertolotti AM, et al. Usefulness of PCR strategies for early diagnoses of Chagas disease reactivation and treatment follow-up in heart transplantation. Am J Transplant. 2007;7(6):1633–40.
94. Kransdorf EP, Zakowski PC, Kobashigawa JA. Chagas disease in solid organ and heart transplantation. Curr Opin Infect Dis. 2014;27(5):418–24.
95. Amato Neto V, Matsubara L, Uip DE, et al. Transplante de coraçao: doador com doença de Chagas e evoluçao do receptor. Rev Hosp Clin Fac Med Univ São Paulo. 1992;47:92–4.
96. Kun H, Moore A, Mascola L, et al. Transmission of *Trypanosoma cruzi* by heart transplantation. Clin Infect Dis. 2009;48:1534–48.
97. Riarte A, Luna C, Sinagra A. Enfermedad de Chagas en el huésped trasplantado. Medicina (Buenos Aires). 2000;60:31–2.
98. Riarte A, Luna C, Sabatiello R, et al. Chagas disease in patients kidney transplants. Clin Infect Dis. 1999;29(3):561–7.
99. Fiorelli A, Stolf N, Honorato R, et al. Later evolution after cardiac transplantation in Chagas disease. Transplant Proc. 2005;37:2793–8.
100. Godoy HL, Guerra CM, Viegas RF, et al. Infections in heart transplant recipients in Brazil: the challenge of Chagas' disease. J Heart Lung Transplant. 2010;29:286–90.
101. Altclas J, Sinagra A, Dictar M, et al. Chagas disease in bone marrow transplantation: an approach to preemptive therapy. Bone Marrow Transpl. 2005;36:123–9.
102. Piron M, Fisa R, Casamitjana N, et al. Development of a real-time PCRassay for *Trypanosoma cruzi* detection in blood samples. Acta Trop. 2007;103:195–200.
103. Duffy T, Bisio M, Altcheh J, et al. Accurate real-time PCR strategy for monitoring bloodstream parasitic loads in Chagas disease patients. PLoS Negl Trop Dis. 2009;3:e419.
104. Cicora F, Paz M, Mos FA, Petroni J, Roberti JE. Belatacept-based immunosuppression in a chagasic adult recipient of en bloc pediatric kidneys. Transplantation. 2014;98(4):e34–5.
105. Guiang KM, Cantey P, Montgomery SP, Ailawadhi S, Qvarnstrom Y, Price T, Blodget E. Reactivation of Chagas disease in a bone marrow transplant patient: case report and review of screening and management. Transpl Infect Dis. 2013;15(6):E264–7.
106. Bestetti RB, Souza TR, Lima MF, Theodoropoulos TA, Cordeiro JA, Burdmann EA. Effects of a mycophenolate mofetil-based immunosuppressive regimen in Chagas' heart transplant recipients. Transplantation. 2007;84(3):441–2.
107. Salvador F, Sánchez-Montalvá A, Sulleiro E, Berastegui C, Jauregui A, Pont T, Los-Arcos I, Len A, Gavaldá J, Molina I. Case report: successful lung transplantation from a donor seropositive for *Trypanosoma cruzi* infection (Chagas disease) to a seronegative recipient. Am J Trop Med Hyg. 2017;97(4):1147–50.

108. Disease Argentine Collaborative Transplant Consortium, Casadei D. Chagas' disease and solid organ transplantation. Chagas' Transplant Proc. 2010;42(9):3354–9.
109. Bern C, Montgomery SP, Herwaldt BL, et al. Evaluation and treatment of Chagas disease in the United States: a systematic review. JAMA. 2007;298(18):2171–81.
110. Godoy HL, Guerra CM, Viegas RF, Dinis RZ, Branco JN, Neto VA, Almeida DR. Infections in heart transplant recipients in Brazil: the challenge of Chagas' disease. J Heart Lung Transplant. 2010;29(3):286–90.
111. Cicora F, Escurra V, Bibolini J, Petroni J, González I, Roberti J. Cerebral trypanosomiasis in a renal transplant recipient. Transpl Infect Dis. 2014;16(5):813–7.
112. dos Santos-Neto LL, Polcheira MF, Castro C, Lima RA, Simaan CK, Correa-Lima FA. *Trypanosoma cruzi* high parasitemia in patient with systemic lupus erythematosus. Rev Soc Bras Med Trop. 2003;36:613–5.
113. Bilate AM, Salemi VM, Ramires FJ, de Brito T, Russo M, Fonseca SG, et al. TNF blockade aggravates experimental chronic Chagas disease cardiomyopathy. Microbes Infect. 2007;9:1104–13.
114. Kroll-Palhares K, Silvério JC, Silva AA, Michailowsky V, Marino AP, Silva NM, et al. TNF/TNFR1 signaling up-regulates CCR5 expression by CD8+ T lymphocytes and promotes heart tissue damage during Trypanosoma cruzi infection: beneficial effects of TNF-a blockade. Mem Inst Oswaldo Cruz. 2008;103:375–85.
115. Pérez AR, Fontanella GH, Nocito AL, Revelli S, Bottasso OA. Short treatment with the tumour necrosis factor-alpha blocker infliximab diminishes chronic chagasic myocarditis in rats without evidence of Trypanosoma cruzi reactivation. Clin Exp Immunol. 2009;157:291–9.
116. Nishioka Sde A. Benznidazole in the primary chemoprophylaxis of the reactivation of Chagas' disease in chronic chagasic patients using corticosteroids at immunosuppressive doses: is there sufficient evidence for recommending its use? Rev Soc Bras Med Trop. 2000;33:83–5.
117. Barousse AP, Costa JA, Eposto M, Laplume H, Segura EL. Chagas disease and immunosuppression. Medicina (Buenos Aires). 1980;40(Suppl 1):17–26.

People Voices

14

Francisco Javier Sancho, Colin J. Forsyth, Claudia Nieto-Sanchez, Esteban G. Baus, and Leonardo de la Torre Ávila

14.1 A Mother Against Chagas Disease

Francisco Javier Sancho

When she was 18 years old, Idalia Isabel (Yaya) donated blood for a relative who had broken his hip in his city, Xalapa, in the state of Veracruz, Mexico. What was supposed to be a routine blood collection ended in complications. And blood tests were performed over and over again until her arm turned purple. A week later, she was on a stretcher in the epidemiology area of the hospital, surrounded by doctors who told her they had discovered that she was suffering from Chagas disease.

"And what is that?" she asked.

"A deadly disease," they told her, straight away and unambiguously.

Now she is 26 years old; when she is asked if she remembers herself being afraid at that time, she answers "No" with an open smile. This is strange. Everybody is

F. J. Sancho
Global Chagas Disease Coalition, Barcelona, Spain
e-mail: javier.sancho@coalicionchagas.org

C. J. Forsyth
Drugs for Neglected Diseases Initiative, Geneva, Switzerland
e-mail: cforsyth@dndi.org

C. Nieto-Sanchez · E. G. Baus
Center for Research on Health in Latin America (CISeAL), College of Exact and Natural Sciences - Pontificia Universidad Católica del Ecuador and Medical Anthropology Unit - Institute of Tropical Medicine, Antwerp, Belgium
e-mail: cnieto@itg.be; EGBAUS@puce.edu.ec

L. de la Torre Ávila (✉)
InSPIRES Project, The Barcelona Institute for Global Health and International Health Service, Hospital Clínic i Provincial de Barcelona, Barcelona, Spain
e-mail: leonardo.delatorrea@isglobal.org

M.-J. Pinazo Delgado, J. Gascon (eds.), *Chagas Disease*,
https://doi.org/10.1007/978-3-030-44054-1_14

afraid of the word *deadly*; moreover, if it is told you when you are 18 years old. But to face scary words, there are other words that do not provoke fear, and Yaya had saved one of them for her: the catch-all term *mother*.

"I only knew that my mother was going to solve everything. I have always trusted her. And I left it up to her."

In fact, Elvira Hernández, a schoolteacher, a widow with three children including Yaya, moved heaven and earth to get an accurate diagnosis and a prompt treatment for her daughter. She had already had the experience of dealing with her other children's severe health complications. It was not the first time. "This *chamaca* is not going to die because of Chagas disease," she promised herself.

Elvira was not satisfied with what some health workers told her. She now says that she applied the methodology she had learnt while working as a teacher, when she had to help students with learning difficulties or behavioral problems: to be patient, devote her time to understand the cause of the situation, and put herself together to find solutions. She surfed the Net and immediately learnt that, although Chagas disease causes more than 7000 deaths per year and affects around eight million people mainly in Latin America but also in non-endemic countries such as the USA or Spain, it is not a synonym of death if it is treated promptly. She also learnt that it is not only transmitted by the kissing bug or *chinche besucona*, as it is called in Mexico, but also through blood transfusions, organ transplants, oral transmission, and even from mother to child during pregnancy or childbirth. This scared her, and she decided to have a blood test immediately. Later, she knew that her daughter possibly had caught the disease through a *vinchuca* because the test result was negative.

It was relevant to know whether there was a medicine for the disease. Elvira read that the two drugs to treat it (benznidazole and nifurtimox), in spite of being ancient and having side effects, are effective especially in children and women of childbearing age but also in adults with a chronic condition who have not generated other complications in vital organs such as the heart, liver, or intestines. She learnt that only 30% out of the people who get the infection might develop the disease. And the word *death* became distant. Even so, she contacted other specialists abroad. She could travel to Argentina, invited to an assembly of affected people; and there she went with Yaya. She met Dr. Sergio Sosa-Estani, who works for the Drugs for Neglected Diseases Initiative (DNDi) and is a member of the Global Chagas Disease Coalition. He indicated prompt treatment and follow-up care.

The treatment produced side effects, but Yaya overcame them and 2 months later she started to feel much better. From 2010, when she was diagnosed, onward her tests have been free from the infection. She became a lawyer. "In fact, I wanted to be a doctor. I was diagnosed when I was about to enroll at university. I was advised to choose a more relaxed university career because the disease was going to burden me. That is why I changed to Law. Today, I am a safe and sound lawyer."

Since then, she helps her mother in the association they founded to accompany other people diagnosed in Mexico. In her Facebook Page, they answered weekly the enquiries of potentially affected people who have not received enough care and do not know where to go in order to get it, as Yaya comments.

Yaya and Elvira are also members of the Asociación Mexicana de Personas Afectadas por Chagas (AMEPACH), which in 2018 hosted the V Assembly of the international federation FINDECHAGAS that gathered some 20 associations of affected people from several endemic and non-endemic countries in Xalapa.

During the Assembly, the associations addressed the strengthening of the cohesion among them and the visibility of this problematic as well as the establishment of collaboration networks among all the actors involved in the fight against Chagas disease. One of their greatest challenges is to fight against words that stigmatize or cause panic when talking about the disease: *an illness of poverty*; *deadly*; *forgotten*; *an illness of migrants*, among others. Chagas disease is nowadays a global health challenge affecting all social classes. However, there are still many gaps, mainly as regard the access of affected people to diagnosis and treatment.

Research and programs developed by some organizations of the Global Chagas Disease Coalition, such as Mundo Sano, DNDi, ISGlobal, CEADES, Ayuda en Acción, or Baylor College, among others, attempt to advance drastically in those aspects. The benznidazole register in Mexico and the USA in 2017 was one of the steps to bring treatment closer to affected people. On the other hand, the Pan American Health Organization (PAHO) has also included the control of the Chagas disease mother-to-child transmission among the priorities for the following years, at the same level as some serious illnesses, such as AIDS, syphilis, or hepatitis B, are approached.

Another gap is related to actual affectation data of this illness around the world. Only in Mexico, for example, some studies estimate that the number of affected people might be around one million, although official data register no more than a thousand positive cases per year mainly detected in blood banks.

Arturo Díaz, 62 years old, is a doctor in the area of internal medicine in Tantoyuca hospital in Veracruz. He attended the affected people's assembly because he became interested in the disease 10 years ago, when he had a patient with cardiomegaly, with neither history of high blood pressure nor other complications. His patient was 32 years old. "I thought his condition could be due to Chagas disease. The patient underwent a rapid test, and he tested positive. Then, we performed more serological tests. If two out of three results are positive, we consider the patient affected by Chagas disease," says Dr. Díaz.

Dr. Díaz comments that, since then, many colleagues have updated their knowledge about Chagas disease to be able to diagnose it promptly. After having assisted more than a hundred cases in diverse states of Mexico, he states that "we must be responsible with this disease. We must pay more attention. Underestimating it means putting many people's lives at risk. In my area of expertise, when we performed a cardiovascular evaluation to a patient referred by a surgeon, we have the chance of investigating whether that patient is affected or not."

Dr. Díaz has participated in numerous first-rate trainings for healthcare workers. But, according to him, much remains to be done; for example, to consider the Chagas disease training in university programs so that health professionals have the precise tools for diagnosis and treatment.

"It is not just a problem concerning health authorities. We cannot and must not expect them to solve it by themselves. Chagas disease, like Dengue fever, is one of the most worrying vector-borne diseases. We should not put the blame on others; instead, we have to help each other, share experiences and knowledge, and look for more help."

During the Assembly of the associations of affected people, the delegates chose the new steering committee of the federation that will work during the following 3 years on behalf of all of them. Precisely, Elvira Hernández's association, AMEPACH, was the one chosen for the presidency. Besides, the associations issued a manifesto asking for and offering collaboration to health authorities, WHO, and PAHO to work on the implementation of Sustainable Development Goals (SDG) for the control and end of neglected diseases.

Elvira's first action was to travel to Mexico City for a public act of presentation of the federation. She will now lead the fight for affected people not only in her country, but also around the world.

"A friend told me that once you become a mother, you become everyone's mother, so I will fight for all affected people in the same the way I fought for my daughter. And I will do it with her help."

14.2 Yerko: A Life Taken Too Soon by Chagas Disease

Colin J. Forsyth

Yerko Alvarez loved three things: his family, his church, and music. He was one of the first people I met when I moved to Santa Cruz de la Sierra, Bolivia in 1996; he was married to my wife's sister, and our children played together at family gatherings. He had a way of putting people at ease and was quick to smile or make a joke, but he was also a devoted and concerned father, and showed a more serious side when looking after his children, for whom he wanted the best possible future. Frequently, my wife's relatives joined together to celebrate holidays, or birthdays, or simply to enjoy a good *churrasco*. These occasions were always highlighted by Yerko playing the guitar and singing the traditional songs of Santa Cruz.

Yerko, one of seven children, grew up in the rural community of El Forestal, approximately a 2-h drive from Santa Cruz. His childhood home, like most homes in the Bolivian countryside at the time, was made of natural materials, with walls of adobe bricks and a thatched roof. In the cracks and crevices of the adobe and within the thatched roof lived kissing bugs, insects known locally as *vinchucas*, a term derived from a Quechua word describing the way they glide down from the roof in the evening to drink the blood of the people sleeping below [1]. Yerko recalled being bitten by *vinchucas* during his childhood, waking at night to find them in his bed, and in the morning seeing the stains on the walls from where the insects, gorged on blood, dragged themselves back to their hiding places. At the time, they were simply considered a nuisance. The children even played with their eggs, which resembled grains of rice. What Yerko did not realize is that as he slept as a child and scratched the areas where he was bitten by the *vinchucas*, a deadly parasite had entered his bloodstream. There were no obvious signs or symptoms to tell him he had been infected with *Trypanosoma cruzi*, the protozoan that causes Chagas disease. Life went on as if nothing had changed.

Later in his childhood Yerko's family moved to the city of Santa Cruz, and he met and married my sister-in-law. They had three children: two boys followed by a girl. My sister-in-law taught English and Yerko worked as an assistant in a pharmacy; they were very active in their church, where Yerko directed the music. Although they had a decent life, Yerko eventually came to regret not having gone to the university; he realized he had the potential to do more and provide a better future for his children. He therefore enrolled in the university in his late 30s and began studying toward a degree in pharmacy. His plan was to become a pharmacist and run his own pharmacy one day.

Despite having commenced his studies at a later age, Yerko did very well, receiving high grades in his courses. But then something went terribly wrong. As part of his job, Yerko delivered medicines to pharmacies in Santa Cruz. One day, as he was walking in the city carrying a box of medicine to make a delivery, he suddenly cried out in pain and collapsed on the street. The people passing by avoided him, unsure of what was wrong. Fortunately, an acquaintance happened to be walking by, recognized Yerko, and called for help. He was rushed to the hospital. Yerko had suffered a stroke.

The physicians were perplexed. How could a man in the prime of his life be stricken with such a serious, unexpected blow to his health? Despite the fact that Yerko lived in an area endemic for Chagas disease, where in many communities over 50% of the adult population was infected with *T. cruzi* [2], Yerko was not tested and his *T. cruzi* infection was not detected. Instead, the doctors assumed that Yerko had acquired trichinosis from eating undercooked pork. Slowly, he recovered and was even able to return to his studies, and it seemed as if the danger had passed.

But it hadn't. As Yerko continued working toward his pharmacy degree, his health, still fragile in the aftermath of the stroke, began to deteriorate further. It started with a persistent cough which stemmed not from a cold, but from his heart's diminished ability to pump blood. Yerko tried improving his diet and exercising regularly, but he still began to feel increasingly tired. His heart had enlarged. He started to experience palpitations, chest pain, and shortness of breath. Even as his health worsened, he continued looking toward the future. He finished his coursework despite his constant pain and fatigue. His eldest son, now in high school, had to physically support Yerko to help him walk into the classroom for the final exam of the last course he needed for his pharmacy degree.

It was around this time that a doctor realized Yerko had heart damage caused by Chagas disease. This doctor tried antiparasitic treatment, but Yerko's health did not improve; the damage to his heart was irreversible. He died in 2012 at the age of 44.

It is always shocking to lose someone so young. In Bolivia, stories are often told of people in their 30s and 40s, in the prime of their health, who suddenly, unexpectedly collapse dead, usually while playing soccer, enjoying time with their families, or walking home from work. Sudden death is a frequent outcome of Chagas disease. Some people even express the notion that a death from Chagas disease is "good" because it happens quickly and unexpectedly.

However, this was not the case for Yerko and thousands like him. Yerko endured a lengthy, slow, and painful decline, in which his family bore witness to his deterioration and suffering. And it could have been avoided. If he had been tested at a younger age and received antiparasitic treatment, killing the infection which triggered his heart damage, he might still be alive today. Even failing that, if he had received a pacemaker or other critical cardiological interventions, he might still be alive today. But he received none of these things because, despite being at high risk for Chagas disease and living in a region with high prevalence, he was not diagnosed and treated in time. Perhaps his physicians did not receive the necessary training on Chagas disease, or perhaps because of the urban setting they did not consider Chagas disease as a potential risk. Or, like many doctors in Bolivia and elsewhere, they may have felt that Chagas disease was untreatable, and thus it was better not to look for it. Some doctors want to protect patients from the anguish of being diagnosed with a fatal disease [3]. Regardless, we now know that earlier detection and treatment would give Yerko and others like him a better chance for survival. Physician awareness is a key part of making this happen.

Every life lost from Chagas disease is tragic, but part of the story which is often untold, and which raw statistics cannot adequately convey, is the empty space that is left in families and communities when someone is taken too soon. When Yerko died, his children (two of whom were still in high school at the time) lost the guidance and love of a father at a critical juncture when they were transitioning to adulthood. His wife lost her partner of over 20 years. The loss of Yerko's income, just when he was on the cusp of starting a promising career, created financial challenges for the family. His church lost an active and dedicated member, and the community lost a future health professional. The sound of his voice, whether in song or laughter, is now conspicuously absent at family gatherings.

The point of Yerko's story is that every death from Chagas disease matters; every loss is irreplaceable. Therefore, it is imperative we continue taking the necessary steps to prevent and diminish the awful toll of this disease. Greater awareness among physicians who see patients at risk is a critical need in all countries impacted by the disease. Systematic screening programs would also help detect the infection in people like Yerko so that they can be treated while there is still a possibility of preventing chronic complications from the disease. While the available drugs are not perfect, and better evidence of benefits and efficacy are still needed, timely treatment does give people with the infection a chance at living longer lives. Every person with the disease deserves this chance.

14.3 Behind Those Walls[1]

Claudia Nieto-Sanchez and Esteban G. Baus

Back in the 1920s, when Teresa's[2] grandparents arrived to the little town in southern Ecuador that she has always known as her home, people did not have access to electricity, water could only be brought to homes by mules, and the roads were nothing more than bridle paths selectively built in areas where produce was cultivated and harvested. Most farmers (or *hortelanos*, as they were called back then) did not own the land where they lived or worked. Instead, they used to work for a *patron* or landowner who allowed them to work his land and take portions of the harvest in exchange for service to him and his family. The portion of land assigned to them would change year after year, meaning that once the terrain had been prepared for the next harvest, the landowner would reclaim it and use it for his own benefit. Everybody had to work: women, children, and of course, men. Raising cattle, feeding the animals, collecting water, cooking, constructing fences, and farming were among the many responsibilities that local families had to assume in exchange for a place to live. They were never paid an actual salary for their work, and it was not unusual for them to receive physical punishment if their work was not done as the *patron* expected.

"Those were tough times" says Teresa. "Imagine how difficult it could be to work that hard and never see a dime for what you do!" Knowing that the plot they had been assigned would be taken away from them, people like Teresa's grandparents would build temporary huts with the resources they had at hand: dirt, manure, hay, and sticks. Those adobe structures would provide shelter for some time and could be easily taken down when needed. With each move, people would normally take with them the adobe blocks that were still in an acceptable condition to use for building the structural base for their new home. Sometimes, those blocks would be accompanied by uninvited pests.

"My grandmother used to tell us that when she moved here with my grandfather, she could see *chinches*[3] everywhere in the adobe, especially in the cracks. She grew up in a city nearby, and as she was trying to keep her house as tidy as possible, she could see *chinches* walking on the walls or along the edges of their beds. There were many: big, small … many! She was so bothered that she used to bring water on her donkey to boil it and throw on them—there was no other way to kill those *bichos*[4]. My grandfather used to get mad at my grandmother because he thought that the water would weaken the adobe and the house would fall apart. After that, she decided to expose beds, blankets, and *jergas*[5] to the sun on regular basis," remembers Teresa.

[1] Editing and English proofreading: Christine Fram.

[2] Real names have been substituted with pseudonyms.

[3] Local name for triatomines.

[4] Local name for insects.

[5] Form of horse blanket used as mattress in rural areas of Latin America.

She used to hear this and other stories from her grandparents when she was a child, almost 40 years ago. "It was like having a TV under the starts every night," she jokes. Together with her younger uncles and aunts, Teresa used to ask them repeatedly to tell the story of their lives—and through that, the story of her own village. "One day, the landowner asked my grandfather to build a stone fence. He spent days working on it. Once he finished and went home to rest, the fence was destroyed by another worker who disliked my grandfather. The day after, the *patron* scolded him and threaten to hit him. My grandfather reacted, and the *patron* kick him out of the *hacienda*. As he went back home to collect his kids and wife, the landowner followed him and asked him to leave immediately. It was very hard for him. That house was a hut, but was all they had at that time." After walking for several days, the family was able to reach another *hacienda*, find a new job, and initiate another cycle of working and moving on.

It was through these stories that Teresa learned how her grandfather and other farmers from the area organized a movement to rebel against their *patron* by indefinitely occupying the land they had worked throughout all those years of exploitation. "They just refused to leave, and after many years of fighting, the landowners decided to sell their land to the farmers," she remembers. Teresa and most of her neighbors are the grandchildren of that generation of rebellious workers who decided to fight for their rights. The spirit of those times survives not only through stories like hers but also through legal and social arrangements that originated during those times of struggle. Ownership of land in the village, for example, is dependent upon a communal title that secures equal use of living and productive spaces for the heirs of the families who settled in the area in the 1960s, when a land reform in Ecuador facilitated legalization of occupied properties.

Presently, in addition to electricity and regular (once a day) *ranchera*[6] transportation using the gravel-surfaced roads surrounding the village, local families are entitled to basic services such as basic education and healthcare. Although an important achievement in legal terms, gaining access to these kinds of services is still challenging for rural populations in Ecuador. This is particularly true for conditions that, as in the case of Chagas disease (CD), are consistently neglected by healthcare providers and systems. "We never learned that there was a relationship between the *chinches* that we usually see coming from the fields and any disease. It was only when researchers from Quito and Ohio came here that we started talking about Chagas," says Teresa.

She is referring to the Healthy Living Initiative (HLI), a transdisciplinary research effort aimed at describing and addressing the socioeconomic dynamics involved in CD transmission in southern Ecuador. After determining that CD was affecting rural communities from Loja province, the research team decided to go beyond the field's usual borders and embrace a different approach: a complex problem must be treated in a complex way, and a transdisciplinary approach was needed to tackle an issue that was simultaneously environmental, economic, social, and political. To respond to this complexity, the HLI was created as a holistic strategy

[6] Open bus.

that mobilizes various stakeholders interested in improving people's lives and aims to understand community dynamics to promote health. One of the challenges to controlling CD was that the members of these rural communities did not see this disease as a priority; other problems that had a greater effect on their daily lives were considered more urgent. Therefore, another aspect of this strategy involves interventions under the HLI umbrella to address these problems and gradually build trust between the community and the participating research institutions.

Information about CD is so difficult to obtain for local residents that they see HLI's presence as an opportunity to ask questions about their own diagnosis. "There is a man from this neighborhood who got diagnosed with CD some years ago. When he learned that this group was coming here, he decided to come all the way from Quito and talk with them [the researchers] about his problem. He lives with that condition because there is no cure for CD. My grandmother also died from cardiac problems. The doctors said that her heart was enlarged, and we think it was the result from all those years being bitten by *chinches*. She was never diagnosed or treated for Chagas, though."

Besides general education concerning the disease itself, most of the work developed by HLI is focused on CD prevention through infrastructure improvement. "The researchers have told us that sometimes there are no symptoms. We used to scratch our skin with both hands when the *chinche* bit us—we did not know that we were infecting ourselves by doing that. *Chinches* leave a welt that becomes hard after scratching and lasts 3 or 4 days. We don't do that anymore. We are more careful now and when we see the *bicho*, we are careful to pick it up with a plastic bag or a piece of paper to kill it," says Teresa.

HLI has also promoted and helped secure basic sanitation improvements for the communities. For instance, HLI was actively involved in securing access to drinking water systems for this community. In 2016, community members organized to improve their existing water collection and distribution system. After receiving a donation from Rotary Club International, each family contributed the 32 weeks of work necessary to bring tubes and material from the source to their homes. In Teresa's words, this was a substantial improvement for local families. "Now we have access to water day and night. Before the new system was completed, we only had intermittent service, and one neighbor would tell the other 'Hey, water is coming,' and we all had to run to collect water for cooking or to wash clothes. Having water on a regular basis has helped us to clean our homes more regularly." Similarly, some residents have put plaster on their walls, which has also helped them to keep insects under control. "Some *chinches* still come from the fields, but since the walls are plastered, we can spot them more easily. Of course, they can still enter the homes through some openings, but much less frequently than in the past."

The struggle to improve living conditions for local families continues. Teresa and other women in the communities interested in handicrafts created a group of artisans to generate additional income for their homes. They spent over a year, making multiple trips to local government offices and using their own resources to advocate for a permanent stand in the Sunday market to offer their products. "It was a long fight, but we finally achieved it. It is not ideal yet because the transportation that comes down here leaves early, so we have to close our stand before the other merchants, but it is worth it because we can have an independent source of income. I love what I do, and I love where I live. But more than anything, I love to be independent."

14.4 The Most Beautiful Photograph Ever Taken

Leonardo de la Torre Ávila

She turned up in our consulting room breastfeeding her 8-month-old daughter, with a heart rate of 35 beats/min, numbers of a heart which should had stopped but kept on giving. Angélica, a 33-year-old Bolivian woman, was waiting for a phone call from the hospital about a pacemaker implantation she needed, but the doctor in charge of her case claimed that she could not wait any longer and requested an immediate hospitalization. The baby went back home with Angélica's sister-in-law who had recently arrived in Spain and insisted in accompanying her to see the doctor because she had seen Angélica got dizzy when walking.

That night, her heart rate decreased to 29 beats/min; some minutes later, the pacemaker was implanted and Angélica saved her life. Because total anesthesia would have not allowed her to breastfeed her baby, she asked to be administered local anesthesia for the procedure. The next day, she came back home to feed her baby; but before that, she decided to tell us her story.

14.4.1 An Endless Road

Her father came from Vilacaya (Potosí) and her mother from a little town close to Punata (Cochabamba). They met in Yapacaní (Santa Cruz) and some of their kids were born there. Then, the family moved to El Chapare, more precisely to Vueltadero, close to 2 de Marzo; some time later, with seven children, they moved to Yacuiba (Tarija) and, after a while, they came back to El Chapare, but this time to Senda Cinco, near Cerro Verde. When she arrived in the city of Santa Cruz, Angélica, the couple's fifth daughter, was less than 17 years old. The endless road would go on to some points in Spain: Lorca (Murcia) and, after that, Sabadell (Catalonia).

In Yacapaní, they lived on the rice farming; in El Chapare, they grew yucca, rice, and oranges, and they had to sell a small store. Because her father got ill, the family had to move to Yacuiba. After her father's death, Angélica, her mother, and her two younger brothers came back to the tropic of Cochabamba where families were coming from all over the country. "I had to move, I lost the school year, I had to start again," she told us; one day, she realized that her mother was not supporting the family, and she dropped out of school. When they were living in Santa Cruz, one of her brothers who was working as a taxi driver told her that there were working positions at the airport. She stopped cleaning houses to start cleaning the huge toilets. Later, when she gained her bosses' confidence, she started cleaning planes. She replenished magazines asking herself whether she was going to be able to fly on one of those planes someday. And that day would come sooner than later.

At that time, Angélica got pregnant but remained silent because she was afraid of her mother's reaction; besides, her boyfriend had disappeared. "I had always been plump, so no one at work realized that I was pregnant." She continued working in silence until her skin paleness was impossible to hide. She was 7 months pregnant when she had the first medical control, and doctors told her that there were "two little hearts" there. When her twins were born, she knew she had to accept her sister's invitation from Murcia: "You have to come; if not, how are you going to support those two?" Angélica left her 2-month-old babies to her mother and got on a plane.

In Lorca, Angélica worked in the harvest of almonds and olives to send the money she earned to Bolivia. Then, she decided to move to Barcelona where live-in maids had better salaries. She says that she moved to another city because she did not want to be controlled by her family. She took care of an elderly woman for 7 years in Sabadell. We asked her if it was there where she met her husband; she answered that she met him again there, but she had met him for the first time in Bolivia when she was 15 years old and he, Saúl, was a widower 14 years older than her with two daughters and was living in Spain. Like most Peruvians, Saúl decided to perform some immigration procedures—in that case, family reunification—in Bolivia because they were more favorable. "He wanted me to work in his house in Peru to take care of his daughters," she tells us smiling. They were in contact by telephone and, later, they met again.

When Angélica got her immigration papers after 8 years working in Spain, she was already living with Saúl. He helped her to make the travel she had wished for so long: "I went back to Bolivia to get my children." "I left them when they were 6 months old," she reminds us. "My mother always told them about me; I always called them." When they met her, they stuck to her as if nothing had happened, as if she had raised them. "They did not want to be far from me, they took me to their school." As I was thinner than when I left, people in town did not believe I was their mother. "Here they are, look," she says while she shows the most beautiful photograph ever taken, the one of that day at school. "I always talk to them; I always tell them that everything I do is for them."

Angélica stayed 6 months in Bolivia, while the house she had saved the money for was being built. She came back to Barcelona and got a job taking care of another elderly woman who was going to die in her arms some time later. Saúl and Angélica were blessed with another pregnancy; she remained silent again to keep her new job until she was 8 months pregnant. She was really exhausted and told her boss, who let her go with great sadness. Two weeks later, Ángela was born.

14.4.2 A Family Fight Against Chagas Disease

Angélica's father's belly swelled up, but she thinks that alcohol killed him. Her mother is now 60 years old and her heart is damaged by Chagas disease. Her elder sister says that Angélica and their two younger brothers could have acquired Chagas disease in El Chapare. One of those brothers was diagnosed in the oil company he works for. According to the family source, the youngest brother should be also tested because he spent his childhood in the same house; but it is not necessary to screen the elder brothers born in Yapacaní. If the Chagas disease transmission was congenital, as her elder sister states, it is because their mother was weak when the younger brothers were born. The time spent in Yacuiba does not generate suspicions. The memory of *vinchucas* is linked to the tropic "where we used to catch them," she says.

However, Angélica felt depressed when she was diagnosed. "I thought that only my mother had been infected by the disease, not me." Then, she understood why she had been so tired during the last 3 years and the severe dizziness she attributed to other causes. A nurse called Amparo gave her the news. They were both alone in the walk-in clinic in Numancia in a gynecological follow-up visit. Angélica was saved because she was pregnant of her daughter Ángela and the Generalitat de Catalunya had implemented some years earlier a Chagas disease screening protocol for Latin American pregnant women. This is the way of saving one life, could be the newspaper headline. Angélica would not have asked to be diagnosed spontaneously. Once her progressive heart damage was detected, she was saved by the doctors who hurried the pacemaker implantation.

14.4.3 Six Lives Ahead

That night in hospital, Angélica picked up the phone. She called a friend with a positive *T. cruzi test* who had not been treated because of work issues and told her: "You have to be treated, you are young. If you get weak in the future, work is not going to save you." Angélica expected that her call from a hospital bed became binding among her trust networks. Before that, she had already encouraged two acquaintances to ask for diagnosis, and one of them was already under treatment in Sabadell.

All these actions emerged when Angélica knew that her heart has been damaged by Chagas disease. She says that people will not ask to be tested if they are not informed about its true risks. "One does not visit the doctor because of Chagas disease; one visits the doctor because of a headache," she admits. Now, she is asking for a change in practices: "Before it is too late, people must consult about Chagas disease. Isn't it true that the disease progresses little by little but it progresses fast when you get weaker? Sometimes, people seem to be stubborn; they know that Chagas disease is affecting others in the area, but anyway they do not visit the doctor. They must not wait for too long!"

Angélica assures that, if she would had known before, she had followed the treatment. When we ask her why she did not activate the alarms when she knew her mother had Chagas disease, she answers that she simply did not believe that she could be infected too. Maybe she was not properly informed. There is another explanation that can be inferred from something that happened to her when she was a child. Because of her excited shouts when she found a barnyard animal that had given birth in the field, she was attacked by a bee swarm. Some of the bees stung her ear, and someone decided to cure the infection by putting alcohol into it no matter how. One thing or the other perforated her eardrum. Her family could not afford a visit to a doctor. The story explains not only why Angélica hears with difficulty, but also why the decision of looking for a doctor in those communities is usually postponed because they have learnt to consider it unaffordable.

But, although the past should not be ignored, it is now turn to look forward. At their table, Angélica and Saúl think every single day not only of the baby who lives with them but also of his two daughters, of Angélica's twins, and even of a young girl Angélica and her mother adopted in Montero, Santa Cruz. She is a goddaughter whose father died because of Acquired Immune Deficiency Syndrome (AIDS) and was abandoned by her mother but finally was cared by a family with lots of love to offer. Six children in total; that is a good reason to look forward!

"You are going to get better, you are going to get better. Only the good die young," Saúl tells Angélica. She smiles; she learnt to start all over again when she was a young girl. She knows that she must help her mother who will probably need a pacemaker too, which is really expensive in Bolivia. She knows that she will have to deal with the twins' biological father. "When he saw their house and how cute they are, he appeared and says he wants to take them to Chile." She knows that tonight she will have to avoid physical exertion and pick up her baby with the other arm. The most crucial event will happen in a few days, when she will know whether she has infected her little daughter with the parasite or not. The baby has already had the ninth-month test, now they are waiting for the results.

Before leaving the hospital, Angélica receives a phone call from a friend and she gets it without hesitation. Her friend works in France, but she has arrived in Spain because of an emergency and she asks for accommodation at Angélica's home for the same night she is going to be released from hospital. "Come home; we can accommodate you for one night," she answers. Sometimes, Angélica helps some of her friends who work as live-in maids when they come out from work and do not have a place to go to. She lets them sleep in her living room. "Saúl does not get angry," she says quietly.

Finally, the medical team say Angélica goodbye and recommend her to take care of herself. A nurse reminds her that, from now onward, she will have to inform that she has a pacemaker at airport controls. Angélica smiles. We imagine her coming back to Bolivia soon. The metal detector will sound when she gets through the security gate, but she will tell her story and will go on happily on her way toward repeating the most beautiful photograph ever taken.

References

1. Bastien JW. The kiss of death: Chagas' disease in the Americas. Salt Lake City, UT: University of Utah Press; 1998.
2. Araujo TC, Medrano-Mercado N. Chagas disease in Bolivia: a brief review of the urban phenomena. Rev Biomed. 2009;2:236–44.
3. Forsyth CJ. Controlled but not cured: structural processes and explanatory models of Chagas disease in tropical Bolivia. Soc Sci Med. 2015;145:7–16.

GPSR Compliance

The European Union's (EU) General Product Safety Regulation (GPSR) is a set of rules that requires consumer products to be safe and our obligations to ensure this.

If you have any concerns about our products, you can contact us on ProductSafety@springernature.com

In case Publisher is established outside the EU, the EU authorized representative is:

Springer Nature Customer Service Center GmbH
Europaplatz 3
69115 Heidelberg, Germany

Batch number: 10837764

Printed by Printforce, the Netherlands